Miroslav Lovrić
McMaster University

VECTOR CALCULUS

STUDENT SOLUTIONS MANUAL

Addison-Wesley Publishers Limited

Don Mills, Ontario • Reading, Massachusetts
Menlo Park, California • New York • Harlow, England
Amsterdam • Bonn • Sydney • Singapore • Tokyo • Madrid
San Juan • Paris • Seoul • Milan • Mexico City • Taipei

PUBLISHER: Ron Doleman
MANAGING EDITOR: Linda Scott
EDITOR: Santo D'Agostino
COVER DESIGN: Anthony Leung
TECHNICAL ART: Nelson Gonzalez
PRODUCTION COORDINATOR: Wendy Moran
MANUFACTURING COORDINATOR: Sharon Latta Paterson

ISBN 0–201–42799–0

Printed and bound in Canada

A B C D E F – MP – 01 00 99 98 97

FOREWORD

This manual provides solutions to odd-numbered exercises and to all computer projects contained in the textbook *Vector Calculus* by Miroslav Lovrić, published by Addison-Wesley Publishers Ltd.

In most cases, all details of a solution are provided. When there is a need, a related theoretical concept or a method (or a formula) is recalled and discussed first. Sometimes, details of the evaluation of definite integrals are skipped and a reader is referred to a table of integrals or is advised to use a numeric method or appropriate software. (The objective of this course is not to master a dozen integration techniques but rather to understand what the integral involved is all about.)

Do not read this manual! It is a lot more beneficial to try to solve an exercise on your own; start it and see how far you can go. If you get stuck, identify the problem first — try to understand why you are having difficulties, and then look up the solution in the manual. This way, you will learn not only what the problem is (or what you have problems with), but also how that particular problem is dealt with. If you just read a solution, most probably you will not recognize the hard part(s) — or, even worse, you might miss the whole point of the exercise.

It is assumed that you have access to a table of integrals and a calculator. You are strongly encouraged to use a graphing calculator and/or appropriate software (such as Maple or Mathematica).

I accept full responsibility for errors in this text and will be grateful to anybody who brings them to my attention. Your comments and suggestions will be greatly appreciated.

I thank Jeff Hooper for proofreading the whole manuscript and the team at Addison-Wesley (Canada) — Ron Doleman and Linda Scott — for all their help. Thanks to Nelson Gonzalez for helping with computer-related issues.

Miroslav Lovrić
Department of Mathematics and Statistics
McMaster University
Hamilton, Ontario, Canada L8S 4K1
e-mail: lovric@icarus.math.mcmaster.ca

February 1997

Contents

CHAPTER 5 INTEGRATION ALONG PATHS

CHAPTER 6 DOUBLE AND TRIPLE INTEGRALS

CHAPTER 7 INTEGRATION OVER SURFACES; PROPERTIES AND APPLICATIONS OF INTEGRALS

CHAPTER 8 CLASSICAL INTEGRATION THEOREMS OF VECTOR CALCULUS

1. VECTORS, MATRICES AND APPLICATIONS

1.1 Vectors

1. By the Triangle Inequality, $\|\mathbf{v}+\mathbf{w}\| \leq \|\mathbf{v}\| + \|\mathbf{w}\|$ for two vectors $\mathbf{v}$ and $\mathbf{w}$. Assume that $\mathbf{v}$ and $\mathbf{w}$ are parallel and have the same direction, i.e., $\mathbf{w} = t\mathbf{v}$ and $t > 0$. In that case,

$$\|\mathbf{v}+\mathbf{w}\| = \|\mathbf{v}+t\mathbf{v}\| = \|(1+t)\mathbf{v}\| = |1+t|\,\|\mathbf{v}\| = (1+t)\,\|\mathbf{v}\|$$

(since $t > 0$, it follows that $1+t > 0$ and so $|1+t| = 1+t$) and

$$\|\mathbf{v}\| + \|\mathbf{w}\| = \|\mathbf{v}\| + \|t\mathbf{v}\| = \|\mathbf{v}\| + |t|\,\|\mathbf{v}\| = (1+t)\|\mathbf{v}\|$$

(since $t > 0$, $|t| = t$). Therefore, to illustrate $\|\mathbf{v}+\mathbf{w}\| = \|\mathbf{v}\| + \|\mathbf{w}\|$ we can choose any pair of parallel vectors with the same direction. For example, $\mathbf{v} = \mathbf{i}$ and $\mathbf{w} = 4\mathbf{i}$.

If $\mathbf{v}$ and $\mathbf{w}$ are non-parallel, then they form a triangle with sides equal to $\|\mathbf{v}\|$, $\|\mathbf{w}\|$ and $\|\mathbf{v}+\mathbf{w}\|$, in which case $\|\mathbf{v}+\mathbf{w}\| < \|\mathbf{v}\| + \|\mathbf{w}\|$. As an example, take $\mathbf{v} = \mathbf{i}$ and $\mathbf{w} = \mathbf{i}+\mathbf{j}$.

Notice that as the angle between two vectors (whose magnitudes are fixed) gets larger and larger, the magnitude of their sum gets smaller and smaller. Therefore, to find $\mathbf{w}$, $\mathbf{v}$ such that $\|\mathbf{v}+\mathbf{w}\| < (\|\mathbf{v}\|+\|\mathbf{w}\|)/2$ we can look for vectors that form a larger angle, like $\mathbf{v} = -\mathbf{i}+\mathbf{j}$ and $\mathbf{w} = \mathbf{i}$. In this case, $\|\mathbf{v}+\mathbf{w}\| = \|\mathbf{j}\| = 1$ and $(\|\mathbf{v}\| + \|\mathbf{w}\|)/2 = (\sqrt{2}+1)/2 > 1$.

3. Assume that $\mathbf{v} = (v_1, v_2) \in \mathbb{R}^2$ (same proof works in any dimension). If $\|\mathbf{v}\| = 0$ then $\sqrt{v_1^2+v_2^2} = 0$ and $v_1^2+v_2^2 = 0$. Therefore, $v_1 = v_2 = 0$ and $\mathbf{v} = \mathbf{0}$. This proves it one way. The other implication is immediate: if $\mathbf{v} = \mathbf{0} = (0,0)$, then $\|\mathbf{v}\| = \sqrt{0^2+0^2} = 0$.

5. If $x = y = 2$, then $r = \sqrt{x^2+y^2} = \sqrt{8}$ and $\tan\theta = 1$. Hence $\theta = \pi/4$ and the point $(2,2)$ is represented in polar coordinates as $(\sqrt{8}, \pi/4)$. For the remaining three points, $r = \sqrt{8}$ as well. For $(-2,2)$ we get $\tan\theta = 2/(-2) = -1$, i.e., $\theta = -\pi/4 + k\pi$ (k is an integer). Since the point $(-2,2)$ is in the second quadrant, $\theta = 3\pi/4$, and $(\sqrt{8}, 3\pi/4)$ are its polar coordinates. Similarly, for the point $(2,-2)$ we get $\tan\theta = -1$ and (the point is in the fourth quadrant) $\theta = 7\pi/4$. Hence $(\sqrt{8}, 7\pi/4)$ are its polar coordinates. Finally, from $x = -2$ and $y = -2$ we get $\tan\theta = 1$ and (we are now in the third quadrant) $\theta = 5\pi/4$. Consequently, $(\sqrt{8}, 5\pi/4)$ are the polar coordinates of $(-2,-2)$.

7. Rewrite the Triangle Inequality as $\|\mathbf{a}+\mathbf{b}\| \leq \|\mathbf{a}\| + \|\mathbf{b}\|$ and substitute $\mathbf{a} = \mathbf{v}-\mathbf{w}$ and $\mathbf{b} = \mathbf{w}$. Thus $\|\mathbf{v}-\mathbf{w}+\mathbf{w}\| \leq \|\mathbf{v}-\mathbf{w}\|+\|\mathbf{w}\|$, so $\|\mathbf{v}\| \leq \|\mathbf{v}-\mathbf{w}\|+\|\mathbf{w}\|$ and $\|\mathbf{v}-\mathbf{w}\| \geq \|\mathbf{v}\|-\|\mathbf{w}\|$. Consider a triangle whose sides are $\mathbf{v}$, $\mathbf{w}$ and $\mathbf{v}-\mathbf{w}$. The above inequality states that the difference of lengths of two sides in a triangle is smaller than the length of the third side. The equality holds in the case when $\mathbf{v}$ and $\mathbf{w}$ are parallel, of the same direction, and $\|\mathbf{v}\| \geq \|\mathbf{w}\|$.

9. If $\mathbf{v}$ and $\mathbf{w}$ are parallel of opposite directions, then $\mathbf{v}+\mathbf{w} = \mathbf{0}$ and hence 0 is the smallest value of $\|\mathbf{v}+\mathbf{w}\|$. By the Triangle Inequality $\|\mathbf{v}+\mathbf{w}\| \leq \|\mathbf{v}\| + \|\mathbf{w}\|$, the largest value of

$\|\mathbf{v}+\mathbf{w}\|$ is $\|\mathbf{v}\|+\|\mathbf{w}\| = 2$ (and this happens when $\mathbf{v}$ and $\mathbf{w}$ are parallel and of the same direction; i.e., when $\mathbf{v}=\mathbf{w}$, since both vectors are of the same length).

11. By definition, $\|\mathbf{v}\| = \sqrt{0^2+2^2+(-1)^2} = \sqrt{5}$.

13. Since $\|\mathbf{w}\| = \sqrt{9+16} = 5$, it follows that

$$\mathbf{v} = (3/5)\mathbf{i} + (4/5)\mathbf{j} \qquad \text{and} \qquad \|\mathbf{v}\| = \sqrt{9/25+16/25} = 1.$$

In general, $\mathbf{v} = \mathbf{w}/\|\mathbf{w}\| = (1/\|\mathbf{w}\|)\mathbf{w}$ has the same direction as $\mathbf{w}$. Its length is 1 (it is called the unit vector in the direction of $\mathbf{w}$), since

$$\|\mathbf{v}\| = \left\|\frac{\mathbf{w}}{\|\mathbf{w}\|}\right\| = \frac{1}{\|\mathbf{w}\|}\|\mathbf{w}\| = 1.$$

15. Let $A(a_1,a_2)$, $B(b_1,b_2)$ be points in $\mathbb{R}^2$. If A and B lie on the same horizontal line, then their distance is $|b_1-a_1|$. Since, in that case, $a_2=b_2$, we can write (recall that $|x|=\sqrt{x^2}$)

$$|b_1-a_1| = \sqrt{(b_1-a_1)^2} = \sqrt{(b_1-a_1)^2+(b_2-a_2)^2}.$$

If A and B lie on the same vertical line, then their distance is $|b_2-a_2|$; since $a_1=b_1$, we obtain the same formula. If A and B do not lie on the same horizontal or the same vertical line, then construct the triangle as shown in Figure 1.1.15(a). It is a right triangle with sides $|b_1-a_1|$ and $|b_2-a_2|$. By the Pythagorean Theorem, its hypotenuse is $d(A,B) = \sqrt{(b_1-a_1)^2+(b_2-a_2)^2}$.

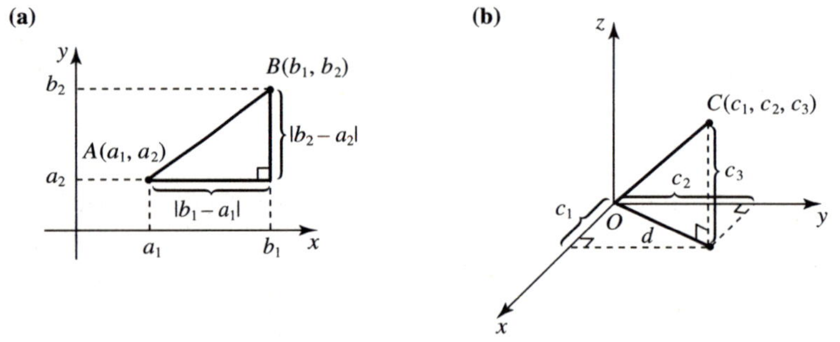

Figure 1.1.15

The proof in $\mathbb{R}^3$ is done in a similar way. Here is an alternative: first, compute the distance $d(O,C)$ from a point $C(c_1,c_2,c_3)$ to the origin, see Figure 1.1.15(b). By the Pythagorean Theorem, $d^2 = c_1^2+c_2^2$ and (again by the Pythagorean Theorem) $d(O,C) = \sqrt{d^2+c_3^2} = \sqrt{c_1^2+c_2^2+c_3^2}$. Now take any points $A(a_1,a_2,a_3)$ and $B(b_1,b_2,b_3)$ in $\mathbb{R}^3$. Take a line segment $\overline{AB}$ and translate it by the vector $-(a_1,a_2,a_3)$, thus obtaining the segment $\overline{OC}$, joining the origin O and the point $C(b_1-a_1,b_2-a_2,b_3-a_3)$. By the formula we have just derived, the length of $\overline{AB}$ = distance from the origin to $C = \sqrt{(b_1-a_1)^2+(b_2-a_2)^2+(b_3-a_3)^2}$.

17. Recall that the representative of a vector $\mathbf{v}=(v_1,v_2,v_3)$ whose tail is at $A(a_1,a_2,a_3)$ is the directed line segment $\overrightarrow{AB}$, where the point B has coordinates $B(a_1+v_1,a_2+v_2,a_3+$

v_3). Let $\mathbf{v} = (0, 2, -1)$. The directed line segment $\overrightarrow{A_1B_1}$, where $A_1(0,1,1)$ and $B_1(0,3,0)$ is the representative that starts at $A_1(0,1,1)$. Similarly, the directed line segments $\overrightarrow{A_2B_2}$, $\overrightarrow{A_3B_3}$ and $\overrightarrow{A_4B_4}$, where $A_2(0,3,0)$, $B_2(0,5,-1)$, $A_3(8,9,-4)$, $B_3(8,11,-5)$, $A_4(10,-1,4)$ and $B_4(10,1,3)$, are the representatives of $\mathbf{v}$ that start at A_2, A_3 and A_4 respectively.

19. We use elementary vector operations. First of all, $\mathbf{a}-2\mathbf{b} = (2\mathbf{i}-\mathbf{j}+\mathbf{k})-2(\mathbf{k}-3\mathbf{i}) = 8\mathbf{i}-\mathbf{j}-\mathbf{k}$. Since $\|\mathbf{c}\| = 2$, it follows that $\mathbf{a} - \mathbf{c}/\|\mathbf{c}\| = (2\mathbf{i} - \mathbf{j} + \mathbf{k}) - 2\mathbf{i}/2 = \mathbf{i} - \mathbf{j} + \mathbf{k}$. Similarly, $3\mathbf{a}+\mathbf{c}-\mathbf{j}+\mathbf{k} = 3(2\mathbf{i}-\mathbf{j}+\mathbf{k})+2\mathbf{i}-\mathbf{j}+\mathbf{k} = 8\mathbf{i}-4\mathbf{j}+4\mathbf{k}$. Since $\mathbf{b}+2\mathbf{a} = (\mathbf{k}-3\mathbf{i})+2(2\mathbf{i}-\mathbf{j}+\mathbf{k}) = \mathbf{i}-2\mathbf{j}+3\mathbf{k}$ and $\|\mathbf{b}+2\mathbf{a}\| = \sqrt{14}$, it follows that the unit vector in the direction of $\mathbf{b}+2\mathbf{a}$ is $(\mathbf{b}+2\mathbf{a})/\|\mathbf{b}+2\mathbf{a}\| = (\mathbf{i}-2\mathbf{j}+3\mathbf{k})/\sqrt{14}$.

21. The vector $-3\mathbf{i}$ has length 3 and it makes an angle of π radians with respect to the positive x-axis. Consequently, its polar form is $-3\mathbf{i} = 3(\cos\pi\,\mathbf{i} + \sin\pi\,\mathbf{j})$. The length of $\mathbf{i}/2-\mathbf{j}$ is $\sqrt{5}/2 \approx 1.11803$. We need the angle between the positive x-axis and $\mathbf{i}/2-\mathbf{j}$ (notice that $\mathbf{i}/2-\mathbf{j}$ is in the fourth quadrant). From $\tan\theta_1 = -1/(1/2) = -2$ it follows that $\theta_1 = \arctan(-2)$, and the required angle is $2\pi + \arctan(-2) \approx 5.17604$ rad. Hence $\mathbf{i}/2-\mathbf{j} \approx 1.11803(\cos 5.17604\,\mathbf{i} + \sin 5.17604\,\mathbf{j})$. Finally, let $\mathbf{v} = \mathbf{i}-4\mathbf{j}$ ($\mathbf{v}$ is in the fourth quadrant). Then $\|\mathbf{v}\| = \sqrt{17} \approx 4.12311$ and $\tan\theta_2 = -4$, so that the angle between the positive x-axis and the vector $\mathbf{v}$ is $2\pi + \arctan(-4) \approx 4.95737$ rad. Therefore, $\mathbf{i}-4\mathbf{j} \approx 4.12311(\cos 4.95737\,\mathbf{i} + \sin 4.95737\,\mathbf{j})$.

23. Let $\mathbf{v} = (v_1, v_2, v_3)$. Since

$$(\alpha+\beta)\mathbf{v} = (\alpha+\beta)(v_1, v_2, v_3) = ((\alpha+\beta)v_1, (\alpha+\beta)v_2, (\alpha+\beta)v_3)$$

and

$$\begin{aligned}\alpha\mathbf{v}+\beta\mathbf{v} &= \alpha(v_1, v_2, v_3) + \beta(v_1, v_2, v_3) = (\alpha v_1, \alpha v_2, \alpha v_3) + (\beta v_1, \beta v_2, \beta v_3)\\ &= (\alpha v_1 + \beta v_1, \alpha v_2 + \beta v_2, \alpha v_3 + \beta v_3),\end{aligned}$$

it follows (by the distributivity of multiplication of real numbers) that $(\alpha+\beta)\mathbf{v} = \alpha\mathbf{v}+\beta\mathbf{v}$. Similarly,

$$(\alpha\beta)\mathbf{v} = (\alpha\beta)(v_1, v_2, v_3) = ((\alpha\beta)v_1, (\alpha\beta)v_2, (\alpha\beta)v_3)$$

and

$$\alpha(\beta\mathbf{v}) = \alpha(\beta(v_1, v_2, v_3)) = \alpha(\beta v_1, \beta v_2, \beta v_3) = (\alpha(\beta v_1), \alpha(\beta v_2), \alpha(\beta v_3)),$$

together with the associativity of multiplication of real numbers prove the second identity.

1.2 Applications in Geometry and Physics

1. The equation of the line is

$$\ell(t) = (1,3) + t\,(3\mathbf{v}) = (1,3) + t(-3,-15) = (1-3t, 3-15t),$$

where $t \in \mathbb{R}$. If we replace $3\mathbf{v}$ by a vector $m\mathbf{v}$ (where $m \neq 0$; clearly, $m\mathbf{v}$ is parallel to $\mathbf{v}$), we get infinitely many parametrizations, one for each non-zero value of m:

$$\ell(t) = (1,3) + t\,m(-1,-5) = (1-tm, 3-5tm),$$

where $t \in \mathbb{R}$.

3. The equation of the line is $\ell(t) = \mathbf{a} + t\mathbf{v}$, where $\mathbf{a} = (1,1)$ and $\mathbf{v}$ is the vector from $(1,1)$ to $(-2,4)$; i.e., $\mathbf{v} = (-3,3)$. Hence $\ell(t) = (1,1) + t(-3,3) = (1-3t, 1+3t)$, where $t \in \mathbb{R}$, parametrizes the given line.

To describe the half-line starting at $(1,1)$ in the direction towards $(-2,4)$, we need all vectors that point in the same direction as $\mathbf{v}$. Hence $\ell(t) = (1-3t, 1+3t)$ with $t \geq 0$ is its parametric representation.

When $t = 0$, $\ell(0) = (1,1)$; for $t = 1$, we get $\ell(1) = (-2,4)$. If $0 < t < 1$, then $\ell(t)$ is a point on the line between $(1,1)$ and $(-2,4)$. Consequently $\ell(t) = (1-3t, 1+3t)$, where $0 \leq t \leq 1$, describes the desired line segment.

5. Choose $\mathbf{a} = (2,1)$ and let $\mathbf{v}$ be the vector from $(2,1)$ to $(-1,5)$; i.e., $\mathbf{v} = (-3,4)$. Then $\ell(t) = (2,1) + t(-3,4)$, $t \in \mathbb{R}$, is a parametric equation of the line containing the given points. The half-line ℓ' "contains" all vectors parallel to $\mathbf{v}$ that are of the same direction as $\mathbf{v}$. Hence $\ell'(t) = (2,1) + t(-3,4)$ with $t \geq 0$ parametrizes ℓ'.

7. The rectangle in question is spanned by $\mathbf{v} = 3\mathbf{i}$ and $\mathbf{w} = \mathbf{j}$; see Figure 1.2.7.

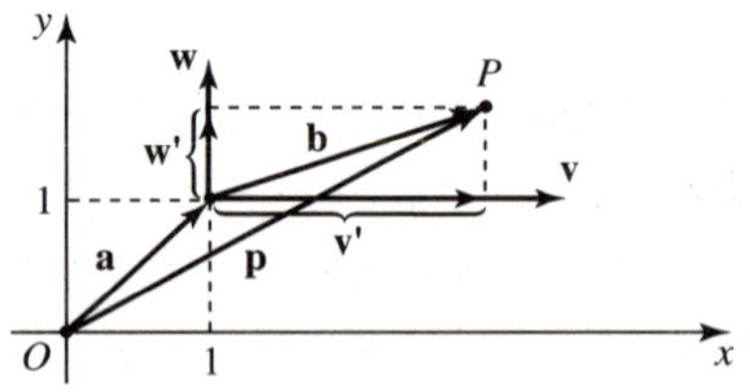

Figure 1.2.7

Let $P(x,y)$ be a point in the rectangle, $\mathbf{b}$ the vector from $(1,1)$ to P and $\mathbf{a}$ the vector from the origin to $(1,1)$. Then ($\mathbf{p}$ is the position vector of P) $\mathbf{p} = \mathbf{a} + \mathbf{b}$, where $\mathbf{b} = \mathbf{v}' + \mathbf{w}'$. Since P belongs to the rectangle, $\mathbf{v}' = \alpha\mathbf{v}$ and $0 \leq \alpha \leq 1$. Similarly, $\mathbf{w}' = \beta\mathbf{w}$ and $0 \leq \beta \leq 1$.

It follows that $\mathbf{b} = \alpha(3\mathbf{i}) + \beta\mathbf{j}$, and therefore $\mathbf{p} = \mathbf{i} + \mathbf{j} + 3\alpha\mathbf{i} + \beta\mathbf{j} = (1+3\alpha)\mathbf{i} + (1+\beta)\mathbf{j}$, $0 \le \alpha, \beta \le 1$, is a vector description of the rectangle.

9. The relative position of the two cars is either of the two differences of the displacement vectors, $\mathbf{v} = \mathbf{i} + 3\mathbf{j} - \mathbf{k} - (\mathbf{j} + 2\mathbf{k}) = \mathbf{i} + 2\mathbf{j} - 3\mathbf{k}$ or $-\mathbf{v}$. The distance between the cars is $\|\mathbf{v}\| = \sqrt{14}$.

11. The particle moves along the line

$$\ell(t) = \textbf{initial point} + \textbf{displacement} = (3,2,4) + t(3,0,-1),$$

$t \ge 0$. It will cross the xy-plane when the z-coordinate of $\ell(t)$ is zero; i.e., when $4 - t = 0$ or $t = 4$. The point where it crosses the plane is $\ell(4) = (15,2,0)$.

13. The information given describes the vector $\mathbf{F}$ in polar form. Hence

$$\mathbf{F} = \|\mathbf{F}\|(\cos(\pi/6)\mathbf{i} + \sin(\pi/6)\mathbf{j}) = 10\left(\tfrac{\sqrt{3}}{2}\mathbf{i} + \tfrac{1}{2}\mathbf{j}\right) = 5\sqrt{3}\mathbf{i} + 5\mathbf{j}.$$

15. The position vectors of the four masses are $\mathbf{0}$, $\mathbf{i} - 2\mathbf{j}$, $\mathbf{i} + \mathbf{j}$ and $(\mathbf{i} - 2\mathbf{j}) + (\mathbf{i} + \mathbf{j}) = 2\mathbf{i} - \mathbf{j}$. By (1.4), (total mass is $M = 8$)

$$\mathbf{c}_M = \tfrac{1}{8}(2(\mathbf{0}) + 2(\mathbf{i} - 2\mathbf{j}) + 2(\mathbf{i} + \mathbf{j}) + 2(2\mathbf{i} - \mathbf{j})) = \tfrac{1}{8}(8\mathbf{i} - 4\mathbf{j}) = \mathbf{i} - \tfrac{1}{2}\mathbf{j}.$$

Therefore, the center of mass of the system is located at $(1, -1/2)$.

17. Place the triangle in the first quadrant (call it the triangle ABC) so that its vertex A is at the origin and the side $\overline{AB}$ lies on the x-axis — thus $A(0,0)$ and $B(2,0)$; see Figure 1.2.17. To find the coordinates of $C(c_1, c_2)$ use the fact that the distances from C to A and from C to B must be 2. Since $d(A,C) = 2$, it follows that $d(A,C)^2 = 4$ and $c_1^2 + c_2^2 = 4$. Similarly, $d(B,C) = 2$ implies that $(c_1 - 2)^2 + c_2^2 = 4$. Subtracting the second equation from the first we get $(c_1 - 2)^2 - c_1^2 = 0$ and $c_1 = 1$. Consequently, $c_2 = \sqrt{3}$ and the coordinates of C are $(1, \sqrt{3})$.

We decided to solve this question so as to practice the distance formula. There are other ways of getting C: for example, by symmetry, the x-coordinate of C must be half-way between 0 and 2; i.e., $c_1 = 1$. The y-coordinate is the height of an equilateral triangle of side 2; i.e., $c_2 = \sqrt{2^2 - 1^2} = \sqrt{3}$ by the Pythagorean Theorem. Alternatively, using trigonometry, $c_1 = 2\cos(\pi/3) = 2(1/2) = 1$ and $c_2 = 2\sin(\pi/3) = 2(\sqrt{3}/2) = 1\sqrt{3}$.

All masses are 1 kg. The position vectors are $\mathbf{r}_A = \mathbf{0}$, $\mathbf{r}_B = 2\mathbf{i}$ and $\mathbf{r}_C = \mathbf{i} + \sqrt{3}\mathbf{j}$. Since the total mass is $M = 3$, we get that

$$\mathbf{c}_M = \tfrac{1}{3}(1(\mathbf{0}) + 1(2\mathbf{i}) + 1(\mathbf{i} + \sqrt{3}\mathbf{j})) = \tfrac{1}{3}(3\mathbf{i} + \sqrt{3}\mathbf{j}) = \mathbf{i} + \tfrac{\sqrt{3}}{3}\mathbf{j}.$$

In words, the center of mass $\mathbf{c}_M$ is located on the axis of symmetry of the triangle (one such axis is the line through C perpendicular to $\overline{AB}$), $\sqrt{3}/3$ units away from the side $\overline{AB}$.

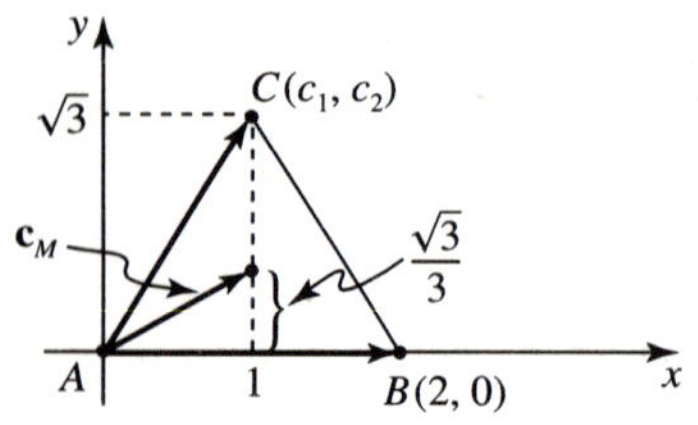

Figure 1.2.17

19. The forces are

$$\mathbf{F}_1 = 5(\cos(\pi/10)\mathbf{i} + \sin(\pi/10)\mathbf{j}) \approx 4.755\mathbf{i} + 1.545\mathbf{j},$$
$$\mathbf{F}_2 = 5(\cos(\pi/5)\mathbf{i} + \sin(\pi/5)\mathbf{j}) \approx 4.045\mathbf{i} + 2.939\mathbf{j},$$
$$\mathbf{F}_3 = 5(\cos(\pi/2)\mathbf{i} + \sin(\pi/2)\mathbf{j}) = 5\mathbf{j}$$

and

$$\mathbf{F}_4 = 5(\cos(13\pi/10)\mathbf{i} + \sin(13\pi/10)\mathbf{j}) \approx -2.939\mathbf{i} - 4.045\mathbf{j}.$$

Their resultant is

$$\mathbf{F} = \mathbf{F}_1 + \mathbf{F}_2 + \mathbf{F}_3 + \mathbf{F}_4 \approx 5.861\mathbf{i} + 5.439\mathbf{j}.$$

1.3 The Dot Product

1. Let $\mathbf{v} = (v_1, \ldots, v_n)$ and $\mathbf{w} = (w_1, \ldots, w_n)$. Then $\mathbf{v} \cdot \mathbf{w} = v_1 w_1 + \ldots + v_n w_n$ and $\mathbf{w} \cdot \mathbf{v} = w_1 v_1 + \ldots + w_n v_n$ are equal due to commutativity of multiplication of real numbers.

3. Let $\mathbf{u} = (u_1, u_2, u_3)$, $\mathbf{v} = (v_1, v_2, v_3)$ and $\mathbf{w} = (w_1, w_2, w_3)$. Then $\mathbf{v} + \mathbf{w} = (v_1 + w_1, v_2 + w_2, v_3 + w_3)$ and

$$\mathbf{u} \cdot (\mathbf{v} + \mathbf{w}) = u_1(v_1 + w_1) + u_2(v_2 + w_2) + u_3(v_3 + w_3).$$

Since

$$\mathbf{u} \cdot \mathbf{v} + \mathbf{u} \cdot \mathbf{w} = u_1 v_1 + u_2 v_2 + u_3 v_3 + u_1 w_1 + u_2 w_2 + u_3 w_3,$$

it follows (by distributivity of multiplication of real numbers) that $\mathbf{u} \cdot (\mathbf{v} + \mathbf{w}) = \mathbf{u} \cdot \mathbf{v} + \mathbf{u} \cdot \mathbf{w}$.

The remaining two identities are proven in a similar way. From $\mathbf{u} = (u_1, u_2, u_3)$ we get $\alpha\mathbf{u} = (\alpha u_1, \alpha u_2, \alpha u_3)$ and

$$(\alpha\mathbf{u}) \cdot \mathbf{v} = (\alpha u_1)v_1 + (\alpha u_2)v_2 + (\alpha u_3)v_3 = \alpha u_1 v_1 + \alpha u_2 v_2 + \alpha u_3 v_3.$$

Similarly,

$$\alpha(\mathbf{u} \cdot \mathbf{v}) = \alpha(u_1 v_1 + u_2 v_2 + u_3 v_3)$$

and the identity $(\alpha\mathbf{u}) \cdot \mathbf{v} = \alpha(\mathbf{u} \cdot \mathbf{v})$ follows from the distributivity of multiplication of real numbers.

To prove the third identity we use the commutativity of the dot product (twice) and the identity that we have just proved:

$$\mathbf{u}\cdot(\alpha\mathbf{v}) = (\alpha\mathbf{v})\cdot\mathbf{u} = \alpha(\mathbf{v}\cdot\mathbf{u}) = \alpha(\mathbf{u}\cdot\mathbf{v}).$$

5. From linear algebra we know that any vector in $\mathbb{R}^3$ can be written as a linear combination of three (non-zero) mutually orthogonal vectors; i.e., $\mathbf{a} = a_{\mathbf{u}}\mathbf{u} + a_{\mathbf{v}}\mathbf{v} + a_{\mathbf{w}}\mathbf{w}$, where $a_{\mathbf{u}}$, $a_{\mathbf{v}}$ and $a_{\mathbf{w}}$ are real numbers. Computing the dot product of $\mathbf{a}$ with $\mathbf{u}$, we get

$$\begin{aligned}\mathbf{a}\cdot\mathbf{u} &= (a_{\mathbf{u}}\mathbf{u} + a_{\mathbf{v}}\mathbf{v} + a_{\mathbf{w}}\mathbf{w})\cdot\mathbf{u}\\ &= a_{\mathbf{u}}\mathbf{u}\cdot\mathbf{u} + a_{\mathbf{v}}\mathbf{v}\cdot\mathbf{u} + a_{\mathbf{w}}\mathbf{w}\cdot\mathbf{u} = a_{\mathbf{u}}\mathbf{u}\cdot\mathbf{u},\end{aligned}$$

since (by orthogonality) the remaining two terms on the right side vanish. Hence $\mathbf{a}\cdot\mathbf{u} = a_{\mathbf{u}}\|\mathbf{u}\|^2$ and $a_{\mathbf{u}} = \mathbf{a}\cdot\mathbf{u}/\|\mathbf{u}\|^2$. The expressions for $a_{\mathbf{v}}$ and $a_{\mathbf{w}}$ are obtained analogously, by computing the dot product of $\mathbf{a}$ with $\mathbf{v}$ and $\mathbf{w}$.

7. Let θ_1 denote the angle at $(0,3,4)$: it is the angle between the vectors from $(0,3,4)$ to $(0,3,0)$ (call it $\mathbf{u}_1$) and from $(0,3,4)$ to $(12,0,5)$ (call it $\mathbf{v}_1$). It follows that $\mathbf{u}_1 = (0,0,-4)$ and $\mathbf{v}_1 = (12,-3,1)$ and

$$\cos\theta_1 = \frac{\mathbf{u}_1\cdot\mathbf{v}_1}{\|\mathbf{u}_1\|\,\|\mathbf{v}_1\|} = \frac{-4}{4\sqrt{154}} = \frac{-1}{\sqrt{154}}$$

and therefore $\theta_1 = \arccos(-1/\sqrt{154}) \approx 1.6515$ rad. Denote by θ_2 the angle at $(0,3,0)$; that is, θ_2 is the angle between the vectors $\mathbf{u}_2 = (0,0,4)$ (from $(0,3,0)$ to $(0,3,4)$) and $\mathbf{v}_2 = (12,-3,5)$ (from $(0,3,0)$ to $(12,0,5)$). We get

$$\cos\theta_2 = \frac{\mathbf{u}_2\cdot\mathbf{v}_2}{\|\mathbf{u}_2\|\,\|\mathbf{v}_2\|} = \frac{20}{4\sqrt{178}}$$

and $\theta_2 = \arccos(5/\sqrt{178}) \approx 1.1867$ rad. Since $\theta_1 + \theta_2 + \theta_3 = \pi$, it follows that $\theta_3 = 0.3034$ rad.

9. Computing the dot product of the given vectors, we get

$$\left(\mathbf{w} - \frac{\mathbf{v}\cdot\mathbf{w}}{\|\mathbf{v}\|^2}\mathbf{v}\right)\cdot\mathbf{v} = \mathbf{w}\cdot\mathbf{v} - \frac{\mathbf{v}\cdot\mathbf{w}}{\|\mathbf{v}\|^2}\mathbf{v}\cdot\mathbf{v} = \mathbf{w}\cdot\mathbf{v} - \frac{\mathbf{v}\cdot\mathbf{w}}{\|\mathbf{v}\|^2}\|\mathbf{v}\|^2 = 0,$$

by the distributivity and the commutativity of the dot product, and the fact that $\mathbf{v}\cdot\mathbf{v} = \|\mathbf{v}\|^2$.

11. Consider the cube spanned by $\mathbf{i}$, $\mathbf{j}$ and $\mathbf{k}$ with one vertex at the origin (as far as angles are concerned, we are free to choose any cube). In that case, the (direction of the) diagonal can be expressed as $\mathbf{i}+\mathbf{j}+\mathbf{k}$, and hence the angle θ between the diagonal and the side represented by $\mathbf{i}$ is

$$\cos\theta = \frac{\mathbf{i}\cdot(\mathbf{i}+\mathbf{j}+\mathbf{k})}{\|\mathbf{i}\|\,\|\mathbf{i}+\mathbf{j}+\mathbf{k}\|} = \frac{1}{\sqrt{3}},$$

i.e., $\theta = \arccos(1/\sqrt{3}) \approx 0.9553$ rad. The remaining two angles (between the diagonal and the sides represented by $\mathbf{j}$ and $\mathbf{k}$) are computed in the same way, and are both equal to θ.

13. The angle $\theta_{\mathbf{v}}$ between $\mathbf{v}$ and $\mathbf{a}$ is

$$\begin{aligned}\cos\theta_{\mathbf{v}} &= \frac{\mathbf{v}\cdot\mathbf{a}}{\|\mathbf{v}\|\,\|\mathbf{a}\|} = \frac{\mathbf{v}\cdot(\|\mathbf{v}\|\mathbf{w}+\|\mathbf{w}\|\mathbf{v})}{\|\mathbf{v}\|\,\|\mathbf{a}\|} = \frac{\|\mathbf{v}\|\mathbf{v}\cdot\mathbf{w}+\|\mathbf{w}\|\mathbf{v}\cdot\mathbf{v}}{\|\mathbf{v}\|\,\|\mathbf{a}\|} \\ &= \frac{\|\mathbf{v}\|(\mathbf{v}\cdot\mathbf{w}+\|\mathbf{v}\|^2)}{\|\mathbf{v}\|\,\|\mathbf{a}\|} = \frac{\mathbf{v}\cdot\mathbf{w}+\|\mathbf{v}\|^2}{\|\mathbf{a}\|},\end{aligned}$$

since $\|\mathbf{v}\| = \|\mathbf{w}\|$. Similarly, the angle $\theta_{\mathbf{w}}$ between $\mathbf{w}$ and $\mathbf{a}$ is computed to be

$$\begin{aligned}\cos\theta_{\mathbf{w}} &= \frac{\mathbf{w}\cdot\mathbf{a}}{\|\mathbf{w}\|\,\|\mathbf{a}\|} = \frac{\mathbf{w}\cdot(\|\mathbf{v}\|\mathbf{w}+\|\mathbf{w}\|\mathbf{v})}{\|\mathbf{w}\|\,\|\mathbf{a}\|} = \frac{\|\mathbf{v}\|\mathbf{w}\cdot\mathbf{w}+\|\mathbf{w}\|\mathbf{w}\cdot\mathbf{v}}{\|\mathbf{w}\|\,\|\mathbf{a}\|} \\ &= \frac{\|\mathbf{w}\|(\|\mathbf{w}\|^2+\mathbf{w}\cdot\mathbf{v})}{\|\mathbf{w}\|\|\mathbf{a}\|} = \frac{\|\mathbf{w}\|^2+\mathbf{w}\cdot\mathbf{v}}{\|\mathbf{a}\|}.\end{aligned}$$

By assumption, $\|\mathbf{v}\| = \|\mathbf{w}\|$, and therefore $\cos\theta_{\mathbf{v}} = \cos\theta_{\mathbf{w}}$, and since (in the definition of the angle between two vectors there is a requirement that $0 \le \theta_{\mathbf{v}}, \theta_{\mathbf{w}} < \pi$) it follows that $\theta_{\mathbf{v}} = \theta_{\mathbf{w}}$.

15. Let $\mathbf{a} = \mathbf{i}+2\mathbf{j}$, $\mathbf{v} = \mathbf{i}+\mathbf{j}$ and $\mathbf{w} = \mathbf{i}-\mathbf{j}$. Vectors $\mathbf{v}$ and $\mathbf{w}$ are orthogonal ($\mathbf{v}\cdot\mathbf{w} = 0$) and hence $\mathbf{a} = a_{\mathbf{v}}\mathbf{v}+a_{\mathbf{w}}\mathbf{w}$, where $a_{\mathbf{v}}\mathbf{v} = pr_{\mathbf{v}}\mathbf{a} = (\mathbf{a}\cdot\mathbf{v}/\|\mathbf{v}\|^2)\mathbf{v}$ and $a_{\mathbf{w}}\mathbf{w} = pr_{\mathbf{w}}\mathbf{a} = (\mathbf{a}\cdot\mathbf{w}/\|\mathbf{w}\|^2)\mathbf{w}$ are the vector orthogonal projections of $\mathbf{a}$ onto $\mathbf{v}$ and $\mathbf{w}$; see (1.9) and the text following it. Since

$$a_{\mathbf{v}}\mathbf{v} = \frac{(\mathbf{i}+2\mathbf{j})\cdot(\mathbf{i}+\mathbf{j})}{\|\mathbf{i}+\mathbf{j}\|^2}(\mathbf{i}+\mathbf{j}) = \tfrac{3}{2}(\mathbf{i}+\mathbf{j})$$

and

$$a_{\mathbf{v}}\mathbf{w} = \frac{(\mathbf{i}+2\mathbf{j})\cdot(\mathbf{i}-\mathbf{j})}{\|\mathbf{i}-\mathbf{j}\|^2}(\mathbf{i}-\mathbf{j}) = -\tfrac{1}{2}(\mathbf{i}-\mathbf{j}),$$

the desired decomposition is

$$\mathbf{a} = \mathbf{i}+2\mathbf{j} = \tfrac{3}{2}(\mathbf{i}+\mathbf{j}) - \tfrac{1}{2}(\mathbf{i}-\mathbf{j}).$$

17. Using the formula of Example 1.24, we get $3(x-0)+0(y+2)-1(z-2) = 0$; i.e., $3x-z+2 = 0$.

19. According to the formula of Example 1.24, we need a point (and we have it — the origin) and a normal vector $\mathbf{n}$. Take $\mathbf{n}$ to be the vector from $(3,2,-1)$ to $(0,0,7)$ (we could have taken the vector in the opposite direction — it would not make any difference). Then $\mathbf{n} = (-3,-2,8)$ and the equation of the plane is $-3(x-0)-2(y-0)+8(z-0) = 0$; i.e., $3x+2y-8z = 0$.

21. Consider a constant force $\mathbf{F}$ moving an object around a triangle ABC: first from A to B, then to C and then back to A. Let $\mathbf{a}$, $\mathbf{b}$ and $\mathbf{c}$ be the vectors represented by the directed line segments $\overrightarrow{AB}$, $\overrightarrow{BC}$ and $\overrightarrow{CA}$ respectively. The work along $\overrightarrow{AB}$ is the dot product of the force and the displacement vector; i.e., it is equal to $\mathbf{F}\cdot\mathbf{a}$. Similarly, the work along the remaining two sides is $\mathbf{F}\cdot\mathbf{b}$ and $\mathbf{F}\cdot\mathbf{c}$, and the total work of $\mathbf{F}$ is $W = \mathbf{F}\cdot\mathbf{a}+\mathbf{F}\cdot\mathbf{b}+\mathbf{F}\cdot\mathbf{c} = \mathbf{F}\cdot(\mathbf{a}+\mathbf{b}+\mathbf{c})$. Since $\mathbf{c} = -\mathbf{b}-\mathbf{a}$, it follows that $W = 0$.

An analogous proof works for any polygon. The key fact that is needed is that if the sides of a polygon are oriented so that the terminal point of one side is the initial point of the neighboring one (i.e., all sides are oriented counterclockwise, or all sides are oriented clockwise), then the vector sum of all directed line segments thus obtained is zero. As above, the work of $\mathbf{F}$ is the dot product of $\mathbf{F}$ and the vector sum of all sides, which is zero.

23. From $\mathbf{u}\cdot\mathbf{v} = (\mathbf{i}+\mathbf{j}-2\mathbf{k})\cdot(2\mathbf{j}+\mathbf{k}) = 2-2 = 0$, $\mathbf{u}\cdot\mathbf{w} = (\mathbf{i}+\mathbf{j}-2\mathbf{k})\cdot(5\mathbf{i}-\mathbf{j}+2\mathbf{k}) = 5-1-4 = 0$ and $\mathbf{v}\cdot\mathbf{w} = (2\mathbf{j}+\mathbf{k})\cdot(5\mathbf{i}-\mathbf{j}+2\mathbf{k}) = -2+2 = 0$, it follows that $\mathbf{u}$, $\mathbf{v}$ and $\mathbf{w}$ are mutually orthogonal. From the three-dimensional version of Theorem 1.7 (the formula appears immediately after the proof; see also Exercise 5) we get (replace $\mathbf{a}$ by $\mathbf{i}$) $\mathbf{i} = i_{\mathbf{u}}\mathbf{u} + i_{\mathbf{v}}\mathbf{v} + i_{\mathbf{w}}\mathbf{w}$, where $i_{\mathbf{u}} = \mathbf{i}\cdot\mathbf{u}/\|\mathbf{u}\|^2 = 1/6$, $i_{\mathbf{v}} = \mathbf{i}\cdot\mathbf{v}/\|\mathbf{v}\|^2 = 0$ and $i_{\mathbf{w}} = \mathbf{i}\cdot\mathbf{w}/\|\mathbf{w}\|^2 = 1/6$. Thus $\mathbf{i} = \mathbf{u}/6 + \mathbf{w}/6$.

25. We will show that any vector $\mathbf{a} \in \mathbb{R}^3$ perpendicular to all three of $\mathbf{u}$, $\mathbf{v}$ and $\mathbf{w}$ must be a zero vector. Since $\mathbf{u}$, $\mathbf{v}$ and $\mathbf{w}$ are mutually orthogonal, it follows that (see Theorem 1.7 and the text immediately following it) $\mathbf{a} = a_{\mathbf{u}}\mathbf{u} + a_{\mathbf{v}}\mathbf{v} + a_{\mathbf{w}}\mathbf{w}$, for some real numbers $a_{\mathbf{u}}$, $a_{\mathbf{v}}$ and $a_{\mathbf{w}}$. Computing the dot product of $\mathbf{a}$ with $\mathbf{u}$, we get $\mathbf{a}\cdot\mathbf{u} = a_{\mathbf{u}}\mathbf{u}\cdot\mathbf{u} + \mathbf{v}\cdot\mathbf{u} + \mathbf{w}\cdot\mathbf{u}$. Since $\mathbf{a}$ must be orthogonal to $\mathbf{u}$ and $\mathbf{u}$, $\mathbf{v}$ and $\mathbf{w}$ are mutually orthogonal, it follows that $\mathbf{a}\cdot\mathbf{u} = 0$, $\mathbf{v}\cdot\mathbf{u} = 0$ and $\mathbf{w}\cdot\mathbf{u} = 0$, and so $0 = a_{\mathbf{u}}\|\mathbf{u}\|^2$ and $a_{\mathbf{u}} = 0$, since $\mathbf{u}$ is a non-zero vector by assumption. Computing the dot product of $\mathbf{a}$ with $\mathbf{v}$ and $\mathbf{w}$ and proceeding as above, we will get $a_{\mathbf{v}} = 0$ and $a_{\mathbf{w}} = 0$, so that $\mathbf{a} = \mathbf{0}$. Therefore, in $\mathbb{R}^3$, there cannot exist four (or more) mutually orthogonal vectors).

1.4 Matrices and Determinants

1. Using elementary matrix operations,

$$2B - 16I_2 = 2\begin{bmatrix} 0 & 5 \\ 4 & 0 \end{bmatrix} - 16\begin{bmatrix} 1 & 0 \\ 0 & 1 \end{bmatrix} = \begin{bmatrix} 0 & 10 \\ 8 & 0 \end{bmatrix} - \begin{bmatrix} 16 & 0 \\ 0 & 16 \end{bmatrix} = \begin{bmatrix} -16 & 10 \\ 8 & -16 \end{bmatrix}.$$

3. Since C is a 3×2 matrix and B a 2×2 matrix, the product CB is defined and is a 3×2 matrix. The product BA is defined, and is a 2×3 matrix. Since CB and BA are not of the same type, the difference $CB - BA$ is not defined.

5. The product AC is a 2×2 matrix, and consequently, the expression $AC + I_2$ is defined. Its value is

$$AC + I_2 = \begin{bmatrix} 2 & -1 & 1 \\ 0 & -5 & 4 \end{bmatrix}\begin{bmatrix} -1 & 0 \\ 1 & 2 \\ 3 & -1 \end{bmatrix} + \begin{bmatrix} 1 & 0 \\ 0 & 1 \end{bmatrix} = \begin{bmatrix} 0 & -3 \\ 7 & -14 \end{bmatrix} + \begin{bmatrix} 1 & 0 \\ 0 & 1 \end{bmatrix} = \begin{bmatrix} 1 & -3 \\ 7 & -13 \end{bmatrix}.$$

7. The product BA is defined, and is equal to the 2×3 matrix

$$BA = \begin{bmatrix} 0 & 5 \\ 4 & 0 \end{bmatrix}\begin{bmatrix} 2 & -1 & 1 \\ 0 & -5 & 4 \end{bmatrix} = \begin{bmatrix} 0 & -25 & 20 \\ 8 & -4 & 4 \end{bmatrix}.$$

Since C is of type 3×2, the product $C(BA)$ is defined. Its value is

$$C(BA) = \begin{bmatrix} -1 & 0 \\ 1 & 2 \\ 3 & -1 \end{bmatrix} \begin{bmatrix} 0 & -25 & 20 \\ 8 & -4 & 4 \end{bmatrix} = \begin{bmatrix} 0 & 25 & -20 \\ 16 & -33 & 28 \\ -8 & -71 & 56 \end{bmatrix}.$$

9. The product AC is of type 2×2, hence $(AC)^2 = (AC)(AC)$ is also of type 2×2. Since B^2 is a 2×2 matrix, the expression $(AC)^2 + 4B^2$ is defined. From

$$AC = \begin{bmatrix} 2 & -1 & 1 \\ 0 & -5 & 4 \end{bmatrix} \begin{bmatrix} -1 & 0 \\ 1 & 2 \\ 3 & -1 \end{bmatrix} = \begin{bmatrix} 0 & -3 \\ 7 & -14 \end{bmatrix}$$

we get

$$(AC)^2 = \begin{bmatrix} 0 & -3 \\ 7 & -14 \end{bmatrix} \begin{bmatrix} 0 & -3 \\ 7 & -14 \end{bmatrix} = \begin{bmatrix} -21 & 42 \\ -98 & 175 \end{bmatrix}.$$

Since

$$B^2 = \begin{bmatrix} 0 & 5 \\ 4 & 0 \end{bmatrix} \begin{bmatrix} 0 & 5 \\ 4 & 0 \end{bmatrix} = \begin{bmatrix} 20 & 0 \\ 0 & 20 \end{bmatrix},$$

it follows that

$$(AC)^2 + 4B^2 = \begin{bmatrix} -21 & 42 \\ -98 & 175 \end{bmatrix} + \begin{bmatrix} 80 & 0 \\ 0 & 80 \end{bmatrix} = \begin{bmatrix} 59 & 42 \\ -98 & 255 \end{bmatrix}.$$

11. We use properties of matrix operations. From $3A - X = I_2 + 4(C - X)$ it follows that $3X = I_2 + 4C - 3A$ and $X = (I_2 + 4C - 3A)/3$. Hence

$$X = \tfrac{1}{3}\left(\begin{bmatrix} 1 & 0 \\ 0 & 1 \end{bmatrix} + \begin{bmatrix} 0 & 4 \\ 4 & 0 \end{bmatrix} - \begin{bmatrix} 6 & -3 \\ 12 & 0 \end{bmatrix}\right) = \tfrac{1}{3}\begin{bmatrix} -5 & 7 \\ -8 & 1 \end{bmatrix} = \begin{bmatrix} -5/3 & 7/3 \\ -8/3 & 1/3 \end{bmatrix}.$$

13. Let $X = \begin{bmatrix} x_{11} & x_{12} \\ x_{21} & x_{22} \end{bmatrix}$. Then $AX = B$ implies

$$\begin{bmatrix} 2 & -1 \\ 4 & 0 \end{bmatrix} \begin{bmatrix} x_{11} & x_{12} \\ x_{21} & x_{22} \end{bmatrix} = \begin{bmatrix} 10 & 1 \\ 0 & 0 \end{bmatrix}$$

and

$$\begin{bmatrix} 2x_{11} - x_{21} & 2x_{12} - x_{22} \\ 4x_{11} & 4x_{12} \end{bmatrix} = \begin{bmatrix} 10 & 1 \\ 0 & 0 \end{bmatrix}.$$

Comparing corresponding entries, we get $2x_{11} - x_{21} = 10$, $2x_{12} - x_{22} = 1$, $4x_{11} = 0$ and $4x_{12} = 0$. It follows that $x_{11} = x_{12} = 0$, and so $x_{21} = -10$ and $x_{22} = -1$, and hence $X = \begin{bmatrix} 0 & 0 \\ -10 & -1 \end{bmatrix}$.

15. The map $\mathbf{F}_C$ is given by $\mathbf{F}_C(\mathbf{v}) = C \cdot \mathbf{v}$, where $\mathbf{v} \in \mathbb{R}^2$. If $\mathbf{v} = (v_1, v_2)$, then

$$\mathbf{F}_C(\mathbf{v}) = \mathbf{F}_C\left(\begin{bmatrix} v_1 \\ v_2 \end{bmatrix}\right) = \begin{bmatrix} 0 & 1 \\ 1 & 0 \end{bmatrix} \begin{bmatrix} v_1 \\ v_2 \end{bmatrix} = \begin{bmatrix} v_2 \\ v_1 \end{bmatrix}.$$

In words, $\mathbf{F}_C$ switches the x- and y-components of a vector. Geometrically, $\mathbf{F}_C$ is a mapping that assigns to a vector its symmetric image with respect to the line $y = x$. Similarly,

$$\mathbf{F}_{I_2}(\mathbf{v}) = \mathbf{F}_{I_2}\left(\begin{bmatrix} v_1 \\ v_2 \end{bmatrix}\right) = \begin{bmatrix} 1 & 0 \\ 0 & 1 \end{bmatrix}\begin{bmatrix} v_1 \\ v_2 \end{bmatrix} = \begin{bmatrix} v_1 \\ v_2 \end{bmatrix};$$

i.e., $\mathbf{F}_{I_2}$ is the identity mapping: it maps a vector to itself.

17. By definition,

$$\det\begin{bmatrix} 3 & 4 \\ 4 & 3 \end{bmatrix} = 3 \cdot 3 - 4 \cdot 4 = -7.$$

19. By definition,

$$\det\begin{bmatrix} \cos\theta & \sin\theta \\ -r\sin\theta & r\cos\theta \end{bmatrix} = r\cos^2\theta + r\sin^2\theta = r.$$

21. By definition,

$$\begin{vmatrix} 2 & 1 & 4 \\ 0 & 6 & 5 \\ 0 & 2 & 0 \end{vmatrix} = 2\begin{vmatrix} 6 & 5 \\ 2 & 0 \end{vmatrix} - 1\begin{vmatrix} 0 & 5 \\ 0 & 0 \end{vmatrix} + 4\begin{vmatrix} 0 & 6 \\ 0 & 2 \end{vmatrix} = 2(-10) - 1(0) + 4(0) = -20.$$

23. Let $A = \begin{bmatrix} a_{11} & a_{12} \\ a_{21} & a_{22} \end{bmatrix}$ and $B = \begin{bmatrix} b_{11} & b_{12} \\ b_{21} & b_{22} \end{bmatrix}$. Then

$$AB = \begin{bmatrix} a_{11} & a_{12} \\ a_{21} & a_{22} \end{bmatrix}\begin{bmatrix} b_{11} & b_{12} \\ b_{21} & b_{22} \end{bmatrix} = \begin{bmatrix} a_{11}b_{11} + a_{12}b_{21} & a_{11}b_{12} + a_{12}b_{22} \\ a_{21}b_{11} + a_{22}b_{21} & a_{21}b_{12} + a_{22}b_{22} \end{bmatrix}$$

and

$$\begin{aligned} \det(AB) &= (a_{11}b_{11} + a_{12}b_{21})(a_{21}b_{12} + a_{22}b_{22}) - (a_{21}b_{11} + a_{22}b_{21})(a_{11}b_{12} + a_{12}b_{22}) \\ &= a_{11}b_{11}a_{21}b_{12} + a_{11}b_{11}a_{22}b_{22} + a_{12}b_{21}a_{21}b_{12} + a_{12}b_{21}a_{22}b_{22} \\ &\quad - a_{21}b_{11}a_{11}b_{12} - a_{21}b_{11}a_{12}b_{22} - a_{22}b_{21}a_{11}b_{12} - a_{22}b_{21}a_{12}b_{22} \\ &= a_{11}b_{11}a_{22}b_{22} + a_{12}b_{21}a_{21}b_{12} - a_{21}b_{11}a_{12}b_{22} - a_{22}b_{21}a_{11}b_{12}. \end{aligned}$$

Since

$$\begin{aligned} \det(A)\det(B) &= (a_{11}a_{22} - a_{21}a_{12})(b_{11}b_{22} - b_{21}b_{12}) \\ &= a_{11}a_{22}b_{11}b_{22} - a_{11}a_{22}b_{21}b_{12} - a_{21}a_{12}b_{11}b_{22} + a_{21}a_{12}b_{21}b_{12}, \end{aligned}$$

it follows that $\det(AB) = \det(A)\det(B)$. In words, the determinant of the product (of two 2×2 matrices) is equal to the product of their determinants (as a matter of fact, the same formula holds for $n \times n$ matrices, $n \geq 2$).

We will now use this fact to prove the second part. Start with $\det(AB) = \det(A)\det(B)$. Since $\det(A)$ and $\det(B)$ are real numbers, it follows that $\det(A)\det(B) = \det(B)\det(A)$, and, again by the above fact, $\det(B)\det(A) = \det(BA)$. Putting it all together, we get $\det(AB) = \det(BA)$.

25. Let $A = \begin{bmatrix} a_{11} & a_{12} \\ a_{21} & a_{22} \end{bmatrix}$. Then $B = \begin{bmatrix} a_{11} & a_{12} \\ \alpha a_{11} & \alpha a_{12} \end{bmatrix}$ and $\det(B) = \alpha a_{11}a_{12} - a_{12}\alpha a_{11} = 0$.

27. Every time we interchange two rows or two columns in a matrix, its determinant changes sign. In order to obtain B from A, we make two interchanges, and therefore $\det(B) = -(-\det(A)) = \det(A)$.

29. Let us look a bit closer at the formula for the determinant of a matrix. Let $A = [a_{ij}]$ be a 3×3 matrix. Then $\det(A)$ is equal to

$$\begin{vmatrix} a_{11} & a_{12} & a_{13} \\ a_{21} & a_{22} & a_{23} \\ a_{31} & a_{32} & a_{33} \end{vmatrix} = a_{11}\begin{vmatrix} a_{22} & a_{23} \\ a_{32} & a_{33} \end{vmatrix} - a_{12}\begin{vmatrix} a_{21} & a_{23} \\ a_{31} & a_{33} \end{vmatrix} + a_{13}\begin{vmatrix} a_{21} & a_{22} \\ a_{31} & a_{32} \end{vmatrix}$$
$$= a_{11}a_{22}a_{33} - a_{11}a_{32}a_{23} - a_{12}a_{21}a_{33} + a_{12}a_{23}a_{31} + a_{13}a_{21}a_{32} - a_{13}a_{31}a_{22}.$$

The expression for $\det(A)$ consists of six terms. Each term contains *exactly one* entry from each row (and exactly one entry from each column) of A. Therefore, if we multiply all elements in one row of A (or in one column of A) by α, each of the six terms will contain the factor α, and therefore $\det(B) = \alpha\det(A)$.

An alternative to this reasoning is a straightforward computation.

31. By the definition of $\mathbf{F}$ and the distributivity of matrix multiplication,

$$\mathbf{F}_{A+B}(\mathbf{v}) = (A+B)\mathbf{v} = A\mathbf{v} + B\mathbf{v} = \mathbf{F}_A(\mathbf{v}) + \mathbf{F}_B(\mathbf{v});$$

i.e., $\mathbf{F}_{A+B} = \mathbf{F}_A + \mathbf{F}_B$. Similarly (using associativity)

$$\mathbf{F}_{2AB}(\mathbf{v}) = (2AB)\mathbf{v} = 2A(B\mathbf{v}) = 2A(\mathbf{F}_B(\mathbf{v})) = 2\mathbf{F}_A(\mathbf{F}_B(\mathbf{v}));$$

i.e., $\mathbf{F}_{2AB} = 2(\mathbf{F}_A \circ \mathbf{F}_B)$, where $\circ$ denotes the composition of functions. Finally,

$$\mathbf{F}_{7B}(\mathbf{v}) = 7B\mathbf{v} = 7\mathbf{F}_B(\mathbf{v}),$$

and so $\mathbf{F}_{7B} = 7\mathbf{F}_B$.

1.5 The Cross Product

1. Using the definition, we get

$$\mathbf{v}\times\mathbf{w}=\begin{vmatrix}\mathbf{i}&\mathbf{j}&\mathbf{k}\\2&0&-3\\1&4&1\end{vmatrix}=\mathbf{i}\begin{vmatrix}0&-3\\4&1\end{vmatrix}-\mathbf{j}\begin{vmatrix}2&-3\\1&1\end{vmatrix}+\mathbf{k}\begin{vmatrix}2&0\\1&4\end{vmatrix}=12\mathbf{i}-5\mathbf{j}+8\mathbf{k}.$$

Similarly,

$$(\mathbf{v}+\mathbf{w})\times\mathbf{k}=\begin{vmatrix}\mathbf{i}&\mathbf{j}&\mathbf{k}\\3&4&-2\\0&0&1\end{vmatrix}=4\mathbf{i}-3\mathbf{j}.$$

By the distributivity of the cross product,

$$(2\mathbf{v}-\mathbf{w})\times(\mathbf{v}+\mathbf{w})=(2\mathbf{v})\times\mathbf{v}-\mathbf{w}\times\mathbf{v}+2\mathbf{v}\times\mathbf{w}-\mathbf{w}\times\mathbf{w}.$$

Since $(2\mathbf{v})\times\mathbf{v}=2\mathbf{v}\times\mathbf{v}=\mathbf{0}$, $\mathbf{w}\times\mathbf{w}=\mathbf{0}$ and $\mathbf{v}\times\mathbf{w}=-\mathbf{w}\times\mathbf{v}$, it follows that

$$(2\mathbf{v}-\mathbf{w})\times(\mathbf{v}+\mathbf{w})=3(\mathbf{v}\times\mathbf{w})=3(12\mathbf{i}-5\mathbf{j}+8\mathbf{k})=36\mathbf{i}-15\mathbf{j}+24\mathbf{k}.$$

Finally,

$$\mathbf{i}\times(\mathbf{v}\times\mathbf{w})=\begin{vmatrix}\mathbf{i}&\mathbf{j}&\mathbf{k}\\1&0&0\\12&-5&8\end{vmatrix}=-8\mathbf{j}-5\mathbf{k}.$$

3. Let $\mathbf{u}=(u_1,u_2,u_3)$, $\mathbf{v}=(v_1,v_2,v_3)$ and $\mathbf{w}=(w_1,w_2,w_3)$. Then $\mathbf{u}+\mathbf{v}=(u_1+v_1,u_2+v_2,u_3+v_3)$ and

$$\begin{aligned}(\mathbf{u}+\mathbf{v})\times\mathbf{w}&=\begin{vmatrix}\mathbf{i}&\mathbf{j}&\mathbf{k}\\u_1+v_1&u_2+v_2&u_3+v_3\\w_1&w_2&w_3\end{vmatrix}\\&=((u_2+v_2)w_3-(u_3+v_3)w_2)\mathbf{i}-((u_1+v_1)w_3-(u_3+v_3)w_1)\mathbf{j}\\&\quad+((u_1+v_1)w_2-(u_2+v_2)w_1)\mathbf{k}.\end{aligned}$$

On the other hand,

$$\begin{aligned}\mathbf{u}\times\mathbf{w}+\mathbf{v}\times\mathbf{w}&=\begin{vmatrix}\mathbf{i}&\mathbf{j}&\mathbf{k}\\u_1&u_2&u_3\\w_1&w_2&w_3\end{vmatrix}+\begin{vmatrix}\mathbf{i}&\mathbf{j}&\mathbf{k}\\v_1&v_2&v_3\\w_1&w_2&w_3\end{vmatrix}\\&=(u_2w_3-u_3w_2)\mathbf{i}-(u_1w_3-u_3w_1)\mathbf{j}+(u_1w_2-u_2w_1)\mathbf{k}\\&\quad+(v_2w_3-v_3w_2)\mathbf{i}-(v_1w_3-v_3w_1)\mathbf{j}+(v_1w_2-v_2w_1)\mathbf{k}.\end{aligned}$$

Comparing the components, we conclude that $(\mathbf{u}+\mathbf{v})\times\mathbf{w}$ and $\mathbf{u}\times\mathbf{w}+\mathbf{v}\times\mathbf{w}$ are equal.

5. All vectors orthogonal to both $\mathbf{i}+\mathbf{j}-2\mathbf{k}$ and $\mathbf{k}$ must be parallel to their cross product $(\mathbf{i}+\mathbf{j}-2\mathbf{k})\times\mathbf{k}=-\mathbf{j}+\mathbf{i}$. Divide $-\mathbf{j}+\mathbf{i}$ by its length to get a unit vector: $(\mathbf{i}-\mathbf{j})/\sqrt{2}$; this is one result. The other one is $-(\mathbf{i}-\mathbf{j})/\sqrt{2}$.

7. Almost any guess for $\mathbf{u}$, $\mathbf{v}$ and $\mathbf{w}$ will work. For example, $\mathbf{u}=\mathbf{i}$, $\mathbf{v}=\mathbf{i}$ and $\mathbf{w}=\mathbf{j}$. Then $(\mathbf{u}\times\mathbf{v})\times\mathbf{w}=(\mathbf{i}\times\mathbf{i})\times\mathbf{j}=\mathbf{0}\times\mathbf{w}=\mathbf{0}$, but $\mathbf{u}\times(\mathbf{v}\times\mathbf{w})=\mathbf{i}\times(\mathbf{i}\times\mathbf{j})=\mathbf{i}\times\mathbf{k}=-\mathbf{j}$.

9. The volume of the parallelepiped is equal to the absolute value of the scalar triple product of given vectors. From

$$(3\mathbf{i}+\mathbf{j}-\mathbf{k})\cdot((-\mathbf{i}-\mathbf{j})\times(\mathbf{j}+\mathbf{k}))=\begin{vmatrix}3 & 1 & -1\\ -1 & -1 & 0\\ 0 & 1 & 1\end{vmatrix}=3(-1)-1(-1)-1(-1)=-1,$$

it follows that the volume is 1.

11. Let $\mathbf{v}$ be the vector from $(0,2)$ to $(3,2)$ (so $\mathbf{v}=3\mathbf{i}$) and $\mathbf{w}$ the vector from $(0,2)$ to $(1,1)$ (so $\mathbf{w}=\mathbf{i}-\mathbf{j}$). The area of the parallelogram spanned by $\mathbf{v}$ and $\mathbf{w}$ is

$$\|\mathbf{v}\times\mathbf{w}\|=\left\|\begin{vmatrix}\mathbf{i} & \mathbf{j} & \mathbf{k}\\ 3 & 0 & 0\\ 1 & -1 & 0\end{vmatrix}\right\|=\|-3\mathbf{k}\|=3.$$

13. Using four given points, we will construct three vectors (all starting at the same point) and compute their scalar triple product. The three vectors (and hence all four points) belong to the same plane if and only if the scalar triple product is zero. Let $\mathbf{u}$ be the vector from $(1,0,0)$ to $(1,2,-1)$ (so $\mathbf{u}=(0,2,-1)$), $\mathbf{v}$ the vector from $(1,0,0)$ to $(0,2,-4)$ (so $\mathbf{v}=(-1,2,-4)$) and $\mathbf{w}$ the vector from $(1,0,0)$ to $(2,-1,0)$ (so $\mathbf{w}=(1,-1,0)$). Then

$$\mathbf{u}\cdot(\mathbf{v}\times\mathbf{w})=\begin{vmatrix}0 & 2 & -1\\ -1 & 2 & -4\\ 1 & -1 & 0\end{vmatrix}=-7,$$

and therefore the four points do not belong to the same plane.

15. Construct the vectors $\mathbf{v}$ (from $(1,0,0)$ to $(-3,2,2)$, so $\mathbf{v}=(-4,2,2)$) and $\mathbf{w}$ (from $(1,0,0)$ to $(3,-1,-1)$, so $\mathbf{w}=(2,-1,-1)$). From $\mathbf{v}\times\mathbf{w}=\mathbf{0}$ it follows that $\mathbf{v}$ and $\mathbf{w}$ are parallel, and since they have the point $(1,0,0)$ in common, all three points must lie on the same line.

17. In order to find an equation of the plane, we need a point and a vector $\mathbf{n}$ normal (perpendicular) to it. We have infinitely many points to choose from, since both ℓ_1 and ℓ_2 lie in the plane. Take, for example, $\ell_2(1)=(3,1,0)$. A vector that is perpendicular to the plane must be perpendicular to both lines; i.e., must be perpendicular to their direction vectors $(-1,0,1)$ and $(3,0,-1)$. Hence

$$\mathbf{n}=\begin{vmatrix}\mathbf{i} & \mathbf{j} & \mathbf{k}\\ -1 & 0 & 1\\ 3 & 0 & -1\end{vmatrix}=2\mathbf{j},$$

and so $0(x-3)+2(y-1)+0(z-0)=0$; i.e., $y=1$ represents the desired equation.

19. A normal vector to the plane in question is

$$\mathbf{n}=\mathbf{v}\times\mathbf{w}=\begin{vmatrix}\mathbf{i} & \mathbf{j} & \mathbf{k}\\ 2 & 4 & 0\\ 0 & 1 & -1\end{vmatrix}=-4\mathbf{i}+2\mathbf{j}+2\mathbf{k}.$$

Since the plane contains the origin, its equation is $-4(x-0)+2(y-0)+2(z-0)=0$; i.e., $2x-y-z=0$. We have to draw the line through $(3,2,0)$ that is perpendicular to the plane and then measure the distance from $(3,2,0)$ to the point of intersection of the line and the plane. The vector $\mathbf{n}$ (or any vector parallel to it, like $2\mathbf{i}-\mathbf{j}-\mathbf{k}$) defines the direction of the line, and hence $\ell(t)=(3,2,0)+t(2,-1,-1)=(3+2t,2-t,-t)$, $t\in\mathbb{R}$, is its parametric equation. Substituting $x=3+2t$, $y=2-t$ and $z=-t$ into the equation of the plane, we get $2(3+2t)-(2-t)+t=0$ and $t=-2/3$. It follows that $\ell(-2/3)=(5/3,8/3,2/3)$ is the point of intersection of the plane and the line $\ell(t)$ normal to it. Finally, the distance $d((3,2,0),(5/3,8/3,2/3))=\sqrt{8/3}$ between $(3,2,0)$ and $(5/3,8/3,2/3)$ is the distance between the point $(3,2,0)$ and the plane.

21. The torque of $\mathbf{F}=2\mathbf{k}$ is given by $\mathbf{T}=\mathbf{r}\times\mathbf{F}$, where $\mathbf{r}$ is the position vector of the point where $\mathbf{F}$ is applied. Since $\|\mathbf{r}\|=1$ and $\|\mathbf{F}\|=2$ it follows that

$$\|\mathbf{T}\|=\|\mathbf{r}\|\,\|\mathbf{F}\|\,\sin\theta=2\sin\theta,$$

where θ is the angle between the position vector of a point and $\mathbf{F}$. The magnitude of $\mathbf{T}$ is the largest when $\theta=\pi/2$; i.e., along the equator (that is, at the points on the intersection of the sphere and the xy-plane). $\|\mathbf{T}\|$ is the smallest when $\theta=0$ or $\theta=\pi$; i.e., on the North Pole and on the South Pole (i.e., at $(0,0,\pm1)$).

23. Since the angular speed of the disk is 10 rad/s, the angular velocity is parallel to the axis of rotation, and so $\mathbf{w}=\pm10\mathbf{k}$. Due to the counterclockwise orientation, we must choose $\mathbf{w}=10\mathbf{k}$ ("right-hand" rule). The tangential velocity vector is $\mathbf{v}=\mathbf{r}\times\mathbf{w}$ ($\mathbf{r}$ is the position vector of a point on the disk) and its magnitude is $\|\mathbf{v}\|=\|\mathbf{r}\|\,\|\mathbf{w}\|\,\sin\theta=10\|\mathbf{r}\|\,\sin\theta$, where θ is the angle between the position vector and the z-axis (that is the axis of rotation). Since the disk lies in the xy-plane, $\theta=\pi/2$ and hence $\|\mathbf{v}\|=10\|\mathbf{r}\|$. Consequently, $\|\mathbf{v}\|$ has the largest value when $\mathbf{r}$ has the largest value (i.e., on the edge of the disk), and the smallest value at the center of the disk.

1.6 Chapter Review

Solution to Computer Project

1. (a) The velocities of the two cars are $\mathbf{v}_A = -50\mathbf{i}$ and $\mathbf{v}_B = 80\mathbf{j}$. Their positions at time t are $\mathbf{p}_A(t) = (10 - 50t)\mathbf{i}$ and $\mathbf{p}_B(t) = 80t\mathbf{j}$.

(b) The relative position of the two cars at time t is

$$\mathbf{p}(t) = \mathbf{p}_A(t) - \mathbf{p}_B(t) = (10 - 50t)\mathbf{i} - 80t\mathbf{j},$$

and their distance is

$$\begin{aligned} d(t) &= \|\mathbf{p}(t)\| = \sqrt{(10-50t)^2 + (80t)^2} \\ &= \sqrt{8900t^2 - 10000t + 100} = 10\sqrt{89t^2 - 10t + 1}. \end{aligned}$$

(c) See Figure 1.6.1 (a).

(d) Looking at the plot, we see that the distance is smallest when $t \approx 0.05$ hours. Using calculus, we will improve this approximation. From

$$d'(t) = 10\frac{178t - 10}{2\sqrt{89t^2 - 10t + 1}} = 0$$

it follows that $178t - 10 = 0$ and $t = 10/178 \approx 0.05618$. Since $d'(t) < 0$ when $t < 10/178$ and $d'(t) > 0$ when $t > 10/178$, we conclude that $t = 10/178$ is a point where $d(t)$ attains its minimum. The minimum distance is $d(10/178) \approx 8.47998$ km.

(e) We have to solve $d(t) = 200$ for t; we can try to read an approximation from the graph, see Figure 1.6.1 (b). From

$$10\sqrt{89t^2 - 10t + 1} = 200$$

we get $89t^2 - 10t - 399 = 0$, and (drop the negative solution) $t \approx 2.17427$ h.

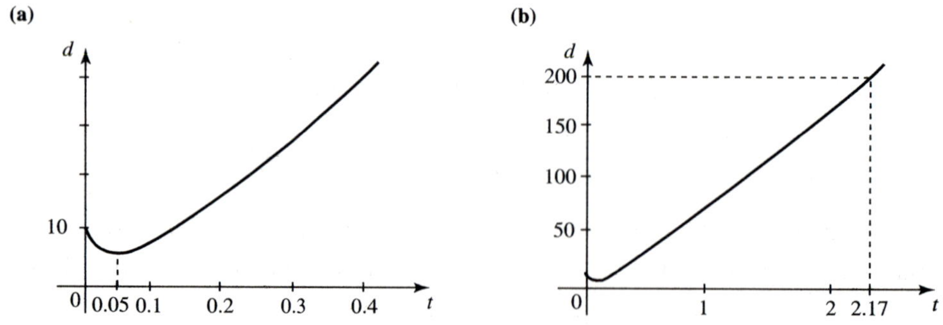

Figure 1.6.1

Solutions to Further Explorations

1. Using the fact that $\|\mathbf{a}\|^2 = \mathbf{a}\cdot\mathbf{a}$ for any vector $\mathbf{a}$, we expand the left side as

$$\begin{aligned}\|\mathbf{v}-\mathbf{w}\|^2 + \|\mathbf{v}+\mathbf{w}\|^2 &= (\mathbf{v}-\mathbf{w})\cdot(\mathbf{v}-\mathbf{w}) + (\mathbf{v}+\mathbf{w})\cdot(\mathbf{v}+\mathbf{w})\\ &= \mathbf{v}\cdot\mathbf{v} - \mathbf{w}\cdot\mathbf{v} - \mathbf{v}\cdot\mathbf{w} + \mathbf{w}\cdot\mathbf{w} + \mathbf{v}\cdot\mathbf{v} + \mathbf{w}\cdot\mathbf{v} + \mathbf{v}\cdot\mathbf{w} + \mathbf{w}\cdot\mathbf{w}.\end{aligned}$$

The four terms containing the dot product of $\mathbf{v}$ and $\mathbf{w}$ cancel. What is left is $2\mathbf{v}\cdot\mathbf{v}+2\mathbf{w}\cdot\mathbf{w} = 2(\|\mathbf{v}\|^2 + \|\mathbf{w}\|^2)$, and we are done.

Consider the parallelogram spanned by $\mathbf{v}$ and $\mathbf{w}$. The vectors $\mathbf{v}-\mathbf{w}$ and $\mathbf{v}+\mathbf{w}$ represent its diagonals; the identity we proved states that the sum of the squares of the lengths of the two diagonals in a parallelogram is equal to the sum of the squares of lengths of its four sides.

3. By the formula for the angle between two vectors, the angle α between $\mathbf{v} = v_1\mathbf{i}+v_2\mathbf{j}+v_3\mathbf{k}$ and the positive x-axis satisfies $\cos\alpha = \mathbf{v}\cdot\mathbf{i}/(\|\mathbf{v}\|\,\|\mathbf{i}\|) = v_1/\|\mathbf{v}\|$. Similarly, the angle β between $\mathbf{v}$ and the positive y-axis satisfies $\cos\beta = \mathbf{v}\cdot\mathbf{j}/(\|\mathbf{v}\|\,\|\mathbf{j}\|) = v_2/\|\mathbf{v}\|$; the angle γ between $\mathbf{v}$ and the positive z-axis satisfies $\cos\gamma = \mathbf{v}\cdot\mathbf{k}/(\|\mathbf{v}\|\,\|\mathbf{k}\|) = v_3/\|\mathbf{v}\|$. Since $\|\mathbf{v}\|^2 = v_1^2+v_2^2+v_3^2$, we get

$$\cos^2\alpha + \cos^2\beta + \cos^2\gamma = \frac{v_1^2}{\|\mathbf{v}\|^2} + \frac{v_2^2}{\|\mathbf{v}\|^2} + \frac{v_3^2}{\|\mathbf{v}\|^2} = \frac{\|\mathbf{v}\|^2}{\|\mathbf{v}\|^2} = 1.$$

Take any vector $\mathbf{v} \in \mathbb{R}^3$; then

$$\begin{aligned}\mathbf{v} = v_1\mathbf{i} + v_2\mathbf{j} + v_3\mathbf{k} &= \|\mathbf{v}\|\left(\frac{v_1}{\|\mathbf{v}\|}\mathbf{i} + \frac{v_2}{\|\mathbf{v}\|}\mathbf{j} + \frac{v_3}{\|\mathbf{v}\|}\mathbf{k}\right)\\ &= \|\mathbf{v}\|(\cos\alpha\,\mathbf{i} + \cos\beta\,\mathbf{j} + \cos\gamma\,\mathbf{k})\end{aligned}$$

is a required representation of a vector in terms of its direction cosines.

5. Most of the work in finding Π is done once we find a normal vector. Using given points, form the vectors $\mathbf{v}$ (from $(1,2,0)$ to $(2,0,4)$, hence $\mathbf{v} = (1,-2,4)$) and $\mathbf{w}$ (from $(1,2,0)$ to $(-1,-1,2)$, hence $\mathbf{w} = (-2,-3,2)$). Since $\mathbf{v}$ and $\mathbf{w}$ belong to Π, a normal to Π is given by their cross product

$$\mathbf{v}\times\mathbf{w} = \begin{vmatrix}\mathbf{i} & \mathbf{j} & \mathbf{k}\\ 1 & -2 & 4\\ -2 & -3 & 2\end{vmatrix} = 8\mathbf{i} - 10\mathbf{j} - 7\mathbf{k}.$$

Consequently, an equation of Π is $8(x-1)-10(y-2)-7(z-0) = 0$; i.e., $8x-10y-7z+12 = 0$.

A normal (call it $\mathbf{n}$) of a plane that is perpendicular to Π must belong to Π (i.e., be parallel to it). Since $\mathbf{v}$ and $\mathbf{w}$ span Π, $\mathbf{n} = t\mathbf{v} + s\mathbf{w}$ for some real numbers s and t. Hence $\mathbf{n} = t(1,-2,4) + s(-2,-3,2) = (t-2s, -2t-3s, 4t+2s)$ and

$$(t-2s)(x-1) + (-2t-3s)(y-2) + (4t+2s)(z-0) = 0;$$

i.e.,

$$(t-2s)x - (2t+3s)y + (4t+2s)z + 3t + 8s = 0,$$

where $t, s \in \mathbb{R}$ (and at least one of s or t is non-zero), represents all planes through $(1,2,0)$ that are perpendicular to Π.

7. Let $A = [a_{ij}]$, $i,j = 1,2,3$. Then

$$B = \begin{bmatrix} a_{11} & a_{12} & a_{13} \\ a_{21} & a_{22} & a_{23} \\ \alpha a_{11} + \beta a_{21} & \alpha a_{12} + \beta a_{22} & \alpha a_{13} + \beta a_{23} \end{bmatrix}$$

and

$$\begin{aligned} \det(B) = a_{11}&(a_{22}(\alpha a_{13} + \beta a_{23}) - (\alpha a_{12} + \beta a_{22})a_{23}) \\ &- a_{12}(a_{21}(\alpha a_{13} + \beta a_{23}) - (\alpha a_{11} + \beta a_{21})a_{23}) \\ &+ a_{13}(a_{21}(\alpha a_{12} + \beta a_{22}) - (\alpha a_{11} + \beta a_{21})a_{22}) \\ = \alpha&(a_{11}a_{22}a_{13} - a_{11}a_{12}a_{23} - a_{12}a_{21}a_{13} + a_{12}a_{11}a_{23} + a_{13}a_{21}a_{12} - a_{13}a_{11}a_{22}) \\ &+ \beta(a_{11}a_{22}a_{23} - a_{11}a_{22}a_{23} - a_{12}a_{21}a_{23} + a_{12}a_{21}a_{23} + a_{13}a_{21}a_{22} - a_{13}a_{21}a_{22}) = 0. \end{aligned}$$

The determinant of B can be interpreted as the scalar triple product of $\mathbf{u} = (a_{11}, a_{12}, a_{13})$, $\mathbf{v} = (a_{21}, a_{22}, a_{23})$, and $\mathbf{w} = (\alpha a_{11} + \beta a_{21}, \alpha a_{12} + \beta a_{22}, \alpha a_{13} + \beta a_{23})$. In that context, the absolute value of $\det(B)$ represents the volume of the parallelepiped spanned by $\mathbf{u}$, $\mathbf{v}$ and $\mathbf{w}$. Since $\mathbf{w} = \alpha\mathbf{u} + \beta\mathbf{v}$, all three vectors belong to the same plane and the volume is zero. Consequently, $\det(B) = 0$.

9. (a) For example,

$$A = \begin{bmatrix} 0 & 2 \\ 0 & 1 \end{bmatrix} \quad \text{and} \quad \mathbf{v} = \begin{bmatrix} 3 \\ 0 \end{bmatrix},$$

or

$$A = \begin{bmatrix} 2 & -1 \\ 4 & -2 \end{bmatrix} \quad \text{and} \quad \mathbf{v} = \begin{bmatrix} 3 \\ 6 \end{bmatrix}$$

satisfy $\mathbf{F}_A(\mathbf{v}) = A \cdot \mathbf{v} = \mathbf{0}$.

(b) Let

$$A = \begin{bmatrix} a_{11} & a_{12} \\ a_{21} & a_{22} \end{bmatrix} \quad \text{and} \quad \mathbf{v} = \begin{bmatrix} v_1 \\ v_2 \end{bmatrix}.$$

Since $\mathbf{F}_A(\mathbf{v}) = \mathbf{0}$ it follows that

$$\begin{bmatrix} a_{11} & a_{12} \\ a_{21} & a_{22} \end{bmatrix} \cdot \begin{bmatrix} v_1 \\ v_2 \end{bmatrix} = \begin{bmatrix} a_{11}v_1 + a_{12}v_2 \\ a_{21}v_1 + a_{22}v_2 \end{bmatrix} = \begin{bmatrix} 0 \\ 0 \end{bmatrix};$$

i.e., $a_{11}v_1 + a_{12}v_2 = 0$ and $a_{21}v_1 + a_{22}v_2 = 0$. We have obtained a system of two equations in two unknowns v_1 and v_2. The matrix of the system is A, and since $\det(A) \neq 0$, the only solution is the trivial solution $v_1 = 0$ and $v_2 = 0$. Hence $\mathbf{v} = \mathbf{0}$.

2. CALCULUS OF FUNCTIONS OF SEVERAL VARIABLES

2.1 Real-Valued and Vector-Valued Functions of Several Variables

1. The function f is a polynomial in two variables, and so its domain is $\mathbb{R}^2$. Since any $c \in \mathbb{R}$ can be obtained as a value of f (for example, $f(0, c+7) = c$), it follows that the range of f is $\mathbb{R}$.

3. Since $\tan a$ is not defined when $a = \pi/2 + k\pi$ (k is an integer), the domain of f consists of all ordered 4-tuples (u, v, t, z) such that $u + v \neq \pi/2 + k\pi$, with no restrictions on t and z (i.e., $t, z \in \mathbb{R}$). Let c be any real number; since $f(0, 0, c^{1/3}, 1) = \tan 0 + (c^{1/3})^3 1^3 = c$, it folows that the range of f is $\mathbb{R}$.

5. The denominator $x^2 + y^2$ is zero only when $x = 0$ and $y = 0$, hence the domain of f is the set $\{(x, y) \mid x \neq 0 \text{ and } y \neq 0\} = \mathbb{R}^2 - \{(0, 0)\}$. From $0 \leq x^2 \leq x^2 + y^2$ it follows that (divide all sides by $x^2 + y^2$) $0 \leq x^2/(x^2 + y^2) \leq 1$ and (multiply by 3) $0 \leq f(x, y) \leq 3$. Consequently, the range of f is $[0, 3]$.

7. Since $\ln a$ is defined for $a > 0$, the domain of $\mathbf{F}$ consists of all ordered pairs (x, y) such that $x > 0$ and $y > 0$. The range of $\ln a$ is $\mathbb{R}$, and so the range of $\mathbf{F}$ consists of all ordered pairs (c, d) such that $c \in \mathbb{R}$ and $d \geq 0$.

9. The absolute value function is defined for all real numbers, and therefore the domain of f is $\mathbb{R}^2$. Since $|a| \geq 0$ for any real number a, it follows that the range of f is $[0, \infty)$.

11. An ordered pair (x, y) belongs to the domain of f if and only if $4 - 4x^2 - y^2 \geq 0$; i.e., if and only if $4x^2 - y^2 \leq 4$. The equation $4x^2 - y^2 = 4$, or $x^2 - y^2/4 = 1$, represents the hyperbola whose asymptotes are $y = \pm 2x$ and the x-intercepts are $x = \pm 1$. This hyperbola divides $\mathbb{R}^2$ in three unbounded regions. The domain is the middle region; i.e., the one that contains the origin. Since $4 - 4x^2 - y^2 = 4 - (4x^2 + y^2) \leq 4$, it follows that $\sqrt{4 - 4x^2 - y^2} \leq 2$. Consequently, the range of f is $[0, 2]$.

13. Since $|x| = 0$ only when $x = 0$, it follows that the domain of f is $\{(x, y) \mid x \neq 0, y \in \mathbb{R}\}$. The fact that $f(1, c) = c$ (for any $c \in \mathbb{R}$) implies that the range is $\mathbb{R}$.

15. Recall that $\mathbf{F}(x, y) = (-y/\sqrt{x^2 + y^2}, x/\sqrt{x^2 + y^2})$. Since $\|\mathbf{F}(x, y)\| = 1$, the range of $\mathbf{F}$ consists of unit vectors. Consider a direction defined by a vector (x_0, y_0); i.e., the direction from $(0, 0)$ to (x_0, y_0). The vector $\mathbf{F}(y_0, -x_0) = (x_0/\sqrt{x_0^2 + y_0^2}, y_0/\sqrt{x_0^2 + y_0^2})$ is a unit vector in that chosen direction. Consequently, unit vectors $\mathbf{F}(x, y)$ for $(x, y) \in \mathbb{R}^2$ point in all directions in $\mathbb{R}^2$, and so the range of $\mathbf{F}$ consists of all unit vectors. Identifying tips of vectors with points in $\mathbb{R}^2$ (in which case their initial points are all located at the origin), we can describe the range of $\mathbf{F}$ as a circle of radius 1 centered at the origin.

17. The distance from (x, y, z) to the origin is given by $d(x, y, z) = \sqrt{x^2 + y^2 + z^2}$. By assumption, the function f is inversely proportional to the square of d; i.e., $f(x, y, z) =$

$k/d^2(x,y,z) = k/(x^2+y^2+z^2)$, for some $k \in \mathbb{R}$, $k \neq 0$. From $f(1,2,3) = 3$ it follows that $3 = k/14$; i.e., $k = 42$. So $f(x,y,z) = 42/(x^2+y^2+z^2)$.

19. The domain of W consists of ordered pairs $(T,v) \in \mathbb{R}^2$, where T is the temperature and v is the wind speed. There is no unique choice for the domain. It is given that $5 \le v \le 45$. However, for the temperature T, we can select an interval that seems reasonable (considering the relevance of the information on the wind chill), for example $-100 \le T \le 60$. Hence, in this case, the domain of W is the set $\{(T,v) \mid -100 \le T \le 60, 5 \le v \le 45\}$.

21. The domain of f is a subset of $\mathbb{R}^3$ defined by $xyz \neq 0$. In other words, it consists of all points $(x,y,z) \in \mathbb{R}^3$ such that $x \neq 0$ and $y \neq 0$ and $z \neq 0$. Geometrically, it represents the space $\mathbb{R}^3$ with the three coordinate planes removed (recall that $x = 0$ represents the yz-plane, $y = 0$ is the xz-plane, and $z = 0$ is the xy-plane).

23. There are no restrictions on r and θ, and so the domain of $\mathbf{F}$ consists of all ordered pairs (r,θ), where $r, \theta \in \mathbb{R}$.

25. The equation $x^2 - y^2 = 1$ represents the hyperbola whose asymptotes are the lines $y = \pm x$ and the x-intercepts are at $(\pm 1, 0)$. This hyprebola divides $\mathbb{R}^2$ into three regions. To find the one(s) that belong to the domain, all we have to do is pick a point in each and check whether the inequality $x^2 - y^2 \ge 1$ is satisfied or not. It follows that the "inside" of the hyperbola (i.e., the regions on the right and on the left) and the hyperbola itself belong to the domain; see Figure 2.1.25. Using set notation, we write the domain as $\{(x,y) \mid x^2 - y^2 \ge 1\}$.

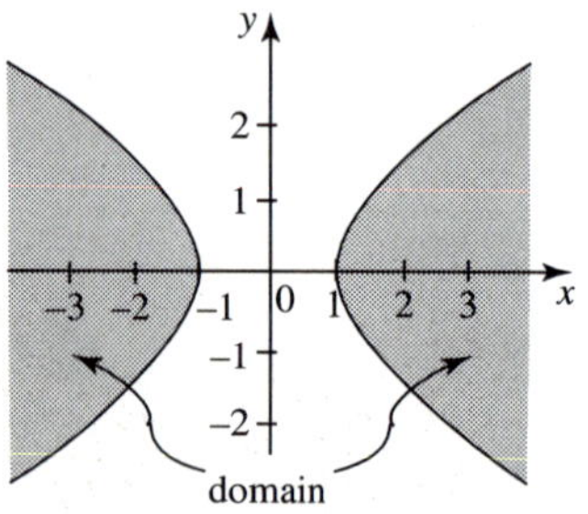

Figure 2.1.25

2.2 Graph of a Function of Several Variables

1. There are no points such that $f(x,y) = c \neq 1$, and $f(x,y) = c = 1$ holds for all (x,y). It follows that there are no level curves for $c \neq 1$ and the "level curve" (now the term *level set* comes in handy) of value $c = 1$ is the xy-plane.

3. From $f(x,y) = 3 - x^2 - y^2 = c$ it follows that $x^2 + y^2 = 3 - c$. Since $x^2 + y^2 \geq 0$, it follows that there are no level curves if $c > 3$. If $c = 3$, then $x^2 + y^2 = 0$ implies that $(x,y) = (0,0)$, i.e., the "level curve" of value $c = 3$ is the origin. If $c < 3$, then the level curve of value c is the circle, centered at the origin, of radius $\sqrt{3-c}$.

5. Since $e^{xy} > 0$, it follows that $f(x,y) = e^{xy}$ has no level curves for $c \leq 0$. If $c > 0$ then $e^{xy} = c$ implies $xy = \ln c$ and hence $y = (\ln c)/x$. If $c > 1$ then $\ln c > 0$ and $y = (\ln c)/x$ is a hyperbola with branches in the first and the third quadrants; if $c < 1$ then $\ln c < 0$ and $y = (\ln c)/x$ is a hyperbola with branches in the second and the fourth quadrants; if $c = 1$ then $y = (\ln 1)/x = 0$ is the x-axis.

7. From $f(x,y) = x/y = c$ it follows that $y = x/c$ and $y \neq 0$. Therefore, the level curve of value c (if $c \neq 0$) is the line of slope $1/c$ going through the origin (the origin is not included since $y \neq 0$). If $c = 0$ then $x/y = 0$ implies that $x = 0$. Consequently, the level curve of value c is the y-axis (again, without the origin).

9. The level curve of value c is defined by $f(x,y) = y - \sin x = c$; i.e., $y = \sin x + c$. It is obtained by moving the graph of $y = \sin x$ up c units if $c \geq 0$ or down $|c|$ units if $c < 0$.

11. The equation $x^2 - y^2 = c$, $c \neq 0$, represents a family of hyperbolas with asymptotes $y = \pm x$.

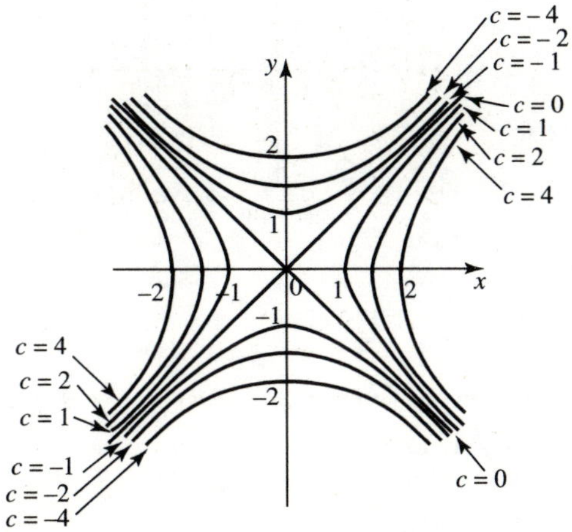

Figure 2.2.11

If $c > 0$, the x-intercepts are at $(\pm\sqrt{c}, 0)$ — therefore all hyperbolas with $c > 0$ are located in the regions to the left and to the right of the origin; see Figure 2.2.11. If $c < 0$, then there are no x-intercepts; the y-intercepts are at $(0, \pm\sqrt{-c})$. Consequently, the hyperbolas are located in the regions above and below the origin, as shown in Figure 2.2.11. If $c = 0$, then $x^2 - y^2 = 0$ implies that $y = \pm x$. The corresponding "level curve" consists of two lines, $y = x$ and $y = -x$, that intersect at the origin (they are also asymptotes of all hyperbolas).

13. From $f(x, y, z) = y - x^2 = c$ it follows that $y = x^2 + c$. In the xy-plane, this is the parabola $y = x^2$ moved up c units if $c \geq 0$ and down $|c|$ units if $c < 0$. Since there are no conditions on z, the surface $y = x^2 + c$ is a parabolic sheet obtained by glueing copies of $y = x^2 + c$ on top of each other (and below each other); see Figure 2.2.13. In other words, $y = x^2 + c$ is a surface all of whose horizontal cross-sections are the same curve, namely the parabola $y = x^2 + c$.

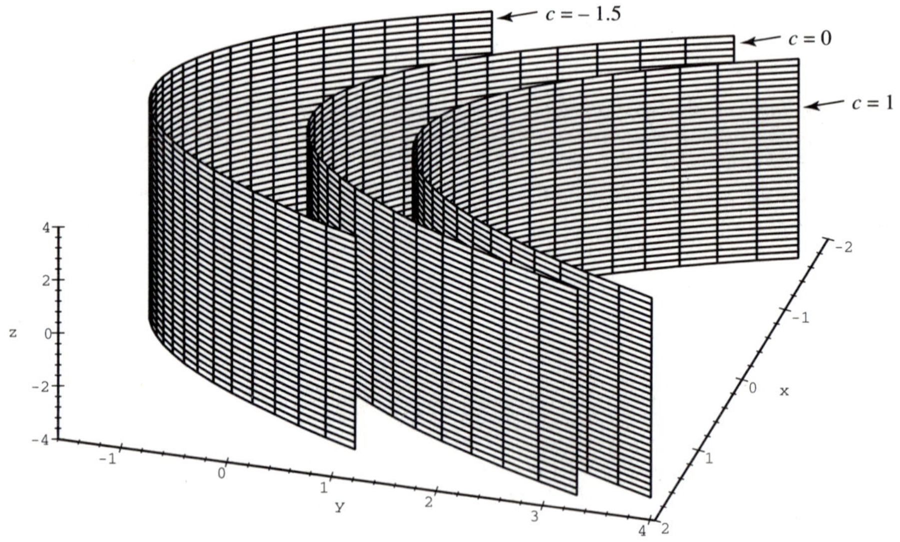

Figure 2.2.13

15. From $f(x, y, z) = c$ it follows that $4x^2 + 4y^2 + z^2 = c - 1$. There are no level surfaces for $c < 1$, since $4x^2 + 4y^2 + z^2 \geq 0$. The level surface of value $c = 1$ degenerates to a point (the origin). If $c > 1$, rewrite the equation as (divide by $c - 1$)

$$\frac{x^2}{(\sqrt{c-1}/2)^2} + \frac{y^2}{(\sqrt{c-1}/2)^2} + \frac{z^2}{(\sqrt{c-1})^2} = 1.$$

This equation represents the ellipsoid with semiaxes $\sqrt{c-1}/2$ (in the x-direction), $\sqrt{c-1}/2$ (in the y-direction) and $\sqrt{c-1}$ (in the z-direction), symmetric with respect to the origin (i.e., its center is at the origin).

17. From $3x + 4 = c$ it follows that $x = (c - 4)/3$. So the level surface of value c is the plane parallel to the yz-plane that crosses the x-axis at $((c - 4)/3, 0, 0)$.

19. From $T(x, y) = c$ it follows that $50/(1 + x^2 + 3y^2) = c$. The left side is always positive, so there are no isotherms for $c \le 0$. If $c > 0$, then

$$x^2 + 3y^2 = \frac{50}{c} - 1.$$

The isotherm of value $c = 50$ is the point $(0,0)$. There are no isotherms for $c > 50$, since in that case $50/c - 1 < 0$. If $0 < c < 50$ then $50/c - 1 > 0$. Let $d = 50/c - 1$, divide the above equation by d and simplify to get

$$\frac{x^2}{d} + \frac{y^2}{(d/3)} = 1.$$

This equation represents the ellipse with semi-axes $d^{1/2} = \left(\frac{50}{c} - 1\right)^{1/2}$ (in the x-direction) and $(d/3)^{1/2} = \left(\frac{50}{3c} - \frac{1}{3}\right)^{1/2}$ (in the y-direction).

21. (a) In two dimensions, the equation $x = a$ represents the line, parallel to the y-axis, going through $(a, 0)$. In three dimensions, it describes the plane, parallel to the yz-plane, going through $(a, 0, 0)$.

(b) In two dimensions, $y = b$ represents the line parallel to the x-axis, going through $(0, b)$. In $\mathbb{R}^3$, it is the plane, parallel to the xz-plane, going through $(0, b, 0)$.

23. (a) Since $x^2 + y^2 + z^2$ is the distance squared from the origin, the equation $x^2 + y^2 + z^2 = a^2$ represents all points whose distance from the origin is a. It is a sphere, centered at the origin, of radius a.

(b) This equation states that (compute the square root of both sides) the distance from a point (x, y, z) to the fixed point (m, n, p) is a. It represents a sphere, centered at (m, n, p), of radius a.

25. Consider the level curves $z = \sqrt{2 - x^2 - y^2} = c$ first. If $c < 0$, there are no level curves. Since $2 - x^2 - y^2 = 2 - (x^2 + y^2) \le 2$, it follows that $z = \sqrt{2 - x^2 - y^2} \le \sqrt{2}$, and so there are no level curves for $c > \sqrt{2}$ either. If $c = \sqrt{2}$, the "level curve" is the origin. If $0 \le c < \sqrt{2}$, then $\sqrt{2 - x^2 - y^2} = c$ implies that $x^2 + y^2 = 2 - c^2$ — consequently, the level curve of value c is a circle of radius $\sqrt{2 - c^2}$.

The trace in the xz-plane is (take $y = 0$) $z = \sqrt{2 - x^2}$; i.e., $x^2 + z^2 = 2$ and $z \ge 0$. It is the upper semi-circle of radius $\sqrt{2}$ in the xz-plane. Similarly, the trace in the yz-plane is (let $x = 0$) $z = \sqrt{2 - y^2}$; it is the upper semi-circle of radius $\sqrt{2}$ in the yz-plane. In words, the surface under investigation is made of circles, smaller ones placed on top of larger circles, from $z = 0$ to $z = \sqrt{2}$. The traces in the xz-plane and in the yz-plane are semi-circles — therefore, the surface in question is the upper hemisphere; (i.e., the part of the sphere centered at the origin of radius $\sqrt{2}$ that lies above and in the xy-plane).

27. The level curve of value c is the circle $x^2 + y^2 = 9$, no matter what c is. The trace in the xz-plane is (substitute $y = 0$) $x^2 = 9$; i.e., $x = \pm 3$ (two vertical lines in the xz-plane). The trace in the yz-plane is $y^2 = 9$; i.e., $y = \pm 3$ (two vertical lines in the yz-plane). The surface is built of circles (of radius 3) stacked vertically (since the traces are vertical lines) on top of each other. It is a cylinder of radius 3, whose axis of (rotational) symmetry is the z-axis.

29. The equation $z = 4$ represents all points whose z-coordinates are 4. It is the plane, parallel to the xy-plane, 4 units above it. The trace in the xz-plane is the horizontal line $z = 4$ and the trace in the yz-plane is the horizontal line $z = 4$.

31. There are no level curves for $c > 2$, since $z = 2 - x^2 - y^2 \leq 2$. The level curve of value $c = 2$ is the point $(0, 0)$. For $c < 2$ the level curve is $2 - x^2 - y^2 = c$, or $x^2 + y^2 = 2 - c$ — it is a circle of radius $\sqrt{2 - c}$. The trace in the xz-plane is the parabola $z = 2 - x^2$, and the trace in the yz-plane is the parabola $z = 2 - y^2$. It follows that the surface is built of circles that decrease in size as the height increases. The circles "decrease" in such a way that the vertical cross-sections (i.e., the traces) are parabolas. The surface is a paraboloid, symmetric with respect to the z-axis, turned upside-down, with the vertex at $(0, 0, 2)$; see Figure 2.2.31.

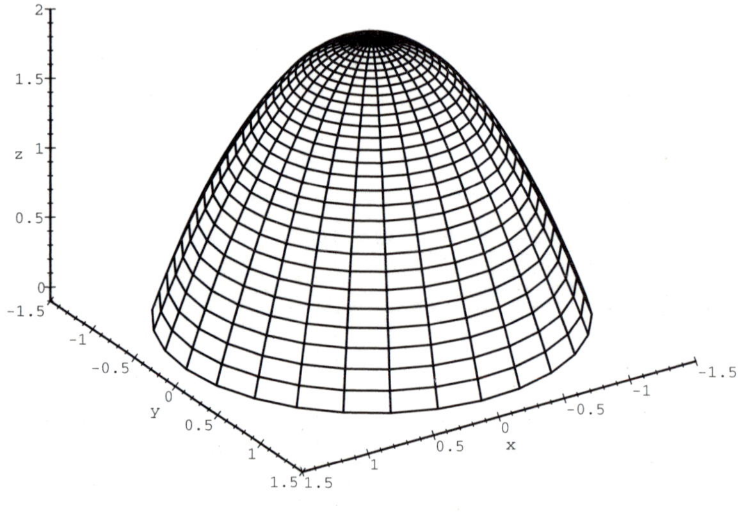

Figure 2.2.31

33. Since $z = \sqrt{x^2 + y^2}$, it follows that the surface in question lies above the xy-plane; it touches it at $(0, 0, 0)$ (since the "level curve" of value $c = 0$ satisfies $\sqrt{x^2 + y^2} = 0$; i.e., $x = 0$ and $y = 0$). If $c > 0$ then $\sqrt{x^2 + y^2} = c$ implies $x^2 + y^2 = c^2$; so the level curve of value c is a circle of radius c. The trace in the xz-plane (substitute $y = 0$) is $z = \sqrt{x^2} = |x|$; i.e., consists of those parts of the lines $z = \pm x$ that lie above the xy-plane. Similarly, the trace in the yz-plane (substitute $x = 0$) is $z = |y|$.

It follows that the surface in question is built of circles that increase in size as the height increases. Since the traces are lines, it is a cone; see Figure 2.2.33.

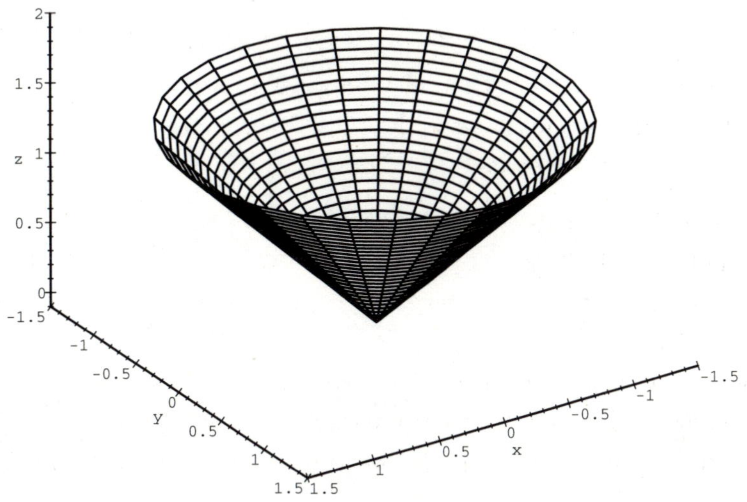

Figure 2.2.33

35. $\mathbf{F}(\mathbf{x}) = \mathbf{x}/\|\mathbf{x}\|$ is the unit vector in the direction of the vector $\mathbf{x}$ (i.e., in the radial direction away from the origin); see Figure 2.2.35.

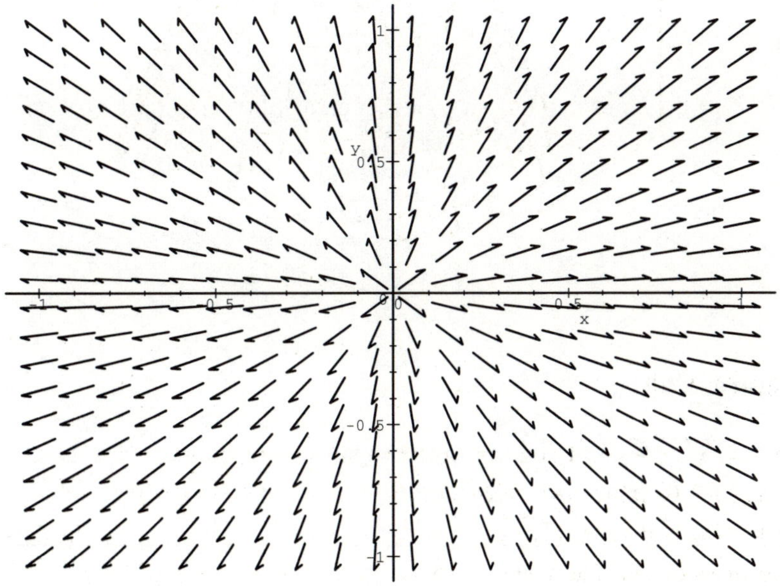

Figure 2.2.35

37. $\mathbf{F}(\mathbf{x})$ is a unit vector from $\mathbf{x}_0$ to $\mathbf{x}$, see Figure 2.2.37. It is the vector field of Exercise 35, translated by the vector $\mathbf{i} + 2\mathbf{j}$, so that the origin gets moved to $(1, 2)$.

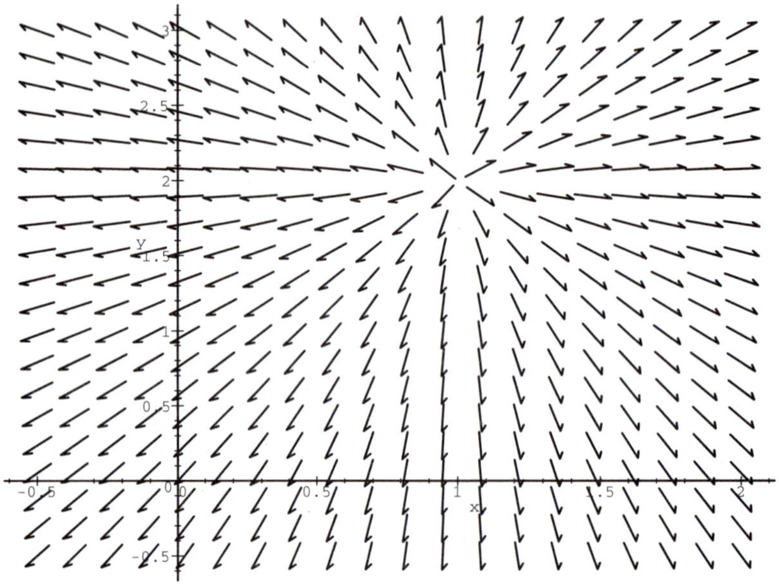

Figure 2.2.37

39. Since $\mathbf{F}(x, y)$ does not depend on y, its value at all points on a vertical line is the same. See Figure 2.2.39.

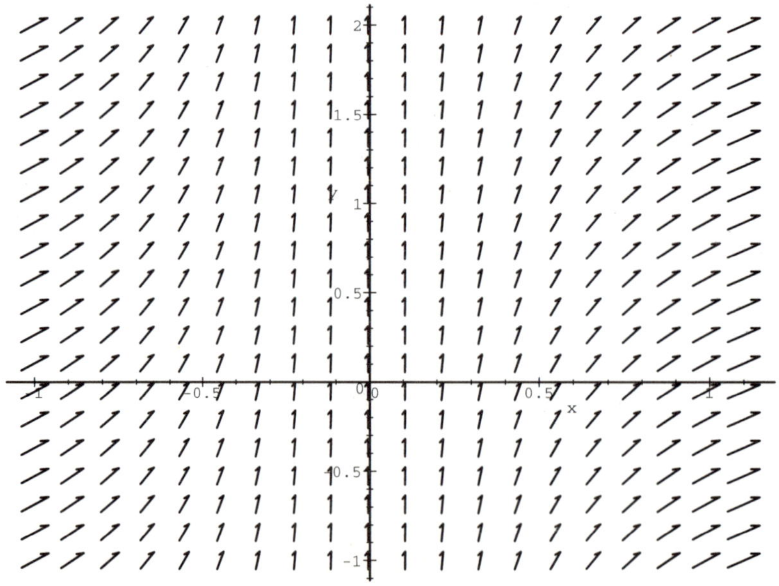

Figure 2.2.39

41. In words, the vector field $\mathbf{F}$ assigns to a vector $x\mathbf{i} + y\mathbf{j}$ the vector $x\mathbf{i} - y\mathbf{j}$, which is its mirror image with respect to the x-axis. See Figure 2.2.41.

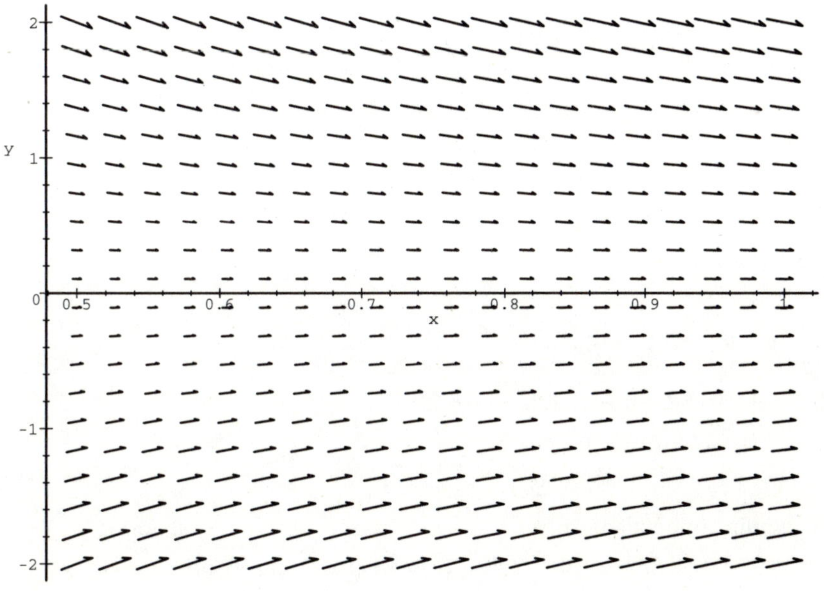

Figure 2.2.41

2.3 Limits and Continuity

1. From $|f(x,y) - 1| < 0.01$ it follows that

$$|e^{-(x^2+y^2)} - 1| < 0.01.$$

Now $-(x^2 + y^2) \leq 0$ implies that $e^{-(x^2+y^2)} \leq 1$; i.e., the quantity inside the absolute value is negative. Therefore, from $|e^{-(x^2+y^2)} - 1| < 0.01$ we get $-(e^{-(x^2+y^2)} - 1) < 0.01$ or $e^{-(x^2+y^2)} > 0.99$. So $-(x^2 + y^2) > \ln 0.99$ and $x^2 + y^2 < -\ln 0.99 \approx 0.01005$. Finally,

$$\sqrt{x^2 + y^2} < \sqrt{-\ln 0.99} \approx 0.10025.$$

It follows that (x, y) must belong to a ball $B((0,0), r)$, with $r < \sqrt{-\ln 0.99} \approx 0.10025$, in order to satisfy $|f(x,y) - 1| < 0.01$.

3. See Figure 2.3.3.

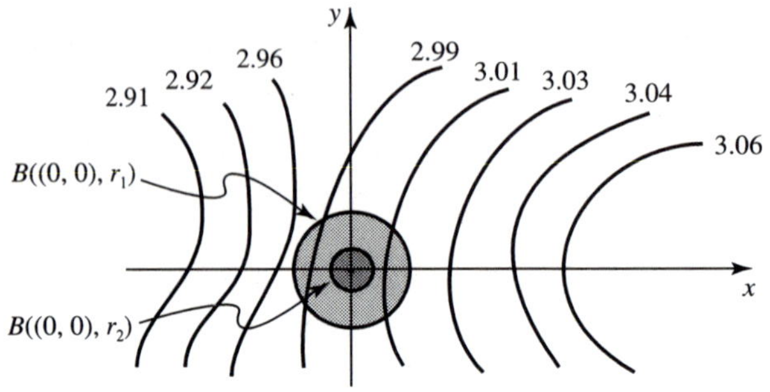

Figure 2.3.3

5. $\lim_{(x,y)\to(0,0)} (1 - y - e^{-x^2-y^2}\cos x) = 1 - 0 - e^0 \cos 0 = 0.$

7. Using the substitution $u = x + y$, we get

$$\lim_{(x,y)\to(0,0)} (x+y)e^{-1/(x+y)} = \lim_{u\to 0} ue^{-1/u}.$$

We have to examine one-sided limits. If $u \to 0^+$, then $-1/u \to -\infty$ and $e^{-1/u} \to 0$. Therefore, $\lim_{u\to 0^+} ue^{-1/u} = 0$. To compute the left limit, use L'Hôpital's Rule:

$$\lim_{u\to 0^-} ue^{-1/u} = \lim_{u\to 0^-} \frac{e^{-1/u}}{1/u} = \lim_{u\to 0^-} \frac{e^{-1/u}(1/u^2)}{-1/u^2} = \lim_{u\to 0^-} -e^{-1/u} = -e^{+\infty} = -\infty.$$

Consequently, $\lim_{u\to 0} ue^{-1/u}$ does not exist.

9. Using polar coordinates, we get

$$\lim_{(x,y)\to(0,0)} \frac{x^2 y}{x^2+y^2} = \lim_{r\to 0} \frac{r^2\cos^2\theta\, r\sin\theta}{r^2} = \lim_{r\to 0} r\cos^2\theta\sin\theta = 0,$$

no matter what θ is.

Alternatively,

$$0 \le \left|\frac{x^2 y}{x^2+y^2} - 0\right| = \frac{x^2|y|}{x^2+y^2} \le \frac{(x^2+y^2)\,|y|}{x^2+y^2} = |y|$$

(the following identities/inequalities were used: $|x^2 y| = |x^2|\cdot|y| = x^2|y|$ and $x^2 \le x^2+y^2$). As $(x,y)\to(0,0)$, $|y|\to 0$, and therefore

$$\lim_{(x,y)\to(0,0)} \frac{x^2 y}{x^2+y^2} = 0.$$

11. Check the approach along the lines $y = mx$:

$$\lim_{(x,y)\to(0,0)} \frac{2x^2-y^2}{x^2+2y^2} = \lim_{x\to 0} \frac{2x^2-m^2x^2}{x^2+2m^2x^2} = \lim_{x\to 0} \frac{2-m^2}{1+2m^2} = \frac{2-m^2}{1+2m^2}.$$

Since the value depends on m (and m is the slope of a line along which we approach the origin) it follows that different approaches give different results. So the limit does not exist.

13. Check the approach along the lines $y = mx$:

$$\lim_{(x,y)\to(0,0)} \frac{2xy}{2x^2+y^2} = \lim_{x\to 0} \frac{2x^2m}{2x^2+m^2x^2} = \lim_{x\to 0} \frac{2m}{2+m^2} = \frac{2m}{2+m^2}.$$

Since the value depends on m (m is the slope of a line along which we approach the origin), the limit does not exist.

15. If $y = mx$ (i.e., we approach the origin along straight lines other than the y-axis), then

$$\lim_{(x,y)\to(0,0)} \frac{x^3y}{x^6+y^2} = \lim_{x\to 0} \frac{x^3mx}{x^6+m^2x^2} = \lim_{x\to 0} \frac{mx^2}{x^4+m^2} = 0,$$

assuming that $m \neq 0$. If $m = 0$, then $y = 0$, and

$$\lim_{(x,y)\to(0,0)} \frac{x^3y}{x^6+y^2} = \lim_{x\to 0} \frac{0}{x^6} = 0.$$

When $x = 0$,

$$\lim_{(x,y)\to(0,0)} \frac{x^3y}{x^6+y^2} = \lim_{y\to 0} \frac{0}{y^2} = 0.$$

This still does not prove anything. As a matter of fact, this limit does not exist, since the approach $y = x^3$ gives

$$\lim_{(x,y)\to(0,0)} \frac{x^3y}{x^6+y^2} = \lim_{(x,y)\to(0,0)} \frac{x^6}{x^6+x^6} = \tfrac{1}{2}.$$

17. There is no difference between analyzing limits of functions of two, three, or more variables. If we use the approach $y = 0$, $z = 0$, we get

$$\lim_{(x,y,z)\to(0,0,0)} \frac{xyz}{x^3+y^3+z^3} = \lim_{x\to 0} \frac{0}{x^3} = 0.$$

However, along $x = y = z$,

$$\lim_{(x,y,z)\to(0,0,0)} \frac{xyz}{x^3+y^3+z^3} = \lim_{x\to 0} \frac{x^3}{3x^3} = \tfrac{1}{3}.$$

Therefore, the limit in question does not exist.

19. $\mathbf{F}$ is a rational function (i.e., a fraction), and is therefore continuous at all points where the denominator is not equal to zero. Since $x^2 + y^2 + z^2 = 0$ implies that $x = 0$, $y = 0$ and $z = 0$, it follows that $\mathbf{F}$ is not continuous only at $(0, 0, 0)$.

21. Both the numerator and the denominator are continuous functions for all x and all y. Since $\sin x \leq 1$ and $\cos x \leq 1$ it follows that $\sin x \cos x \leq 1$ and hence the denominator

$3 - \sin x \cos x$ is never equal to zero. So the function $f(x, y)$ is continuous for all $x, y \in \mathbb{R}$.

23. Both the numerator and the denominator are continuous functions for all $\mathbf{x} \in \mathbb{R}^m$. If $\|\mathbf{x} - \mathbf{x}_0\| = 0$, then $\mathbf{x} = \mathbf{x}_0$; i.e., f is not continuous at $\mathbf{x} = \mathbf{x}_0$ (and is continuous if $\mathbf{x} \neq \mathbf{x}_0$).

25. The limit of f does not exist, since along $y = 0$

$$\lim_{(x,y)\to(0,0)} \frac{xy^3}{x^2 + y^6} = \lim_{x\to 0} \frac{0}{x^2} = 0,$$

but along $x = y^3$

$$\lim_{(x,y)\to(0,0)} \frac{xy^3}{x^2 + y^6} = \lim_{y\to 0} \frac{y^6}{2y^6} = \tfrac{1}{2}.$$

Therefore, it is impossible to define $f(0,0)$ so as to make f continuous at $(0,0)$.

27. The limit of f does not exist, since along $y = 0$,

$$\lim_{(x,y)\to(0,0)} \frac{y^2 + 3|y|}{|x| + 3|y|} = \lim_{x\to 0} \frac{0}{|x|} = 0$$

but along $y = x$, $x > 0$,

$$\lim_{(x,y)\to(0,0)} \frac{y^2 + 3|y|}{|x| + 3|y|} = \lim_{x\to 0} \frac{x^2 + 3x}{4x} = \tfrac{3}{4}.$$

Therefore, it is not possible to assign a numerical value to $f(0,0)$ so as to make f continuous at $(0,0)$.

29. The condition $xy \neq 0$ implies that $x \neq 0$ and $y \neq 0$. Hence U is obtained from $\mathbb{R}^2$ by removing the x-axis and the y-axis. All points in U are interior points. The boundary of U consists of the x-axis and the y-axis.

31. Let $\mathbf{x} = (x_1, \ldots, x_m)$ and $\mathbf{a} = (a_1, \ldots, a_m)$. The dot product function $f(\mathbf{x}) = \mathbf{x} \cdot \mathbf{a} = x_1 a_1 + \ldots + x_m a_m$ is a first-degree (i.e., linear) polynomial in $x_1, \ldots, x_m$ (with coefficients $a_1, \ldots, a_m$), and therefore continuous.

33. Since $\lim_{x\to 0} (\sin x / x) = 1$, it follows that

$$\lim_{(x,y)\to(0,2)} \mathbf{F}(x, y) = \lim_{(x,y)\to(0,2)} \left(\frac{y \sin x}{x}, ye^x \right) = (2, 2).$$

So, if we define $\mathbf{F}(0,2) = (2,2)$, we will make $\mathbf{F}$ continuous at $(0,2)$.

35. From $\lim_{y\to 0} (\cos y - 1)/y = 0$, it follows that

$$\lim_{(x,y)\to(1,0)} \mathbf{F}(x, y) = \lim_{(x,y)\to(1,0)} \left(\sin(x + y), \frac{\cos y - 1}{xy}, e^{xy} \right) = (\sin 1, 0, 1).$$

Therefore, defining $\mathbf{F}(1,0) = (\sin 1, 0, 1)$ will make $\mathbf{F}$ continuous at $(1,0)$.

2.4 Derivatives

1. The set U is the region in $\mathbb{R}^2$ between the circles of radii $\sqrt{2}$ and $\sqrt{3}$ centered at the origin. Since U does not contain the circles, all points in U are interior points. Consequently, U is open in $\mathbb{R}^2$.

3. U is the set of all points that belong to the line $x+y=2$. No ball centered at a point on that line is completely contained in the line, and therefore U is not open in $\mathbb{R}^2$.

5. The set U consists of the four octants where either all of the coordinates x, y and z are positive, or two are negative and the third one is positive. Since the possibility that the product xyz is zero is not allowed, the three coordinate planes do not belong to U. In other words, all points in U are interior points, and therefore, U is open in $\mathbb{R}^3$.

7. Keeping y fixed, we get $f_x = yx^{y-1} + y/x$; keeping x fixed, $f_y = x^y \ln x + \ln x$.

9. By the chain rule,

$$f_x = \frac{1}{x+y+z^2} \cdot 1 = \frac{1}{x+y+z^2}.$$

Similarly,

$$f_z = \frac{1}{x+y+z^2} \cdot 2z = \frac{2z}{x+y+z^2}.$$

11. Using the quotient rule,

$$f_x = \frac{3x^2(x^2-y^2) - 2x(x^3+y^3)}{(x^2-y^2)^2} = \frac{x^4 - 3x^2y^2 - 2xy^3}{(x^2-y^2)^2}.$$

Similarly,

$$f_y = \frac{3y^2(x^2-y^2) - (-2y)(x^3+y^3)}{(x^2-y^2)^2} = \frac{3x^2y^2 + 2yx^3 - y^4}{(x^2-y^2)^2}.$$

13. Using the product rule and the quotient rule, we get

$$f_x = e^{xy} y \cos x \sin y + e^{xy}(-\sin x) \sin y = e^{xy} \sin y (y \cos x - \sin x)$$

and

$$f_y = e^{xy} x \cos x \sin y + e^{xy} \cos x \cos y = e^{xy} \cos x (x \sin y + \cos y).$$

15. Since $f(x_1, \ldots, x_m) = (x_1^2 + \cdots + x_m^2)^{1/2}$, it follows that

$$\frac{\partial f}{\partial x_i} = \tfrac{1}{2}(x_1^2 + \cdots + x_m^2)^{-1/2}\, 2x_i = \frac{x_i}{\sqrt{x_1^2 + \cdots + x_m^2}},$$

for $i = 1, \ldots, m$.

17. By the Fundamental Theorem of Calculus,

$$f_x = \frac{\partial}{\partial x}\left(\int_0^x te^{-t^2}\, dt \right) = xe^{-x^2}.$$

Since $\int_0^x te^{-t^2}\,dt$ is a function of x only, it follows that $f_y = 0$.

19. Using the "prime" notation (′) for the derivatives of f and g, we get $z_x = f'(x)+0 = f'(x)$, since $g(y)$ is constant when x is viewed as a variable. Similarly, $z_y = 0 + g'(y) = g'(y)$.

21. Keeping y constant (so that $g(y)$ is constant as well), we get $z_x = f'(x)/g(y)$. Keeping x fixed and thinking of z as $z = f(x)g(y)^{-1}$, we get

$$z_y = f(x)(-1)g(y)^{-2}g'(y) = -\frac{f(x)g'(y)}{g(y)^2}.$$

23. We have to compute the rates of change of z with respect to x and y at the given point. From

$$z_x(2,1,11) = -2(x-3)\Big|_{(2,1,11)} = 2$$

and

$$z_y(2,1,11) = -8(y-2)^3\Big|_{(2,1,11)} = 8$$

it follows that the hill is steeper in the y (i.e., northern) direction.

25. The function in question is $f(x,y) = -xe^{-x^2-2y^2}$.

(a) By the chain rule,

$$f_y(2,3) = -xe^{-x^2-2y^2}(-4y)\Big|_{(2,3)} = 4xye^{-x^2-2y^2}\Big|_{(2,3)} = 24e^{-22}.$$

(b) Substituting $x = 2$ into the formula for $z = f(x,y)$, we get the equation $z = -2e^{-4-2y^2}$ of the curve that is the intersection of the graph of f and $x = 2$. Its slope at $y = 3$ is

$$z'(3) = 8ye^{-4-2y^2}\Big|_{y=3} = 24e^{-22}.$$

(c) The partial derivative of $f(x,y)$ with respect to y at $(2,3)$ is the slope of the tangent line (at the point where $y = 3$) to the curve $z = f(2,y)$ that is the intersection of the graph of $z = f(x,y)$ and the vertical plane $x = 2$.

27. The derivative of $\mathbf{F}$ at $(0,0)$ is the 3×2 matrix

$$D\mathbf{F}(0,0) = \begin{bmatrix} 0 & 1 \\ 1 & 0 \\ 0 & 0 \end{bmatrix}_{at\ (0,0)} = \begin{bmatrix} 0 & 1 \\ 1 & 0 \\ 0 & 0 \end{bmatrix}.$$

29. The derivative of $\mathbf{F}$ at $(1,1,0)$ is the 2×3 matrix

$$D\mathbf{F}(1,1,0) = \begin{bmatrix} 2x/(x^2+y^2+z^2) & 2y/(x^2+y^2+z^2) & 2z/(x^2+y^2+z^2) \\ 2y & 2x & 1 \end{bmatrix}_{at\ (1,1,0)}$$
$$= \begin{bmatrix} 1 & 1 & 0 \\ 2 & 2 & 1 \end{bmatrix}.$$

31. Since $f(x, y, z) = x^2 + y^2 + z^2$, it follows that

$$Df = [\,2x \quad 2y \quad 2z\,]_{(a_1,a_2,a_3)} = [\,2a_1 \quad 2a_2 \quad 2a_3\,].$$

33. By definition, the gradient of f at $(0,0)$ is

$$\nabla f(0,0) = Df(0,0) = [\,f_x \quad f_y\,]_{at\ (0,0)}\,.$$

Since

$$f_x(0,0) = \cos(x^2+y^2) - (x-y)\cdot 2x\sin(x^2+y^2)\Big|_{(0,0)} = 1$$

and

$$f_y(0,0) = -\cos(x^2+y^2) - (x-y)\cdot 2y\sin(x^2+y^2)\Big|_{(0,0)} = -1,$$

it follows that $\nabla f(0,0) = [\,1 \quad -1\,]$.

35. Let $\mathbf{x} = (x, y, z)$. Then $f(\mathbf{x}) = \sqrt{x^2+y^2+z^2}$ and (by the chain rule)

$$\nabla f(\mathbf{x}) = Df(\mathbf{x}) = \left[\,\frac{x}{\sqrt{x^2+y^2+z^2}} \quad \frac{y}{\sqrt{x^2+y^2+z^2}} \quad \frac{z}{\sqrt{x^2+y^2+z^2}}\,\right].$$

Hence (interpreting $\nabla f(\mathbf{x})$ as a vector, rather than a matrix),

$$\nabla f(\mathbf{x}) = \frac{1}{\sqrt{x^2+y^2+z^2}}[\,x \quad y \quad z\,] = \mathbf{x}/\|\mathbf{x}\|.$$

The domain of ∇f is $\mathbb{R}^3 - \{(0,0,0)\}$.

37. From $f(\mathbf{a}) = f(2,-1) = \ln 4$,

$$f_x(2,-1) = \frac{3}{3x+2y}\Bigg|_{(2,-1)} = \tfrac{3}{4} \quad \text{and} \quad f_y(2,-1) = \frac{2}{3x+2y}\Bigg|_{(2,-1)} = \tfrac{1}{2}$$

we get

$$L_{(2,-1)}(x,y) = \ln 4 + \tfrac{3}{4}(x-2) + \tfrac{1}{2}(y+1).$$

39. From $f(3,2) = 8$,

$$f_x(3,2) = 2x - y\Big|_{(3,2)} = 4 \quad \text{and} \quad f_y(3,2) = -x + y\Big|_{(3,2)} = -1,$$

we get

$$L_{(3,2)}(x,y) = 8 + 4(x-3) - (y-2) = 4x - y - 2.$$

41. From $f(\mathbf{a}) = f(0,1,1) = \sqrt{2}$,

$$f_x(0,1,1) = \frac{x}{\sqrt{x^2+y^2+z^2}}\Bigg|_{(0,1,1)} = 0,$$

$$f_y(0,1,1) = \frac{y}{\sqrt{x^2+y^2+z^2}}\Bigg|_{(0,1,1)} = \frac{1}{\sqrt{2}}$$

and

$$f_z(0,1,1) = \frac{z}{\sqrt{x^2+y^2+z^2}}\bigg|_{(0,1,1)} = \frac{1}{\sqrt{2}}$$

we get

$$L_{(0,1,1)}(x,y,z) = \sqrt{2} + \frac{1}{\sqrt{2}}(y-1) + \frac{1}{\sqrt{2}}(z-1) = \frac{y+z}{\sqrt{2}}.$$

43. Let $f(x,y) = xy(x+y)^{-1}$. In order to verify the given approximation we use the fact that $f(x,y) \approx L_{(2,3)}(x,y)$ for (x,y) close to $(2,3)$. From $f(2,3) = 6/5$,

$$f_x(2,3) = \frac{y(x+y)-xy}{(x+y)^2}\bigg|_{(2,3)} = \frac{y^2}{(x+y)^2}\bigg|_{(2,3)} = \frac{9}{25}$$

and

$$f_y(2,3) = \frac{x(x+y)-xy}{(x+y)^2}\bigg|_{(2,3)} = \frac{x^2}{(x+y)^2}\bigg|_{(2,3)} = \frac{4}{25}$$

it follows that

$$\frac{xy}{x+y} = f(x,y) \approx L_{(2,3)}(x,y) = \tfrac{6}{5} + \tfrac{9}{25}(x-2) + \tfrac{4}{25}(y-3).$$

45. Let $f(x,y) = \sqrt{x^3+y^3}$; we are asked to approximate $f(0.99, 2.02)$. Consider the linear approximation of f at $(1,2)$ (notice that $(1,2)$ is near $(0.99, 2.02)$). From $f(1,2) = \sqrt{9} = 3$,

$$f_x(1,2) = \frac{3x^2}{2\sqrt{x^3+y^3}}\bigg|_{(1,2)} = \tfrac{1}{2}$$

and

$$f_x(1,2) = \frac{3y^2}{2\sqrt{x^3+y^3}}\bigg|_{(1,2)} = 2,$$

it follows that

$$L_{(1,2)}(x,y) = 3 + \tfrac{1}{2}(x-1) + 2(y-2).$$

Recall that "linear approximation" means that $f(x,y) \approx L_{(1,2)}(x,y)$ for points *near* $(1,2)$. In particular,

$$f(0.99, 2.02) \approx L_{(1,2)}(0.99, 2.02) = 3 + \tfrac{1}{2}(-0.01) + 2(0.02) = 3.035.$$

The calculator value is 3.0352441 (last digit(s) could differ, depending on the precision of the calculator).

47. Let $f(x,y) = x\ln y$. We will approximate $f(7.95, 1.02)$ using the linear approximation of f at $(8,1)$. From $f(8,1) = 8\ln 1 = 0$, $f_x(8,1) = \ln y\big|_{(8,1)} = 0$ and $f_y(8,1) = (x/y)\big|_{(8,1)} = 8$ it follows that $L_{(8,1)}(x,y) = 8(y-1)$.

Since $f(x,y) \approx L_{(8,1)}(x,y)$ for (x,y) near $(8,1)$, it follows that

$$f(7.95, 1.02) = 7.95\ln 1.02 \approx L_{(8,1)}(7.95, 1.02) = 8(1.02-1) = 0.16.$$

Therefore, $7.95 \ln 1.02 \approx 0.16$. The calculator value is 0.1574309.

49. Let $f(x,y) = \int_x^y e^{-t^2} dt$; then $\int_{0.995}^{1.02} e^{-t^2} dt = f(0.995, 1.02)$. We will use the fact that $f(0.995, 1.02) \approx L_{(1,1)}(0.995, 1.02)$, where $L_{(1,1)}(x,y)$ is the linear approximation of $f(x,y)$ at $(1,1)$. So we have to compute $L_{(1,1)}(x,y)$ first.

The value of f at $(1,1)$ is $f(1,1) = \int_1^1 e^{-t^2} dt = 0$. From the Fundamental Theorem of Calculus we get

$$f_x = \frac{\partial}{\partial x}\left(\int_x^y e^{-t^2}\, dt\right) = \frac{\partial}{\partial x}\left(-\int_y^x e^{-t^2}\, dt\right) = -e^{-x^2}, \qquad \text{hence} \qquad f_x(1,1) = -e^{-1}.$$

Similarly,

$$f_y = \frac{\partial}{\partial y}\left(\int_x^y e^{-t^2}\, dt\right) = e^{-y^2}, \qquad \text{and hence} \qquad f_y(1,1) = e^{-1}.$$

It follows that

$$L_{(1,1)}(x,y) = -e^{-1}(x-1) + e^{-1}(y-1) = e^{-1}(y-x).$$

Therefore, $f(0.995, 1.02) \approx L_{(1,1)}(0.995, 1.02) = 0.025e^{-1}$. In order to check this approximation, one has to use a numerical integration method.

51. From $\Delta f = f(a + \Delta x, b + \Delta y) - f(a,b)$, with $a = 0$, $b = 1$, $\Delta x = 0.3$ and $\Delta y = 0.01$ we get

$$\Delta f = f(0.3, 1.01) - f(0,1) = e^{0.3} - 1.01e^{1.01} - (1 - e) = 0.2950836.$$

On the other hand, from $df = f_x(a,b)\Delta x + f_y(a,b)\Delta y$ we get

$$df = e^x\Big|_{(0,1)}(0.3) - (e^y + ye^y)\Big|_{(0,1)}(0.01) = 0.3 - 0.02e \approx 0.2456344.$$

Comparing the two, we see that $\Delta f \approx df$.

53. From

$$\Delta f = f(a + \Delta x, b + \Delta y, c + \Delta z) - f(a,b,c) = f(0.98, 2.01, -0.98) - f(1,2,-1)$$

we get $\Delta f \approx 2.919616 - 3 = -0.080384$. On the other hand, from $df = f_x(a,b,c)\Delta x + f_y(a,b,c)\Delta y + f_z(a,b,c)\Delta z$ we get

$$\begin{aligned} df &= (2xy - yz)\Big|_{(1,2,-1)}(-0.02) + (x^2 - xz)\Big|_{(1,2,-1)}(0.01) - (-xy + 3z^2)\Big|_{(1,2,-1)}(0.02) \\ &= 6(-0.02) + 2(0.01) + 1(0.02) \approx -0.08. \end{aligned}$$

Comparing the results, we see that $\Delta f \approx df$.

55. The volume of a cylinder is given by $V(r,h) = \pi r^2 h$, where r is the radius and h the height. The error ΔV is equal to the difference of the measured volume and the exact volume. Therefore, $\Delta V = V(r + \Delta r, h + \Delta h) - V(r,h)$. Since ΔV is approximately equal to the differential dV, it follows that

$$\Delta V \approx dV = 2\pi r h \Delta r + \pi r^2 \Delta h.$$

The error in h is 1.5%, hence $\Delta h = 0.015h$. Similarly, $\Delta r = 0.015r$ and so

$$\Delta V \approx 2\pi r h(0.015r) + \pi r^2(0.015h) = 0.045\pi r^2 h = 0.045V.$$

Consequently, the error in computing the volume is approximately 4.5%.

57. The area of a rectangle is $A = \ell w$, where ℓ is its length and w is its width. From $\Delta A \approx dA$ it follows that

$$\Delta A \approx A_\ell \Delta \ell + A_w \Delta w = w\Delta \ell + \ell \Delta w.$$

The error in ℓ is 2%, hence $\Delta \ell = 0.02\ell$; the error in w is 3%, hence $\Delta w = 0.03w$. Therefore

$$\Delta A \approx w(0.02\ell) + \ell(0.03w) = 0.05w\ell = 0.05A.$$

It follows that the error in computing the area is approximately 5%.

59. Since

$$R(R_1, R_2, R_3) = \left(\frac{1}{R_1} + \frac{1}{R_2} + \frac{1}{R_3}\right)^{-1} = \frac{R_1R_2R_3}{R_2R_3 + R_1R_3 + R_1R_2},$$

it follows that

$$\frac{\partial R}{\partial R_1} = \frac{R_2R_3(R_2R_3 + R_1R_3 + R_1R_2) - R_1R_2R_3(R_2 + R_3)}{(R_2R_3 + R_1R_3 + R_1R_2)^2} = \frac{R_2^2R_3^2}{(R_2R_3 + R_1R_3 + R_1R_2)^2}.$$

By symmetry,

$$\frac{\partial R}{\partial R_2} = \frac{R_1^2R_3^2}{(R_2R_3 + R_1R_3 + R_1R_2)^2} \quad \text{and} \quad \frac{\partial R}{\partial R_3} = \frac{R_1^2R_2^2}{(R_2R_3 + R_1R_3 + R_1R_2)^2}.$$

We need the values of these partial derivatives when $R_1 = 10$, $R_2 = 10$ and $R_3 = 20$;

$$\frac{\partial R}{\partial R_1}(10, 10, 20) = \frac{(100)(400)}{(200 + 200 + 100)^2} = \frac{40000}{250000} = \frac{4}{25}.$$

Similarly,

$$\frac{\partial R}{\partial R_2}(10, 10, 20) = \frac{4}{25} \quad \text{and} \quad \frac{\partial R}{\partial R_3}(10, 10, 20) = \frac{1}{25}.$$

It follows that

$$\Delta R \approx dR = \tfrac{4}{25}(1) + \tfrac{4}{25}(2) + \tfrac{1}{25}(-1) = \tfrac{11}{25} = 0.44.$$

The calculator value is

$$R(11, 12, 19) - R(10, 10, 20) = 4.4077329 - 4 = 0.4077329$$

61. From $\mathbf{c}(t) = (t^2, 1/t)$ we get the velocity $\mathbf{c}'(t) = (2t, -1/t^2)$ and the acceleration $\mathbf{c}''(t) = (2, 2/t^3)$ at t.

At $t = 2$, $\mathbf{c}'(2) = (4, -1/4)$ and $\mathbf{c}''(2) = (2, 1/4)$. Similarly, $\mathbf{c}'(1) = (2, -1)$, $\mathbf{c}''(1) = (2, 2)$, $\mathbf{c}'(1/10) = (1/5, -100)$ and $\mathbf{c}''(1/10) = (2, 2000)$.

Now $\|\mathbf{c}'(t)\| = \sqrt{4t^2 + t^{-4}}$. As $t \to 0$, $\|\mathbf{c}'(t)\| \to \infty$. Similarly, $\|\mathbf{c}''(t)\| = \sqrt{4 + 4t^{-6}}$, and $\|\mathbf{c}''(t)\| \to \infty$ as $t \to 0$.

63. We are actually asked to compute the linear approximation of $z = 6 - x^2 - y^2$ at $(1, 2)$. From $z(1, 2) = 1$, $z_x(1, 2) = -2x\Big|_{(1,2)} = -2$, and $z_y(1, 2) = -2y\Big|_{(1,2)} = -4$, we get

$$L_{(1,2)}(x, y) = 1 - 2(x - 1) - 4(y - 2) = -2x - 4y + 11;$$

so the equation of the tangent plane is $z = -2x - 4y + 11$.

65. The partial derivative f_x is given by $f_x(x, y) = 2xy/(x^2 + y^2)$ if $(x, y) \neq (0, 0)$. By definition,

$$f_x(0,0) = \lim_{h \to 0} \frac{f(h,0) - f(0,0)}{h} = \lim_{h \to 0} \frac{0}{h} = 0,$$

and hence f_x is defined for all (x, y). We will show that $\lim_{(x,y)\to(0,0)} f_x(x, y)$ does not exist, and therefore f_x is not continuous at $(0,0)$ (by definition, it should be equal to $f_x(0,0) = 0$ if f_x were to be continuous).

If $y = mx$ then

$$\lim_{(x,y)\to(0,0)} \frac{2xy}{x^2 + y^2} = \lim_{x \to 0} \frac{2mx^2}{x^2 + m^2x^2} = \frac{2m}{1 + m^2}.$$

Since the result depends on m, it depends on the way we approach $(0,0)$. Consequently, the limit does not exist.

67. The components of $\mathbf{F}(x, y) = (x + y^2, 2xy)$ are $F_1 = x + y^2$ and $F_2 = 2xy$. All partial derivatives of the components of $\mathbf{F}$ exist at $(0,0)$:

$$\frac{\partial F_1}{\partial x}(0,0) = 1, \quad \frac{\partial F_1}{\partial y}(0,0) = 2y\Big|_{(0,0)} = 0,$$

$$\frac{\partial F_2}{\partial x}(0,0) = 2y\Big|_{(0,0)} = 0 \quad \text{and} \quad \frac{\partial F_2}{\partial y}(0,0) = 2x\Big|_{(0,0)} = 0.$$

It follows that

$$D\mathbf{F}(0,0) = \begin{bmatrix} 1 & 0 \\ 0 & 0 \end{bmatrix}$$

and

$$D\mathbf{F}(0,0) \cdot (x, y) = \begin{bmatrix} 1 & 0 \\ 0 & 0 \end{bmatrix} \begin{bmatrix} x \\ y \end{bmatrix} = \begin{bmatrix} x \\ 0 \end{bmatrix}.$$

Now

$$\lim_{(x,y)\to(0,0)} \frac{\|\mathbf{F}(x,y) - \mathbf{F}(0,0) - D\mathbf{F}(0,0) \cdot (x,y)\|}{\|(x,y)\|} = \lim_{(x,y)\to(0,0)} \frac{\|(x + y^2, 2xy) - (x, 0)\|}{\|(x,y)\|}$$

$$= \lim_{(x,y)\to(0,0)} \frac{\|(y^2, 2xy)\|}{\|(x,y)\|} = \lim_{(x,y)\to(0,0)} \frac{\sqrt{y^4 + 4x^2y^2}}{\sqrt{x^2 + y^2}}.$$

Passing to polar coordinates, we get

$$\lim_{(x,y)\to(0,0)} \frac{\sqrt{y^4 + 4x^2y^2}}{\sqrt{x^2 + y^2}} = \lim_{r \to 0} \frac{\sqrt{r^4 \sin^4\theta + 4r^4 \sin^2\theta \cos^2\theta}}{r}$$

$$= \lim_{r \to 0} r\sqrt{\sin^4\theta + 4\sin^2\theta\cos^2\theta} = 0.$$

By definition, it follows that $\mathbf{F}$ is differentiable at $(0,0)$.

2.5 Properties of Derivatives

1. Write $f = g(u, v, w, z)$, where $u = x^2y$, $v = 2x + 5y$, $w = x$ and $z = y$. By the chain rule, $f_x = (\partial g/\partial u)2xy + (\partial g/\partial v)2 + (\partial g/\partial w)1 + (\partial g/\partial z)0$, where $\partial g/\partial u$, $\partial g/\partial v$, $\partial g/\partial w$ and $\partial g/\partial z$ are evaluated at $(x^2y, 2x+5y, x, y)$. Using the notation D_ig for the derivative of g with respect to its i-th variable, we write

$$f_x = D_1g \cdot 2xy + D_2g \cdot 2 + D_3g \cdot 1 + D_4g \cdot 0 = 2xyD_1g + 2D_2g + D_3g,$$

where D_1g, D_2g, etc. are evaluated at $(x^2y, 2x + 5y, x, y)$. Similarly,

$$f_y = D_1g \cdot x^2 + D_2g \cdot 5 + D_3g \cdot 0 + D_4g \cdot 1 = x^2D_1g + 5D_2g + D_4g.$$

3. We use D_if for the derivative of f with respect to its i-th variable, $i = 1, 2, 3$. By the chain rule,

$$\begin{aligned} F_x &= D_1f(h(x), g(y), k(x,y))\, h'(x) + D_2f(h(x), g(y), k(x,y)) \cdot 0 \\ &\quad + D_3f(h(x), g(y), k(x,y))\, \frac{\partial k(x,y)}{\partial x} \\ &= D_1f(h(x), g(y), k(x,y))\, h'(x) + D_3f(h(x), g(y), k(x,y))\, \frac{\partial k(x,y)}{\partial x} \end{aligned}$$

and

$$\begin{aligned} F_y &= D_1f(h(x), g(y), k(x,y)) \cdot 0 + D_2f(h(x), g(y), k(x,y))\, g'(y) \\ &\quad + D_3f(h(x), g(y), k(x,y))\, \frac{\partial k(x,y)}{\partial y} \\ &= D_2f(h(x), g(y), k(x,y))\, g'(y) + D_3f(h(x), g(y), k(x,y))\, \frac{\partial k(x,y)}{\partial y}. \end{aligned}$$

Here is an alternative computation: $F(x,y)\colon \mathbb{R}^2 \to \mathbb{R}$ can be thought of as a composition $F(x,y) = f(G(x,y))$ of the function $G\colon \mathbb{R}^2 \to \mathbb{R}^3$ given by $G(x,y) = (h(x), g(y), k(x,y))$ and the function $f\colon \mathbb{R}^3 \to \mathbb{R}$. By the chain rule, $DF(x,y) = Df(G(x,y)) \cdot DG(x,y)$, where

$$DF(x,y) = [\, F_x(x,y) \quad F_y(x,y)\,]$$

and

$$\begin{aligned} &Df(G(x,y)) \cdot DG(x,y) \\ &\quad = [\, D_1f(G(x,y)) \quad D_2f(G(x,y)) \quad D_3f(G(x,y))\,] \cdot \begin{bmatrix} h'(x) & 0 \\ 0 & g'(y) \\ \partial k(x,y)/\partial x & \partial k(x,y)/\partial y \end{bmatrix} \\ &\quad = \Big[D_1f(G(x,y))\, h'(x) + D_3f(G(x,y))\, \frac{\partial k(x,y)}{\partial x} \\ &\qquad\qquad D_2f(G(x,y))\, g'(y) + D_3f(G(x,y))\, \frac{\partial k(x,y)}{\partial y} \Big]. \end{aligned}$$

Comparing the corresponding entries in $DF(x,y)$ and $Df(G(x,y)) \cdot DG(x,y)$ and replacing $G(x,y)$ by $(h(x), g(y), k(x,y))$, we obtain the expressions for F_x and F_y.

5. By the product rule,

$$\begin{aligned}\nabla(fg)(x,y) &= g(x,y)\nabla f(x,y) + f(x,y)\nabla g(x,y)\\ &= \ln(xy)\,[\,2x+y \quad x\,] + (x^2+xy)\,[\,1/x \quad 1/y\,]\\ &= [\,(2x+y)\ln(xy) + x + y \quad x\ln(xy) + (x^2+xy)/y\,].\end{aligned}$$

By the quotient rule,

$$\begin{aligned}D\left(\frac{f}{g}\right)(2,2) &= \frac{g(2,2)\nabla f(2,2) - f(2,2)\nabla g(2,2)}{g(2,2)^2}\\ &= \frac{\ln 4\,[\,2x+y \quad x\,]_{at\ (2,2)} - 8\,[\,1/x \quad 1/y\,]_{at\ (2,2)}}{(\ln 4)^2}\\ &= \frac{\ln 4\,[\,6 \quad 2\,] - 8\,[\,1/2 \quad 1/2\,]}{(\ln 4)^2}\\ &= [\,(6\ln 4 - 4)/(\ln 4)^2 \quad (2\ln 4 - 4)/(\ln 4)^2\,] \approx [\,2.24672 \quad -0.63867\,].\end{aligned}$$

7. Let $g(x,y,z) = x$. Then

$$D\,(f/x)\,(x,y,z) = D\,(f/g)\,(x,y,z) = \frac{g(x,y,z)\,Df(x,y,z) - f(x,y,z)\,Dg(x,y,z)}{g(x,y,z)^2},$$

and so

$$\begin{aligned}D\,(f/g)\,(1,\pi,-1) &= \frac{1\,Df(1,\pi,-1) - (-2)\,Dg(1,\pi,-1)}{1}\\ &= Df(1,\pi,-1) + 2Dg(1,\pi,-1).\end{aligned}$$

From

$$Df(1,\pi,-1) = [\,2x \quad z\cos(yz) \quad y\cos(yz)\,]_{at\ (1,\pi,-1)} = [\,2 \quad 1 \quad -\pi\,]$$

and

$$Dg(1,\pi,-1) = [\,1 \quad 0 \quad 0\,]_{at\ (1,\pi,-1)} = [\,1 \quad 0 \quad 0\,]$$

we get

$$\begin{aligned}D\,(f/x)\,(1,\pi,-1) &= D\,(f/g)\,(1,\pi,-1)\\ &= [\,2 \quad 1 \quad -\pi\,] + 2\,[\,1 \quad 0 \quad 0\,] = [\,4 \quad 1 \quad -\pi\,].\end{aligned}$$

By the product rule,

$$D(x^2yf)(2,0,1) = f(2,0,1)\,D(x^2y)(2,0,1) + x^2y\Big|_{(2,0,1)}\,Df(2,0,1)$$

(if this notation seems inconvenient, introduce a function $g(x,y,z) = x^2y$ and differentiate gf instead of x^2yf; that's how we did the first part: we used f/g with $g(x,y,z) = x$ instead of f/x). Since

$$D(x^2y)(2,0,1) = [\,2xy \quad x^2 \quad 0\,]_{at\ (2,0,1)} = [\,0 \quad 4 \quad 0\,]$$

and

$$Df(2,0,1) = [\,2x \quad z\cos(yz) \quad y\cos(yz)\,]_{at\ (2,0,1)} = [\,4 \quad 1 \quad 0\,],$$

it follows that

$$D(x^2yf)(2,0,1) = 1 \cdot [0 \quad 4 \quad 0] + 0 \cdot [4 \quad 1 \quad 0] = [0 \quad 4 \quad 0].$$

9. From

$$\frac{\partial w}{\partial \rho} = \frac{\partial f}{\partial x}\frac{\partial x}{\partial \rho} + \frac{\partial f}{\partial y}\frac{\partial y}{\partial \rho} + \frac{\partial f}{\partial z}\frac{\partial z}{\partial \rho}$$

we get

$$\frac{\partial w}{\partial \rho} = \frac{\partial f}{\partial x}\sin\phi\cos\theta + \frac{\partial f}{\partial y}\sin\phi\sin\theta + \frac{\partial f}{\partial z}\cos\phi.$$

Similarly,

$$\begin{aligned}\frac{\partial w}{\partial \theta} &= \frac{\partial f}{\partial x}\frac{\partial x}{\partial \theta} + \frac{\partial f}{\partial y}\frac{\partial y}{\partial \theta} + \frac{\partial f}{\partial z}\frac{\partial z}{\partial \theta}\\ &= \frac{\partial f}{\partial x}(-\rho\sin\phi\sin\theta) + \frac{\partial f}{\partial y}\rho\sin\phi\cos\theta + \frac{\partial f}{\partial z}\cdot 0\\ &= -\rho\sin\phi\left(\frac{\partial f}{\partial x}\sin\theta - \frac{\partial f}{\partial y}\cos\theta\right)\end{aligned}$$

and

$$\begin{aligned}\frac{\partial w}{\partial \phi} &= \frac{\partial f}{\partial x}\frac{\partial x}{\partial \phi} + \frac{\partial f}{\partial y}\frac{\partial y}{\partial \phi} + \frac{\partial f}{\partial z}\frac{\partial z}{\partial \phi}\\ &= \frac{\partial f}{\partial x}(\rho\cos\phi\cos\theta) + \frac{\partial f}{\partial y}\rho\cos\phi\sin\theta - \frac{\partial f}{\partial z}\rho\sin\phi.\end{aligned}$$

11. From

$$(\mathbf{v}\times\mathbf{w})(t) = \begin{vmatrix}\mathbf{i} & \mathbf{j} & \mathbf{k}\\ t^3 & 0 & te^t\\ 0 & -2t & 0\end{vmatrix} = 2t^2e^t\mathbf{i} - 2t^4\mathbf{k}$$

we get

$$(\mathbf{v}\times\mathbf{w})'(t) = (4te^t + 2t^2e^t)\mathbf{i} - 8t^3\mathbf{k} = 2te^t(2+t)\mathbf{i} - 8t^3\mathbf{k}.$$

Using the product rule,

$$\begin{aligned}(\mathbf{v}\times\mathbf{w})'(t) &= \mathbf{v}'(t)\times\mathbf{w}(t) + \mathbf{v}(t)\times\mathbf{w}'(t)\\ &= \begin{vmatrix}\mathbf{i} & \mathbf{j} & \mathbf{k}\\ 3t^2 & 0 & e^t + te^t\\ 0 & -2t & 0\end{vmatrix} + \begin{vmatrix}\mathbf{i} & \mathbf{j} & \mathbf{k}\\ t^3 & 0 & te^t\\ 0 & -2 & 0\end{vmatrix}\\ &= 2t(e^t + te^t)\mathbf{i} - 6t^3\mathbf{k} + 2te^t\mathbf{i} - 2t^3\mathbf{k} = 2te^t(2+t)\mathbf{i} - 8t^3\mathbf{k}.\end{aligned}$$

13. By the Chain Rule, $D(g\circ\mathbf{F})(0,0) = Dg(\mathbf{F}(0,0))\cdot D\mathbf{F}(0,0)$. Since

$$Dg(\mathbf{F}(0,0)) = Dg(1,0,1) = [w \quad 2v \quad u]_{at\ (1,0,1)} = [1 \quad 0 \quad 1]$$

and

$$D\mathbf{F}(0,0) = \begin{bmatrix}e^x & 0\\ y & x\\ 0 & e^y\end{bmatrix}_{at\ (0,0)} = \begin{bmatrix}1 & 0\\ 0 & 0\\ 0 & 1\end{bmatrix},$$

it follows that

$$D(g\circ \mathbf{F})(0,0) = [1 \quad 0 \quad 1]\cdot \begin{bmatrix} 1 & 0 \\ 0 & 0 \\ 0 & 1 \end{bmatrix} = [1 \quad 1].$$

Alternatively, $(g\circ \mathbf{F})(x,y) = g(\mathbf{F}(x,y)) = g(e^x, xy, e^y) = e^x e^y + x^2y^2$, and

$$D(g\circ \mathbf{F})(0,0) = [\, e^x e^y + 2xy^2 \quad e^x e^y + 2x^2 y\,]_{at\ (0,0)} = [1 \quad 1].$$

15. Since $w = f(x, g(x,z), z)$, it follows that (using $D_i f$ for the partial derivative of f with respect to its i-th variable)

$$\frac{\partial w}{\partial x} = D_1 f\frac{\partial x}{\partial x} + D_2 f\frac{\partial g(x,z)}{\partial x} + D_3 f\frac{\partial z}{\partial x} = D_1 f + D_2 f\frac{\partial g(x,z)}{\partial x}.$$

Similarly,

$$\frac{\partial w}{\partial z} = D_2 f\frac{\partial g(x,z)}{\partial z} + D_3 f,$$

where all partial derivatives $D_i f$, $i = 1,2,3$, are computed at $(x, g(x,z), z)$.

17. Let $a = [a_{ij}]$, $i,j = 1,2$, and $\mathbf{x} = (x,y)$. Then

$$\mathbf{F}(\mathbf{x}) = A\cdot \mathbf{x} = \begin{bmatrix} a_{11} & a_{12} \\ a_{21} & a_{22} \end{bmatrix}\cdot \begin{bmatrix} x \\ y \end{bmatrix} = \begin{bmatrix} a_{11}x + a_{12}y \\ a_{21}x + a_{22}y \end{bmatrix},$$

and therefore

$$D\mathbf{F}(\mathbf{x}) = \begin{bmatrix} a_{11} & a_{12} \\ a_{21} & a_{22} \end{bmatrix} = A.$$

The components $a_{11}x + a_{12}y$ and $a_{21}x + a_{22}y$ of $\mathbf{F}$ are polynomials, and so have partial derivatives at all points. It remains to show that (let $\mathbf{x}_0 = (x_0, y_0)$)

$$\lim_{\mathbf{x}\to\mathbf{x}_0} \frac{\|\mathbf{F}(\mathbf{x}) - \mathbf{F}(\mathbf{x}_0) - D\mathbf{F}(\mathbf{x})(\mathbf{x}-\mathbf{x}_0)\|}{\|\mathbf{x}-\mathbf{x}_0\|} = 0.$$

The numerator of this expression is equal to

$$\|A\cdot\mathbf{x} - A\cdot\mathbf{x}_0 - A\cdot(\mathbf{x}-\mathbf{x}_0)\| = \|\mathbf{0}\| = 0,$$

by the distributivity property of matrix multiplication. Therefore,

$$\lim_{\mathbf{x}\to\mathbf{x}_0} \frac{\|\mathbf{F}(\mathbf{x}) - \mathbf{F}(\mathbf{x}_0) - D\mathbf{F}(\mathbf{x})(\mathbf{x}-\mathbf{x}_0)\|}{\|\mathbf{x}-\mathbf{x}_0\|} = \lim_{\mathbf{x}\to\mathbf{x}_0} \frac{0}{\|\mathbf{x}-\mathbf{x}_0\|} = 0,$$

and so $\mathbf{F}$ is differentiable at $\mathbf{x}_0$.

Comment: this solution is provided as an exercise in using the definition of differentiability (Definition 2.12). There is an easier way: since the components of $\mathbf{F}$ are polynomials, their partial derivatives are continuous (i.e., $\mathbf{F}$ is C^1). By Theorem 2.5 in Section 2.4, it follows that $\mathbf{F}$ is differentiable.

19. By the chain rule,

$$\frac{df}{dt} = \frac{\partial f}{\partial x}\frac{dx}{dt} + \frac{\partial f}{\partial y}\frac{dy}{dt}.$$

In order to compute dx/dt and dy/dt we have to use implicit differentiation. From $x^3+tx=8$ it follows that

$$3x^2\frac{dx}{dt}+x+t\frac{dx}{dt}=0 \quad \text{and hence} \quad \frac{dx}{dt}=\frac{-x}{3x^2+t}.$$

Similarly, $ye^y=t$ implies

$$e^y\frac{dy}{dt}+ye^y\frac{dy}{dt}=1 \quad \text{and hence} \quad \frac{dy}{dt}=\frac{1}{e^y(1+y)}.$$

It follows that

$$\frac{df}{dt}=3x^2y\,\frac{-x}{3x^2+t}+x^3\,\frac{1}{e^y(1+y)}.$$

If $t=0$, then $x^3=8$ implies $x=2$; the equation $ye^y=0$ implies $y=0$. Therefore, $(df/dt)(0)=0+8/1=8$.

2.6 Higher Order Partial Derivatives

1. Write z as $z=e^{xy}+2\ln x+3\ln y$. Then $z_x=ye^{xy}+2/x$, $z_y=xe^{xy}+3/y$, $z_{xx}=y^2e^{xy}-2/x^2$, $z_{yy}=x^2e^{xy}-3/y^2$, and $z_{xy}=z_{yx}=e^{xy}(1+xy)$.

3. Using the chain rule and the product rule over and over again, we get

$$z_x=\tfrac{5}{2}(x^2+y^2)^{3/2}\,2x=5x(x^2+y^2)^{3/2}$$

and

$$\begin{aligned} z_{xx}&=5(x^2+y^2)^{3/2}+5x\,\tfrac{3}{2}(x^2+y^2)^{1/2}\,2x\\ &=5(x^2+y^2)^{3/2}+15x^2(x^2+y^2)^{1/2}=5(x^2+y^2)^{1/2}(4x^2+y^2).\end{aligned}$$

By symmetry (interchanging x and y),

$$z_y=5y(x^2+y^2)^{3/2} \quad \text{and} \quad z_{yy}=5(x^2+y^2)^{1/2}(x^2+4y^2).$$

Finally,

$$z_{xy}=z_{yx}=5x\,\tfrac{3}{2}(x^2+y^2)^{1/2}\,2y=15xy(x^2+y^2)^{1/2}.$$

5. Using the chain rule and the product rule, we get

$$z_x=2\sin(x+y)\cos(x+y)=\sin(2x+2y),$$

using the formula $2\sin\theta\cos\theta=\sin 2\theta$ with $\theta=x+y$. So $z_{xx}=2\cos(2x+2y)$, and, by symmetry, $z_{yy}=2\cos(2x+2y)$. Finally, $z_{xy}=z_{yx}=2\cos(2x+2y)$.

7. Using $'$ to denote the derivatives of f and g, we get $z_x=f'(ax+by)\,a+g'(ax/y)\,(a/y)$ and

$$z_{xx}=f''(ax+by)\,a\,a+g''(ax/y)\frac{a}{y}\frac{a}{y}=a^2\left(f''(ax+by)+\frac{g''(ax/y)}{y^2}\right).$$

Similarly, $z_y = f'(ax+by)\,b + g'(ax/y)\,(-ax/y^2)$ and

$$z_{yy} = f''(ax+by)\,b\,b + g''(ax/y)\frac{-ax}{y^2}\frac{-ax}{y^2} + g'(ax/y)\frac{2ax}{y^3}$$
$$= b^2 f''(ax+by) + \frac{a^2x^2 g''(ax/y)}{y^4} + \frac{2axg'(ax/y)}{y^3}.$$

Starting with z_x, and computing the derivative with respect to y, we get

$$z_{xy} = f''(ax+by)\,b\,a + g''(ax/y)\frac{-ax}{y^2}\frac{a}{y} + g'(ax/y)\frac{-a}{y^2}$$
$$= abf''(ax+by) - \frac{a^2xg''(ax/y)}{y^3} - \frac{ag'(ax/y)}{y^2}.$$

Similarly, $z_{yx} = z_{xy}$.

9. The function w is of class C^k, for any k, so we can compute partial derivatives in any order. Since the first term has no z, start with $w_z = 4x^3y^2z^3$. Now $w_{zx} = 12x^2y^2z^3$, $w_{zxx} = 24xy^2z^3$ and $w_{xyzx} = w_{zxxy} = 48xyz^3$.

11. (a) From $u = x^3 - 3xy^2$ it follows that $u_x = 3x^2 - 3y^2$ and $u_y = -6xy$. The partial derivatives of $v = 3x^2y - y^3$ are $v_x = 6xy$ and $v_y = 3x^2 - 3y^2$. Clearly, $u_x = v_y$ and $u_y = -v_x$.

(b) From $u = e^x \cos y$ it follows that $u_x = e^x \cos y$ and $u_y = -e^x \sin y$. From $v = e^x \sin y$ we get $v_x = e^x \sin y$ and $v_y = e^x \cos y$. It follows that u and v satisfy the Cauchy-Riemann equations.

13. A straightforward computation gives $z_x = e^y + ye^x$, $z_{xx} = ye^x$ and $z_{xxx} = ye^x$. Similarly, $z_y = xe^y + e^x$, $z_{yy} = z_{yyy} = xe^y$, and it follows that

$$z_{xxx} + z_{yyy} = xe^y + ye^x = z.$$

Next, we compute the right side. From $z_{xy} = z_{yx} = e^y + e^x$ it follows that $z_{xyy} = e^y$, $z_{yxx} = e^x$ and

$$xz_{xyy} + yz_{yxx} = xe^y + ye^x = z;$$

so both sides are equal to z.

15. Let $f(x,y,z)$ be a C^2 function. It has three partial derivatives in which both derivatives are done with respect to the same variable: f_{xx}, f_{yy} and f_{zz}. There are three other distinct partial derivatives: $f_{xy} = f_{yx}$, $f_{xz} = f_{zx}$ and $f_{yz} = f_{zy}$. Therefore, a C^2 function of three variables has six different second-order partial derivatives.

In general, let $f(x_1, \ldots, x_m)$ be a C^2 function. Organize all second-order partial derivatives in a matrix (call it Hf), so that the ij-th entry of Hf is $\partial^2 f/\partial x_i \partial x_j$. The entries on the main diagonal of Hf are $\partial^2 f/\partial x_1^2, \ldots, \partial^2 f/\partial x_m^2$. If $i \neq j$, then $\partial^2 f/\partial x_i \partial x_j = \partial^2 f/\partial x_j \partial x_i$, since f is C^2. Notice that one of these two derivatives is above the main diagonal, and the other (symmetric with respect to the diagonal) is below it. As a matter of fact, every $\partial^2 f/\partial x_i \partial x_j$ above the diagonal has its matching pair $\partial^2 f/\partial x_j \partial x_i$ below the diagonal (and vice versa). Hence all distinct partial derivatives are either on the diagonal or above it (or on the diagonal and below it). There are m entries on the diagonal, and a total of $m^2 - m$ non-diagonal entries. Half of them are above (and half below) the diagonal; hence the total number of distinct second-order partial derivatives is

$$m + \tfrac{1}{2}(m^2 - m) = \tfrac{1}{2}m^2 + \tfrac{1}{2}m = \tfrac{1}{2}m(m+1).$$

17. Let f', f'', g' and g'' denote the derivatives of f and g. Then $u_x = f + xf' + yg'$, where (here and in what follows) f, g and their derivatives are computed at $x+y$). Computing the derivatives of u_x with respect to x and y, we get $u_{xx} = f' + f' + xf'' + yg'' = 2f' + xf'' + yg''$ and $u_{xy} = f' + xf'' + g' + yg''$. Similarly, $u_y = xf' + g + yg'$ and $u_{yy} = xf'' + g' + g' + yg'' = 2g' + xf'' + yg''$. It follows that

$$u_{xx} - 2u_{xy} + u_{yy} = 2f' + xf'' + yg'' - 2(f' + xf'' + g' + yg'') + 2g' + xf'' + yg'' = 0.$$

19. The domain of z consists of all points (x, y) such that $x^2 + y^2 > 0$, i.e., is the set $\mathbb{R}^2 - \{(0,0)\}$. From $z_x = 2x/(x^2 + y^2)$ we get

$$z_{xx} = \frac{2(x^2+y^2) - 2x\,2x}{(x^2+y^2)^2} = \frac{2y^2 - 2x^2}{(x^2+y^2)^2}.$$

By symmetry, $z_{yy} = (2x^2 - 2y^2)/(x^2+y^2)^2$ and therefore $z_{xx} + z_{yy} = 0$ whenever $(x, y) \neq (0,0)$.

21. Write R as $R = (1/R_1 + 1/R_2)^{-1} = R_1R_2/(R_1 + R_2)$. Then

$$\frac{\partial R}{\partial R_1} = \frac{R_2(R_1+R_2) - R_1R_2}{(R_1+R_2)^2} = \frac{R_2^2}{(R_1+R_2)^2} = R_2^2(R_1+R_2)^{-2}$$

and

$$\frac{\partial^2 R}{\partial R_1^2} = R_2^2(-2)(R_1+R_2)^{-3} = \frac{-2R_2^2}{(R_1+R_2)^3}.$$

By symmetry (interchange R_1 and R_2),

$$\frac{\partial^2 R}{\partial R_2^2} = \frac{-2R_1^2}{(R_1+R_2)^3},$$

and therefore

$$\frac{\partial^2 R}{\partial R_1^2}\frac{\partial^2 R}{\partial R_2^2} = \frac{4R_1^2R_2^2}{(R_1+R_2)^6} = 4\left(\frac{R_1R_2}{R_1+R_2}\right)^2 \frac{1}{(R_1+R_2)^4} = 4R^2\frac{1}{(R_1+R_2)^4}.$$

23. From $V = -GMm(x^2+y^2+z^2)^{-1/2}$ we get, by the chain rule, $V_x = GMmx(x^2+y^2+z^2)^{-3/2}$. Proceed using the product rule and the chain rule:

$$\begin{aligned} V_{xx} &= \frac{GMm(x^2+y^2+z^2)^{3/2} - GMmx\,\frac{3}{2}(x^2+y^2+z^2)^{1/2}\,2x}{(x^2+y^2+z^2)^3} \\ &= \frac{GMm(x^2+y^2+z^2)^{1/2}(x^2+y^2+z^2-3x^2)}{(x^2+y^2+z^2)^3} = \frac{GMm(y^2+z^2-2x^2)}{(x^2+y^2+z^2)^{5/2}}. \end{aligned}$$

By symmetry,

$$V_{yy} = \frac{GMm(x^2+z^2-2y^2)}{(x^2+y^2+z^2)^{5/2}} \quad \text{and} \quad V_{zz} = \frac{GMm(x^2+y^2-2z^2)}{(x^2+y^2+z^2)^{5/2}}.$$

Adding up V_{xx}, V_{yy} and V_{zz} we get zero. Notice that the computation works whenever $(x, y, z) \neq (0, 0, 0)$.

2.7 Cylindrical and Spherical Coordinate Systems

1. The point $(-4, 0)$ in the xy-plane has polar coordinates $(4, \pi)$, and hence the cylindrical coordinates of $(-4, 0, 0)$ are $(4, \pi, 0)$. The cylindrical coordinates of $(0, 0, 3)$ are $(0, \theta, 3)$ for any θ, $0 \leq \theta < 2\pi$, since $(0, 0)$ is represented in polar coordinates as $(0, \theta)$ for any θ. The polar coordinates of $(0, 2)$ are $(2, \pi/2)$; so the cylindrical coordinates of $(0, 2, 4)$ are $(2, \pi/2, 4)$. Finally, the polar coordinates of $(2, -3)$ are $r = \sqrt{13} \approx 3.60555$ and $\theta = \arctan(-3/2) + 2\pi \approx 5.30039$; therefore, the point $(2, -3, -1)$ is represented as $(3.60555, 5.30039, -1)$ in cylindrical coordinates.

3. View T as the composition $T_2 \circ T_1$ of the maps $T_1(r, \theta, z) = (2r, \theta, z)$ and $T_2(r, \theta, z) = (r, \theta + \pi, z)$. The map T_1 does not change the z-coordinate, and so the height of the image of the cube by T_1 will remain the same. Since T_1 doubles the distance in the xy-plane, it maps the given cube of side a onto a rectangular box whose sides are $2a$ and the height is a. The map T_2 now rotates a point in the counterclockwise sense for π rad about the z-axis. It follows that the map T maps the given cube onto a rectangular box whose sides are $2a$ and the height is a and then rotates it for π rad about the z-axis

5. In cylindrical coordinates, $x^2+y^2 = r^2$, and so the paraboloid $z = 4-x^2-y^2$ is represented as $z = 4-r^2$. The equation of the plane $x+2y-z = 0$ is transformed to $r\cos\theta+2r\sin\theta-z = 0$ in cylindrical coordinates. In spherical coordinates, $x^2 + y^2 = \rho^2 \sin^2\phi$, and so $\rho\cos\phi = 4-\rho^2\sin^2\phi$ represents the paraboloid $z = 4-x^2-y^2$. The plane $x+2y-z = 0$ is represented as $\rho\sin\phi\cos\theta + 2\rho\sin\phi\sin\theta - \rho\cos\phi = 0$, or as $\sin\phi\cos\theta + 2\sin\phi\sin\theta - \cos\phi = 0$.

7. The coordinate surface $r = C$ (C is a constant) is the cylinder $x^2 + y^2 = C$ of radius C (if $C > 0$) whose axis of rotation is the z-axis. The "coordinate surface" $r = 0$ is the z-axis. The coordinate surface $\theta = C$ is the plane perpendicular to the xy-plane that contains the z-axis and whose intersection with the xy-plane makes the angle θ with respect to the positive x-axis. The coordinate surface $z = C$ is the plane parallel to the xy-plane, C units above it if $C \geq 0$ and $|C|$ units below it if $C < 0$.

The coordinate curve $r = C_1$, $\theta = C_2$ (C_1 and C_2 are constants) is the line perpendicular to the xy-plane, crossing it at the point with polar coordinates (C_1, C_2). The coordinate curve $r = C_1$, $z = C_2$ is the circle of radius C_1 in the plane $z = C_2$ with the center at $(0, 0, C_2)$. The coordinate curve $\theta = C_1$, $z = C_2$ is the line in the plane $z = C_2$ crossing the z-axis at $(0, 0, C_2)$. When that line is projected onto the xy-plane, it makes the angle θ with respect to the positive x-axis.

9. We must solve equations (2.36) for $\mathbf{i}$, $\mathbf{j}$ and $\mathbf{k}$:

$$\mathbf{e}_\rho = \sin\phi\cos\theta\mathbf{i} + \sin\phi\sin\theta\mathbf{j} + \cos\phi\mathbf{k} \tag{2.36a}$$

$$\mathbf{e}_\theta = -\sin\theta\mathbf{i} + \cos\theta\mathbf{j} \tag{2.36b}$$

$$\mathbf{e}_\phi = \cos\phi\cos\theta\mathbf{i} + \cos\phi\sin\theta\mathbf{j} - \sin\phi\mathbf{k}. \tag{2.36c}$$

Multiply (2.36a) by $\sin\phi$, (2.36c) by $\cos\phi$ and add them up, to get

$$\sin\phi\mathbf{e}_\rho + \cos\phi\mathbf{e}_\phi = \cos\theta\mathbf{i} + \sin\theta\mathbf{j}.$$

Now multiply this identity by $\cos\theta$:

$$\cos\theta\sin\phi\mathbf{e}_\rho + \cos\theta\cos\phi\mathbf{e}_\phi = \cos^2\theta\mathbf{i} + \sin\theta\cos\theta\mathbf{j};$$

multiply (2.36b) by $-\sin\theta$:

$$-\sin\theta\mathbf{e}_\theta = \sin^2\theta\mathbf{i} - \sin\theta\cos\theta\mathbf{j},$$

and add up the two, thus obtaining

$$\mathbf{i} = \cos\theta\sin\phi\mathbf{e}_\rho + \cos\theta\cos\phi\mathbf{e}_\phi - \sin\theta\mathbf{e}_\theta.$$

From (2.36b) it follows that

$$\mathbf{j} = \frac{1}{\cos\theta}\mathbf{e}_\theta + \frac{\sin\theta}{\cos\theta}\mathbf{i} = \sin\phi\sin\theta\mathbf{e}_\rho + \cos\phi\sin\theta\mathbf{e}_\phi + \cos\theta\mathbf{e}_\theta.$$

Similarly, from either (2.36a) or (2.36c) we get that $\mathbf{k} = \cos\phi\mathbf{e}_\rho - \sin\phi\mathbf{e}_\phi$.

11. In cylindrical coordinates, using (2.33), we get

$$\mathbf{F}(r,\theta,z) = \cos\theta\mathbf{e}_r - \sin\theta\mathbf{e}_\theta + \sin\theta\mathbf{e}_r + \cos\theta\mathbf{e}_\theta = (\cos\theta + \sin\theta)\mathbf{e}_r + (\cos\theta - \sin\theta)\mathbf{e}_\theta.$$

To express F in spherical coordinates, we use formulas derived in Exercise 9, and get

$$\begin{aligned}\mathbf{F}(\rho,\theta,\phi) &= \sin\phi\cos\theta\mathbf{e}_\rho - \sin\theta\mathbf{e}_\theta + \cos\phi\cos\theta\mathbf{e}_\phi + \sin\phi\sin\theta\mathbf{e}_\rho + \cos\phi\sin\theta\mathbf{e}_\phi + \cos\theta\mathbf{e}_\theta \\ &= \sin\phi(\cos\theta + \sin\theta)\mathbf{e}_\rho + (\cos\theta - \sin\theta)\mathbf{e}_\theta + \cos\phi(\cos\theta + \sin\theta)\mathbf{e}_\phi.\end{aligned}$$

13. In cylindrical coordinates, using (2.33), we get

$$\mathbf{F}(r,\theta,z) = r\sin\theta(\cos\theta\mathbf{e}_r - \sin\theta\mathbf{e}_\theta) - r\cos\theta(\sin\theta\mathbf{e}_r + \cos\theta\mathbf{e}_\theta) = -r\mathbf{e}_\theta.$$

To express $\mathbf{F}$ in spherical coordinates, we use formulas derived in Exercise 9, and get

$$\begin{aligned}\mathbf{F}(\rho,\theta,\phi) =& \rho\sin\phi\sin\theta(\sin\phi\cos\theta\mathbf{e}_\rho - \sin\theta\mathbf{e}_\theta + \cos\phi\cos\theta\mathbf{e}_\phi) \\ &- \rho\sin\phi\cos\theta(\sin\phi\sin\theta\mathbf{e}_\rho + \cos\theta\mathbf{e}_\theta + \cos\phi\sin\theta\mathbf{e}_\phi) \\ =& -\rho\sin\phi\mathbf{e}_\theta.\end{aligned}$$

15. From (2.36), using the product and the chain rules, we get

$$\frac{d\mathbf{e}_\theta}{dt} = -\cos\theta\frac{d\theta}{dt}\mathbf{i} - \sin\theta\frac{d\theta}{dt}\mathbf{j} = -(\cos\theta\mathbf{i} + \sin\theta\mathbf{j})\frac{d\theta}{dt}.$$

To express $d\mathbf{e}_\theta/dt$ in terms of unit vectors $\mathbf{e}_r$, $\mathbf{e}_\theta$ and $\mathbf{e}_\phi$ we use the formulas of Exercise 9; thus

$$\frac{d\mathbf{e}_\theta}{dt} = -\big[\cos\theta(\sin\phi\cos\theta\mathbf{e}_\rho - \sin\theta\mathbf{e}_\theta + \cos\phi\cos\theta\mathbf{e}_\phi)$$

$$+\sin\theta(\sin\phi\sin\theta\mathbf{e}_\rho+\cos\theta\mathbf{e}_\theta+\cos\phi\sin\theta\mathbf{e}_\phi)]\frac{d\theta}{dt}$$

$$=-(\sin\phi\mathbf{e}_\rho+\cos\phi\mathbf{e}_\phi)\frac{d\theta}{dt}.$$

Similarly,

$$\begin{aligned}\frac{d\mathbf{e}_\phi}{dt}&=\frac{d}{dt}(\cos\phi\cos\theta\mathbf{i}+\cos\phi\sin\theta\mathbf{j}-\sin\phi\mathbf{k})\\&=-\sin\phi\frac{d\phi}{dt}\cos\theta\mathbf{i}-\cos\phi\sin\theta\frac{d\theta}{dt}\mathbf{i}-\sin\phi\frac{d\phi}{dt}\sin\theta\mathbf{j}+\cos\phi\cos\theta\frac{d\theta}{dt}\mathbf{j}-\cos\phi\frac{d\phi}{dt}\mathbf{k}\\&=-(\sin\phi\cos\theta\mathbf{i}+\sin\phi\sin\theta\mathbf{j}+\cos\phi\mathbf{k})\frac{d\phi}{dt}+\cos\phi(-\sin\theta\mathbf{i}+\cos\theta\mathbf{j})\frac{d\theta}{dt}\\&=-\frac{d\phi}{dt}\mathbf{e}_\rho+\cos\phi\frac{d\theta}{dt}\mathbf{e}_\theta.\end{aligned}$$

17. The differential of $x=r\cos\theta$ is $dx=\cos\theta dr-r\sin\theta d\theta$, and the differential of $y=r\sin\theta$ is $dy=\sin\theta dr+r\cos\theta d\theta$. Hence $ds^2=dx^2+dy^2+dz^2=(\cos\theta dr-r\sin\theta d\theta)^2+(\sin\theta dr+r\cos\theta d\theta)^2+dz^2=dr^2+r^2d\theta^2+dz^2$.

Another way of computing ds^2 is the following: in cylindrical coordinates, $\mathbf{r}=x\mathbf{i}+y\mathbf{j}+z\mathbf{k}=r\cos\theta\mathbf{i}+r\sin\theta\mathbf{j}+z\mathbf{k}$. The differential $d\mathbf{r}$ of $\mathbf{r}$ is

$$\begin{aligned}d\mathbf{r}&=\left(\frac{\partial\mathbf{r}}{\partial r}\right)dr+\left(\frac{\partial\mathbf{r}}{\partial\theta}\right)d\theta+\left(\frac{\partial\mathbf{r}}{\partial z}\right)dz\\&=(\cos\theta\mathbf{i}+\sin\theta\mathbf{j})dr+(-r\sin\theta\mathbf{i}+r\cos\theta\mathbf{j})d\theta+\mathbf{k}dz\\&=(\cos\theta dr-r\sin\theta d\theta)\mathbf{i}+(\sin\theta dr+r\cos\theta d\theta)\mathbf{j}+dz\mathbf{k},\end{aligned}$$

and so $d\mathbf{r}\cdot d\mathbf{r}=dr^2+r^2d\theta^2+dz^2$.

The differentials in the case of spherical coordinates are computed as follows: from $x=\rho\sin\phi\cos\theta$ we get

$$dx=\sin\phi\cos\theta d\rho+\rho\cos\phi\cos\theta d\phi-\rho\sin\phi\sin\theta d\theta;$$

$y=\rho\sin\phi\sin\theta$ implies

$$dy=\sin\phi\sin\theta d\rho+\rho\cos\phi\sin\theta d\phi+\rho\sin\phi\cos\theta d\theta,$$

and $z=\rho\cos\phi$ implies

$$dz=\cos\phi d\rho-\rho\sin\phi d\phi.$$

Thus

$$\begin{aligned}ds^2&=dx^2+dy^2+dz^2\\&=(\sin\phi\cos\theta d\rho+\rho\cos\phi\cos\theta d\phi-\rho\sin\phi\sin\theta d\theta)^2\\&\quad+(\sin\phi\sin\theta d\rho+\rho\cos\phi\sin\theta d\phi+\rho\sin\phi\cos\theta d\theta)^2\\&\quad+(\cos\phi d\rho-\rho\sin\phi d\phi)^2.\end{aligned}$$

After simplifying, we get $=d\rho^2+\rho^2\sin^2\phi d\theta^2+\rho^2 d\phi^2$.

2.8 General Curvilinear Coordinate Systems

1. (a) The position vector is $\mathbf{r} = \frac{1}{2}(u^2 - v^2)\mathbf{i} + uv\mathbf{j} + z\mathbf{k}$. It follows that $\partial\mathbf{r}/\partial u = u\mathbf{i} + v\mathbf{j}$, $h_u = \|\partial\mathbf{r}/\partial u\| = \sqrt{u^2+v^2}$ and

$$\mathbf{e}_u = \frac{\partial\mathbf{r}}{\partial u}\Big/\left\|\frac{\partial\mathbf{r}}{\partial u}\right\| = \frac{u}{\sqrt{u^2+v^2}}\mathbf{i} + \frac{v}{\sqrt{u^2+v^2}}\mathbf{j}.$$

Similarly, $\partial\mathbf{r}/\partial v = -v\mathbf{i} + u\mathbf{j}$, $h_v = \|\partial\mathbf{r}/\partial v\| = \sqrt{u^2+v^2}$ and

$$\mathbf{e}_v = \frac{\partial\mathbf{r}}{\partial v}\Big/\left\|\frac{\partial\mathbf{r}}{\partial v}\right\| = \frac{-v}{\sqrt{u^2+v^2}}\mathbf{i} + \frac{u}{\sqrt{u^2+v^2}}\mathbf{j}.$$

Finally, $\partial\mathbf{r}/\partial z = \mathbf{k}$, $h_z = \|\partial\mathbf{r}/\partial z\| = 1$ and $\mathbf{e}_z = \mathbf{k}$.

(b) The fact that the coordinate system is orthogonal follows from $\mathbf{e}_u \cdot \mathbf{e}_v = 0$, $\mathbf{e}_u \cdot \mathbf{e}_z = 0$ and $\mathbf{e}_v \cdot \mathbf{e}_z = 0$.

(c) By (a) and (b), the vectors $\mathbf{e}_u$, $\mathbf{e}_v$ and $\mathbf{e}_z$ are orthonormal. Hence $\mathbf{F} = F_u\mathbf{e}_u + F_v\mathbf{e}_v + F_z\mathbf{e}_z$, where

$$F_u = \mathbf{F}\cdot\mathbf{e}_u = \left(\tfrac{1}{2}(u^2-v^2)\mathbf{i} + u^2v^2\mathbf{j} - 2uv\mathbf{k}\right)\cdot\left(\frac{u}{\sqrt{u^2+v^2}}\mathbf{i} + \frac{v}{\sqrt{u^2+v^2}}\mathbf{j}\right)$$
$$= \frac{\frac{1}{2}u(u^2-v^2) + u^2v^3}{\sqrt{u^2+v^2}}.$$

Similarly,

$$F_v = \mathbf{F}\cdot\mathbf{e}_v = \frac{-\frac{1}{2}v(u^2-v^2) + u^3v^2}{\sqrt{u^2+v^2}}$$

and $F_z = \mathbf{F}\cdot\mathbf{e}_z = -2uv$.

(d) Let us consider $u = C$ (C is a constant) first. Substitute $u = C$ into the equations for x and y, thus getting $x = \frac{1}{2}(C^2 - v^2)$, $y = Cv$; now eliminate v, thus getting $y^2 = C^4 - 2C^2x$. It follows that $u = C$ is a parabolic sheet perpendicular to the xy-plane, whose trace in the xy-plane is the parabola $y^2 = C^4 - 2C^2x$. Similarly, we obtain that the coordinate surface $v = C$ is the parabolic sheet $y^2 = C^4 + 2C^2x$, and the coordinate surface $z = C$ is the plane parallel to the xy-plane, crossing the z-axis at $(0, 0, C)$.

(e) The z-curve $u = C_1$ and $v = C_2$ (C_1 and C_2 are constants) is the intersection of the vertical planes $x = \frac{1}{2}(C_1^2 - C_2^2)$ and $y = C_1C_2$. Hence it is the vertical line that crosses the xy-plane at the point $(\frac{1}{2}(C_1^2 - C_2^2), C_1C_2)$. The u-curve $v = C_2$ and $z = C_3$ (C_2 and C_3 are constants) is the parabola $y^2 = 2C_2^2x + C_2^4$ in the plane $z = C_3$. The v-curve $u = C_1$ and $z = C_3$ (C_1 and C_3 are constants) is the parabola $y^2 = -2C_1^2x + C_1^4$ in the plane $z = C_3$.

3. (a) The position vector is

$$\mathbf{r} = a\sinh\xi\sin\eta\cos\phi\mathbf{i} + a\sinh\xi\sin\eta\sin\phi\mathbf{j} + a\cosh\xi\cos\eta\mathbf{k}.$$

It follows that

$$\frac{\partial\mathbf{r}}{\partial\xi} = a\cosh\xi\sin\eta\cos\phi\mathbf{i} + a\cosh\xi\sin\eta\sin\phi\mathbf{j} + a\sinh\xi\cos\eta\mathbf{k},$$

$$\begin{aligned} h_\xi = \left\| \frac{\partial \mathbf{r}}{\partial \xi} \right\| &= \left(a^2 \cosh^2 \xi \sin^2 \eta \cos^2 \phi + a^2 \cosh^2 \xi \sin^2 \eta \sin^2 \phi + a^2 \sinh^2 \xi \cos^2 \eta\right)^{1/2} \\ &= a\sqrt{\cosh^2 \xi \sin^2 \eta + \sinh^2 \xi \cos^2 \eta} \\ &= a\sqrt{(1 + \sinh^2 \xi) \sin^2 \eta + \sinh^2 \xi (1 - \sin^2 \eta)} \\ &= a\sqrt{\sinh^2 \xi + \sin^2 \eta}, \end{aligned}$$

and hence

$$\mathbf{e}_\xi = \frac{\cosh \xi \sin \eta \cos \phi \mathbf{i} + \cosh \xi \sin \eta \sin \phi \mathbf{j} + \sinh \xi \cos \eta \mathbf{k}}{\sqrt{\sinh^2 \xi + \sin^2 \eta}}.$$

Similarly, $h_\eta = a\sqrt{\sinh^2 \xi + \sin^2 \eta}$ and

$$\mathbf{e}_\eta = \frac{\sinh \xi \cos \eta \cos \phi \mathbf{i} + \sinh \xi \cos \eta \sin \phi \mathbf{j} - \cosh \xi \sin \eta \mathbf{k}}{\sqrt{\sinh^2 \xi + \sin^2 \eta}}.$$

Finally, $h_\phi = a \sinh \xi \sin \eta$, $\mathbf{e}_\phi = -\sin \phi \mathbf{i} + \cos \phi \mathbf{j}$.

(b) We have to check that the dot products of unit vectors from (a) are zero. For example,

$$\begin{aligned} \mathbf{e}_\xi \cdot \mathbf{e}_\phi &= \frac{\cosh \xi \sin \eta \cos \phi \mathbf{i} + \cosh \xi \sin \eta \sin \phi \mathbf{j} + \sinh \xi \cos \eta \mathbf{k}}{\sqrt{\sinh^2 \xi + \sin^2 \eta}} \cdot (-\sin \phi \mathbf{i} + \cos \phi \mathbf{j}) \\ &= \frac{\cosh \xi \sin \eta \cos \phi (-\sin \phi)}{\sqrt{\sinh^2 \xi + \sin^2 \eta}} + \frac{\cosh \xi \sin \eta \sin \phi \cos \phi}{\sqrt{\sinh^2 \xi + \sin^2 \eta}} = 0. \end{aligned}$$

The remaining two dot products are straightforward as well.

(c) From

$$\begin{aligned} \sqrt{x^2 + y^2} &= \sqrt{\sinh^2 \xi \sin^2 \eta \cos^2 \phi + \sinh^2 \xi \sin^2 \eta \sin^2 \phi} \\ &= |\sinh \xi \sin \eta| = \sinh \xi \sin \eta \end{aligned}$$

(since $\xi \geq 0$ and $0 \leq \eta \leq \pi$ guarantee that both factors are positive) it follows that $\mathbf{F} = \sinh \xi \sin \eta \mathbf{k}$. The vectors $\mathbf{e}_\xi$, $\mathbf{e}_\eta$ and $\mathbf{e}_\phi$ are orthonormal; hence $\mathbf{F} = F_\xi \mathbf{e}_\xi + F_\eta \mathbf{e}_\eta + F_\phi \mathbf{e}_\phi$, where

$$\begin{aligned} F_\xi &= \mathbf{F} \cdot \mathbf{e}_\xi \\ &= \sinh \xi \sin \eta \mathbf{k} \cdot \frac{\cosh \xi \sin \eta \cos \phi \mathbf{i} + \cosh \xi \sin \eta \sin \phi \mathbf{j} + \sinh \xi \cos \eta \mathbf{k}}{\sqrt{\sinh^2 \xi + \sin^2 \eta}} \\ &= \frac{\sinh^2 \xi \sin \eta \cos \eta}{\sqrt{\sinh^2 \xi + \sin^2 \eta}}. \end{aligned}$$

Similarly,

$$F_\eta = \mathbf{F} \cdot \mathbf{e}_\eta = \frac{-\sinh \xi \cosh \xi \sin^2 \eta}{\sqrt{\sinh^2 \xi + \sin^2 \eta}}$$

and $F_\phi = \mathbf{F} \cdot \mathbf{e}_\phi = 0$.

(d) Substitute $\phi = C$ into the expressions for x and y to get $x = a \sinh \xi \sin \eta \cos C$ and $y = a \sinh \xi \sin \eta \sin C$; multiply the first equation by $\sin C$, the second by $\cos C$ and subtract, thus getting $(\sin C)x - (\cos C)y = 0$; i.e., $y = (\tan C)x$. It follows that the coordinate surface $\phi = C$ is the plane perpendicular to the xy-plane that crosses it along the line of slope

$\tan C$. If $\cos C = 0$ then $x = 0$, so it is the yz-plane. Substituting $\xi = C$ we get $x = a\sinh C \sin\eta\cos\phi$, $y = a\sinh C\sin\eta\sin\phi$ and $z = a\cosh C\cos\eta$, and therefore (notice that $x^2 + y^2 = a^2\sinh^2 C\sin^2\eta$)

$$\begin{aligned}(x^2+y^2)\cosh^2 C + z^2\sinh^2 C &= a^2\cosh^2 C\sinh^2 C\sin^2\eta + a^2\cosh^2 C\sinh^2 C\cos^2\eta\\ &= a^2\cosh^2 C\sinh^2 C.\end{aligned}$$

Hence

$$x^2\cosh^2 C + y^2\cosh^2 C + z^2\sinh^2 C = a^2\cosh^2 C\sinh^2 C$$

and it follows that the coordinate surface $\xi = C$ is the ellipsoid (of revolution)

$$\frac{x^2}{\sinh^2 C} + \frac{y^2}{\sinh^2 C} + \frac{z^2}{\cosh^2 C} = a^2$$

(provided that $C \neq 0$; if $C = 0$, the "coordinate surface" degenerates to the z-axis). Similarly, we get that the coordinate surface $\eta = C$ is the hyperboloid (of revolution)

$$\frac{x^2}{\sin^2 C} + \frac{y^2}{\sin^2 C} - \frac{z^2}{\cos^2 C} = -a^2;$$

if $\sin C = 0$, the "coordinate surface" degenerates to the part of the z-axis, and if $\cos C = 0$ it is the xy-plane.

5. (a) The position vector is

$$\mathbf{r} = \frac{a\sinh v}{\cosh v - \cos u}\mathbf{i} + \frac{a\sin u}{\cosh v - \cos u}\mathbf{j} + z\mathbf{k}.$$

It follows that

$$\frac{\partial\mathbf{r}}{\partial u} = -\frac{a\sinh v\sin u}{(\cosh v - \cos u)^2}\mathbf{i} + \frac{a(\cosh v\cos u - 1)}{(\cosh v - \cos u)^2}\mathbf{j},$$

and

$$h_u = \left\|\frac{\partial\mathbf{r}}{\partial u}\right\| = \frac{a}{\cosh v - \cos u},$$

and hence

$$\mathbf{e}_u = \frac{-\sinh v\sin u\mathbf{i} + (\cosh v\cos u - 1)\mathbf{j}}{\cosh v - \cos u}.$$

Similarly,

$$\frac{\partial\mathbf{r}}{\partial v} = -\frac{a(\cosh v\cos u - 1)}{(\cosh v - \cos u)^2}\mathbf{i} - \frac{a\sinh v\sin u}{(\cosh v - \cos u)^2}\mathbf{j},$$

and

$$h_v = \left\|\frac{\partial\mathbf{r}}{\partial v}\right\| = \frac{a}{\cosh v - \cos u},$$

and hence

$$\mathbf{e}_v = \frac{-(\cosh v\cos u - 1)\mathbf{i} - \sinh v\sin u\mathbf{j}}{\cosh v - \cos u}.$$

Since $\partial\mathbf{r}/\partial z = \mathbf{k}$, it follows that $h_z = 1$ and $\mathbf{e}_z = \mathbf{k}$.

(b) Since the $\mathbf{k}$ component of $\mathbf{e}_u$ and $\mathbf{e}_v$ is zero, it follows that $\mathbf{e}_u \cdot \mathbf{e}_z = \mathbf{e}_v \cdot \mathbf{e}_z = 0$. Clearly, $\mathbf{e}_u \cdot \mathbf{e}_v = 0$.

(c) Vectors $\mathbf{e}_u$, $\mathbf{e}_v$ and $\mathbf{e}_z$ are orthonormal. Hence $\mathbf{F} = F_u\mathbf{e}_u + F_v\mathbf{e}_v + F_z\mathbf{e}_z$, where

$$F_u = \mathbf{F} \cdot \mathbf{e}_u = z\mathbf{i} \cdot \left(\frac{-\sinh v \sin u\mathbf{i} + (\cosh v \cos u - 1)\mathbf{j}}{\cosh v - \cos u} \right) = \frac{-z \sinh v \sin u}{\cosh v - \cos u}.$$

Similarly,

$$F_v = \mathbf{F} \cdot \mathbf{e}_v = \frac{-z(\cosh v \cos u - 1)}{\cosh v - \cos u}$$

and $F_z = \mathbf{F} \cdot \mathbf{e}_z = z\mathbf{i} \cdot \mathbf{k} = 0$.

2.9 Chapter Review

Solutions to Computer Projects

1. (a) The only requirement is that $x^2 + y^2 \neq 0$. Thus the domain of f is $\mathbb{R}^2 - \{(0,0)\}$.

(b) The function $g_1(x)$ is given by

$$g_1(x) = f(x,x) = \frac{x \sin x + x^3 + 2x}{\sqrt{2x^2}} = \frac{x \sin x + x^3 + 2x}{|x|\sqrt{2}}.$$

From the plot (see Figure 2.9.1 (a)) we see that $\lim_{x\to 0} g_1(x)$ does not exist.

(c) The function $g_2(x)$ is given by

$$g_2(x) = f(x,x^2) = \frac{x \sin x^2 + x^3 + 2x^2}{\sqrt{x^2 + x^4}} = \frac{x \sin x^2 + x^3 + 2x^2}{|x|\sqrt{1 + x^2}}.$$

From the plot (see Figure 2.9.1 (a)) we see that $\lim_{x\to 0} g_2(x) \approx 0$.

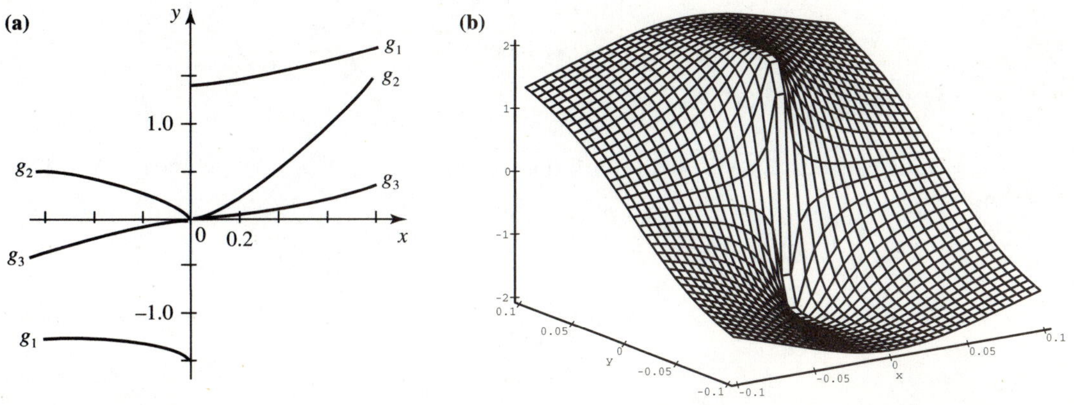

Figure 2.9.1

(d) The function $g_3(x)$ is given by

$$g_3(x) = f(x,0) = \frac{x^3}{\sqrt{x^2}} = \frac{x^3}{|x|}.$$

From the plot (see Figure 2.9.1 (a)) we see that $\lim_{x\to 0} g_3(x) \approx 0$.

(e) The limit $\lim_{(x,y)\to(0,0)} f(x,y)$ does not exist because along $y = x$ (see $g_1(x)$) the limit at 0 does not exist.

(f) See Figure 2.9.1 (b); it suggests that $\lim_{(x,y)\to(0,0)} f(x,y) \approx 0$, which we proved cannot be right.

2. (a) The level curve that goes through $(1,0)$ has the value $c_1 = f(1,0) = 2$. The remaining level curves are computed similarly: for $(0.9,0)$, $c_2 = f(0.9,0) = 2/0.9^2 \approx 2.46914$; for $(0.95,0)$, $c_3 = f(0.95,0) = 2/0.95^2 \approx 2.21607$; for $(1.05,0)$, $c_4 = f(1.05,0) = 2/1.05^2 \approx 1.81406$; for $(1.1,0)$, $c_5 = f(1.1,0) = 2/1.1^2 \approx 1.65289$. Thus the required level curves have equations $f(x,y) = c_i$, $i = 1,\ldots,5$.

(b) See Figure 2.9.2.

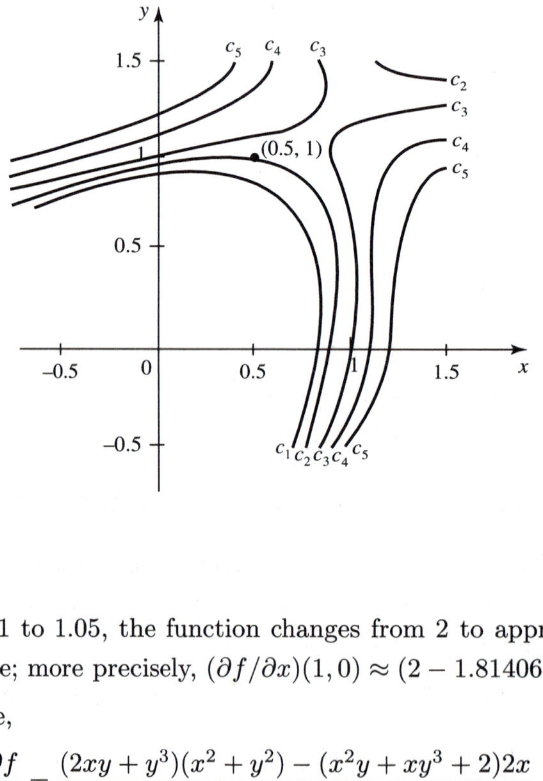

Figure 2.9.2

(c) As x changes from 1 to 1.05, the function changes from 2 to approximately 1.81406; so $(\partial f/\partial x)(1,0)$ is negative; more precisely, $(\partial f/\partial x)(1,0) \approx (2 - 1.81406)/0.05 \approx -3.71882$.

(d) By the quotient rule,

$$\frac{\partial f}{\partial x} = \frac{(2xy + y^3)(x^2 + y^2) - (x^2y + xy^3 + 2)2x}{(x^2 + y^2)^2},$$

and so $\partial f/\partial x(1,0) = -4$.

(e) The function g is given by $g(x) = f(x,0) = 2/x^2$. Its slope at $x = 1$ is

$$g'(1) = -4x^{-3}\Big|_1 = -4.$$

(f) The two are equal, by the geometric interpretation of a partial derivative as the slope of a tangent.

(g) The point $(1/2, 1)$ is very close to the level curve $f(x, y) = c_2$. It takes longer to reach the level curve $f(x, y) = c_3$ from $(1/2, 1)$ if we "walk" in the positive x-direction than if we "walk" in the positive y-direction. In other words, at $(1/2, 2)$, $f(x, y)$ changes faster in the y-direction than in the x-direction. Consequently, $(\partial f/\partial y)(1/2, 2) > (\partial f/\partial x)(1/2, 2)$.

3. (a) From $(x - y)^2 \geq 0$ it follows that $x^2 + y^2 - 2xy \geq 0$, and so $2xy \leq x^2 + y^2$ and $xy/(x^2+y^2) \leq 1/2$. Similarly, $(x+y)^2 \geq 0$ implies $x^2+y^2+2xy \geq 0$ and $xy/(x^2+y^2) \geq -1/2$. Combining the two inequalities, we get

$$-\tfrac{1}{2} \leq \frac{xy}{x^2 + y^2} \leq \tfrac{1}{2}.$$

(b) See Figure 2.9.3 (a).

(c) See Figure 2.9.3 (b); as we will see in part (d), this picture is misleading.

(d) From $f(x, y) = xy/(x^2 + y^2) = c$ it follows that $x^2 + y^2 = xy/c$. On the other hand, simplifying

$$\left(x - \frac{y}{2c}\right)^2 + \left(1 - \frac{1}{4c^2}\right) y^2 = 0$$

we get

$$x^2 - \frac{xy}{c} + \frac{y^2}{4c^2} + y^2 - \frac{y^2}{4c^2} = 0;$$

i.e., $x^2 + y^2 - xy/c = 0$.

Write $(x - y/2c)^2 + (1 - 1/4c^2)y^2 = 0$ as

$$\left(x - \frac{y}{2c}\right)^2 = \left(\frac{1}{4c^2} - 1\right) y^2$$

and compute the square root of both sides (since $-1/2 \leq f(x, y) \leq 1/2$, it follows that $-1/2 \leq c \leq 1/2$ (and we also need $c \neq 0$), and so $1/4c^2 - 1 \geq 0$), thus getting

$$x - ay = \pm by, \qquad \text{where } a = \frac{1}{2c} \text{ and } b = \sqrt{\frac{1}{4c^2} - 1}.$$

Consequently, a level curve of f consists of a pair of lines $y = x/(a \pm b)$ through the origin (compare with Figure 2.9.3 (b)!). For example, when $c = 0.1$, the level curve consists of the lines ($a = 1/0.2 = 5$ and $b = (1/0.04 - 1)^{1/2} = \sqrt{24}$)

$$y = x/(5 \pm \sqrt{24}); \qquad \text{i.e., } y = 0.10102x \text{ and } y = 9.89898x.$$

One more: take $c = 0.49$; then $a = 1/0.98 = 1.02041$ and $b = (1/(4(0.49)^2) - 1)^{1/2} = 0.20306$. So the level curve $f(x, y) = 0.49$ consists of the lines

$$y = x/(1.02041 \pm 0.20306); \text{i.e., } y = 0.81735x \text{ and } y = 1.22347x.$$

The surface in question is built of lines that are parallel to the xy-plane and intersect the z-axis at various heights between $-1/2$ and $1/2$. Since different approaches to $(0, 0)$ give different results, the limit of $f(x, y)$ as $(x, y) \to (0, 0)$ does not exist.

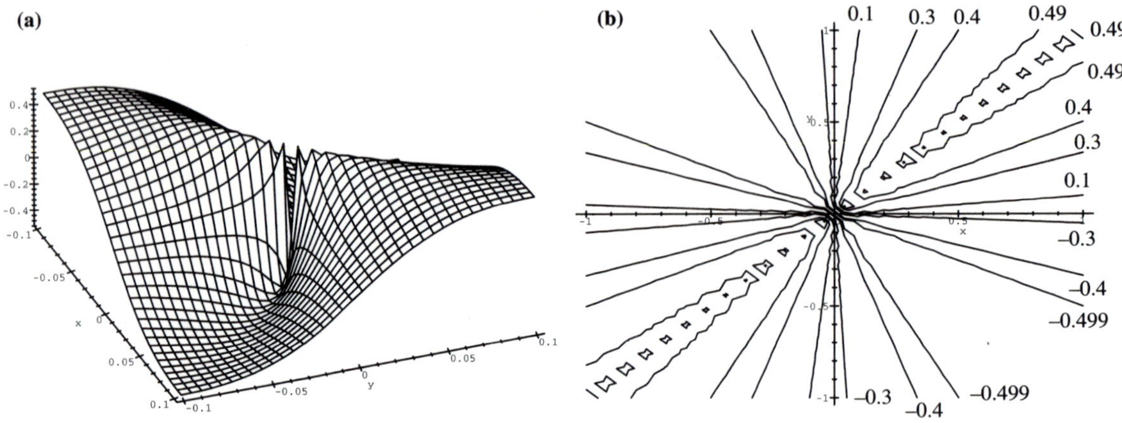

Figure 2.9.3

Solutions to Further Explorations

1. Using $x = r\cos\theta$ and $y = r\sin\theta$ we get

$$\lim_{(x,y)\to(0,0)} f(x,y) = \lim_{r\to 0} \frac{r^5\cos\theta\sin^4\theta}{r^2\cos^2\theta + r^6\sin^6\theta} = \lim_{r\to 0}\frac{r^3\cos\theta\sin^4\theta}{\cos^2\theta + r^4\sin^6\theta} = 0.$$

So, by defining $f(0,0) = 0$, we will make f continuous at $(0,0)$.

3. Since $f(c) \neq 0$, we can select an interval $(f(c)-\epsilon, f(c)+\epsilon)$, where $\epsilon > 0$, that does not contain 0. By definition of continuity of f at c, there must be a number $\delta > 0$ such that, whenever $|x - c| < \delta$, the value $f(x)$ lies within $(f(c)-\epsilon, f(c)+\epsilon)$; i.e., $f(x) \neq 0$. In other words, for any x in the interval $(c-\delta, c+\delta)$ the value $f(x) \neq 0$.

Now assume that $f\colon U \subseteq \mathbb{R}^m \to \mathbb{R}$ is continuous at $\mathbf{a} \in U$ and that $f(\mathbf{a}) \neq 0$. An argument very similar to the one above shows that there is an open ball $B(\mathbf{a}, r)$ centered at $\mathbf{a}$ such that $f(\mathbf{x}) \neq 0$ for all $\mathbf{x}$ in $B(\mathbf{a}, r)$.

5. Let $f(x,y) = 1 - (x^2+y^2-9)^2 = c$; i.e., $(x^2+y^2-9)^2 = 1-c$. It follows that there are no level curves if $c > 1$. If $c = 1$, then $x^2+y^2-9 = 0$ and the level curve is the circle centered at the origin of radius 3. If $c < 1$, then $x^2+y^2 = 9 \pm \sqrt{1-c}$. As long as $c > -80$, $1-c < 81$ and $\sqrt{1-c} < 9$. Therefore, for $-80 < c < 1$, the level curve consists of two circles $x^2+y^2 = 9+\sqrt{1-c}$ and $x^2+y^2 = 9-\sqrt{1-c}$. When $c = -80$, we get $x^2+y^2 = 9 \pm 9$; i.e., the level curve consists of a circle of radius $\sqrt{18}$ and the point $(0,0)$. Finally, when $c < -80$, the level curve is the circle $x^2+y^2 = 9+\sqrt{1-c}$.

7. (a) The rate of change is given by

$$\frac{\partial T}{\partial x}(3/2, 2) = 30e^{-2t}\tfrac{\pi}{2}\cos(\pi x/2)\Big|_{(3/2,2)} = 15\pi e^{-4}\cos(3\pi/4) \approx -0.6103.$$

When $t = 2$, we get the curve $y(x) = T(x,2) = 30e^{-4}\sin(\pi x/2)$. It is a sine curve of period 4 and amplitude $30e^{-4} \approx 0.54947$; see Figure 2.9.7 (a). The graph shows the temperature at points on the rod. For example, at the points $x = 0$, $x = 2$ and $x = 4$ the temperature is zero.

The rod is warmest at $x = 1$ and coldest at $x = 3$. The fact that $(\partial T/\partial x)(3/2, 2) \approx -0.6103$ means that, as we move along the rod from $x = 3/2$ forward, we will experience a decrease in temperature of approximately 0.6103 degrees per unit distance.

(b) The required rate of change is

$$\frac{\partial T}{\partial t}(3/2, 2) = -60e^{-2t}\sin(\pi x/2)\Big|_{(3/2,2)} = -60e^{-4}\sin(3\pi/4) \approx -0.77707.$$

When $x = 3/2$, we get the curve $y(t) = T(3/2, t) = 30e^{-2t}\sin(3\pi/4) = 15\sqrt{2}e^{-2t}$; see Figure 2.9.7 (b). It is an exponentially decreasing function and represents the way the temperature changes at the location $x = 3/2$ on the rod. As $t \to \infty$, the temperature approaches 0. The fact that $(\partial T/\partial t)(3/2, 2) \approx -0.77707$ means that, at $t = 2$, the temperature decreases at a rate of approximately 0.77707 degrees per unit time.

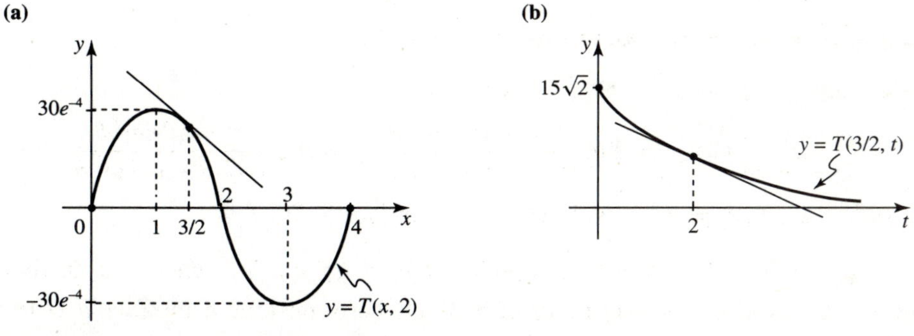

Figure 2.9.7

9. Substitute $x = y = 0$ into the assumption for f to get $|f(0,0)| \le 0$; so $f(0,0) = 0$. Consider the partial derivative

$$\frac{\partial f}{\partial x}(0,0) = \lim_{h\to 0}\frac{f(h,0) - f(0,0)}{h} = \lim_{h\to 0}\frac{f(h,0)}{h}.$$

By assumption, $|f(h,0)| \le h^2 + 0^2 = h^2$, and hence

$$0 \le \frac{|f(h,0)|}{h} \le \frac{h^2}{h} = h,$$

and therefore $|f(h,0)|/h \to 0$ as $h \to 0$. So $\lim_{h\to 0} f(h,0)/h = 0$ and therefore $(\partial f/\partial x)(0,0) = 0$. Similarly, $(\partial f/\partial y)(0,0) = 0$.

To complete the proof, we must show that

$$\lim_{(x,y)\to(0,0)} \frac{|f(x,y) - f(0,0) - Df(0,0)\cdot(x,y)|}{\sqrt{x^2+y^2}} = 0.$$

But $Df(0,0) = [\partial f/\partial x \quad \partial f/\partial y] = [0 \quad 0]$, so that the numerator is equal to $|f(x,y)|$. Using the assumption $|f(x,y)| \le x^2 + y^2$ again, we get

$$0 \le \frac{|f(x,y)|}{\sqrt{x^2+y^2}} \le \frac{x^2+y^2}{\sqrt{x^2+y^2}} = \sqrt{x^2+y^2},$$

and, as $x \to 0$ and $y \to 0$, $|f(x,y)|/\sqrt{x^2+y^2} \to 0$. It follows that f is differentiable at $(0,0)$.

11. (a) Let $f(x,y) = (xy)^{4/5}$. By definition,

$$f_x(0,0) = \lim_{h\to 0} \frac{f(h,0)-f(0,0)}{h} = \lim_{h\to 0} \frac{0-0}{h} = 0;$$

i.e., f_x exists at $(0,0)$. By symmetry, $f_y(0,0) = 0$ and so $Df(0,0) = [\partial f/\partial x \quad \partial f/\partial y] = [0 \quad 0]$ and

$$\begin{aligned}\lim_{(x,y)\to(0,0)} &= \frac{|f(x,y)-f(0,0)-Df(0,0)\cdot(x,y)|}{\sqrt{x^2+y^2}} \\ &= \lim_{(x,y)\to(0,0)} \frac{(xy)^{4/5}}{\sqrt{x^2+y^2}} \\ &= \lim_{r\to 0} \frac{(r^2\sin\theta\cos\theta)^{4/5}}{r} = \lim_{r\to 0} r^{3/5}(\sin\theta\cos\theta)^{4/5} = 0.\end{aligned}$$

So f is differentiable at $(0,0)$.

(b) Yes, since a function differentiable at $(0,0)$ must also be continuous at $(0,0)$.

(c) Yes, see (b).

(d) In order to show that f_x is continuous at $(0,0)$ we have to prove that

$$0 = f_x(0,0) = \lim_{(x,y)\to(0,0)} f_x(x,y).$$

Whenever $x \neq 0$, $f_x(x,y) = 4x^{-1/5}y^{4/5}/5 = 4(y^4/x)^{1/5}/5$. Choosing the approach $y = x^{1/4}$, we get that $\lim_{(x,y)\to(0,0)} f_x(x,y) = \lim_{(x,y)\to(0,0)} 4(1^{1/5})/5 = 4/5$. Consequently, f_x cannot be continuous at $(0,0)$.

13. By definition,

$$f_x(0,0) = \lim_{h\to 0} \frac{f(h,0)-f(0,0)}{h} = \lim_{h\to 0} \frac{1-0}{h} = \pm\infty,$$

and so f_x does not exist at $(0,0)$. (Notice that

$$f_y(0,0) = \lim_{h\to 0} \frac{f(0,h)-f(0,0)}{h} = \lim_{h\to 0} \frac{h-0}{h} = 1,$$

and hence f_y exists at $(0,0)$.) It follows that f is not differentiable at $(0,0)$.

(b) No conclusion on the continuity of f at $(0,0)$ can be made from (a).

(c) We have to check whether $\lim_{(x,y)\to(0,0)} f(x,y) = f(0,0) = 0$ holds. Examining the approach $y = 0$, we get

$$\lim_{(x,y)\to(0,0)} f(x,y) = \lim_{(x,y)\to(0,0)} 1 = 1.$$

So f is not continuous at $(0,0)$.

(d) No, since $f_x(0,0)$ is not defined.

15. By the chain rule,

$$
\begin{aligned}
\frac{\partial F}{\partial t} &= \frac{\partial F}{\partial m}\frac{\partial m}{\partial t} + \frac{\partial F}{\partial r}\frac{\partial r}{\partial t} \\
&= gR^2(R+r)^{-2}\frac{\partial m}{\partial t} + mgR^2(-2)(R+r)^{-3}\frac{\partial r}{\partial t} \\
&= \frac{gR^2}{(R+r)^2}\frac{\partial m}{\partial t} - \frac{2mgR^2}{(R+r)^3}\frac{\partial r}{\partial t}.
\end{aligned}
$$

3. VECTOR-VALUED FUNCTIONS OF ONE VARIABLE

3.1 Curves in $\mathbf{R}^2$ and $\mathbf{R}^3$

1. Let $\mathbf{a} = (3, 1, -2)$ and let $\mathbf{v}$ be the vector from $(3, 1, -2)$ to $(0, 5, 0)$. It follows that $\mathbf{v} = (-3, 4, 2)$ and the required equation is

$$\mathbf{c}(t) = \mathbf{a} + t\mathbf{v} = (3, 1, -2) + t(-3, 4, 2) = (3 - 3t, 1 + 4t, -2 + 2t),$$

where $t \in \mathbb{R}$.

3. There are infinitely many answers. For example, $\mathbf{c}(t) = (\sqrt{5}\cos t, \sqrt{5}\sin t)$, $t \in [0, 2\pi]$; or, $\mathbf{c}(t) = (\sqrt{5}\sin(2t), \sqrt{5}\cos(2t))$, $t \in [0, \pi]$.

5. The parametrization $\overline{\mathbf{c}} = (3\cos t, \sin t)$, $t \in [0, 2\pi]$ gives the right semi-axes, but the ellipse is centered at $(0, 0)$. So we move it to $(-2, -1)$:

$$\mathbf{c}(t) = (-2, -1) + (3\cos t, \sin t) = (-2 + 3\cos t, -1 + \sin t), \qquad t \in [0, 2\pi].$$

7. Let $y = t$; then $x = 1 + t^2$ and a parametrization is $\mathbf{c}(t) = (1 + t^2, t)$, $0 \le t \le 1$.

9. Let $x = t$; then $y^{2/3} = 1 - t^{2/3}$ and $y = \pm(1 - t^{2/3})^{3/2}$. Since $x^{2/3} + y^{2/3} = 1$ implies that $-1 \le x \le 1$ (and $-1 \le y \le 1$, but we do not need it here), it follows that $-1 \le t \le 1$. So

$$\mathbf{c}(t) = (t, \pm(1 - t^{2/3})^{3/2}), \qquad -1 \le t \le 1,$$

is one possible parametrization.

There is a more elegant way to get a parametrization: let $x = \cos^3 t$ and $y = \sin^3 t$; then $x^{2/3} + y^{2/3} = (\cos^3 t)^{2/3} + (\sin^3 t)^{2/3} = 1$! It follows that we can parametrize the given curve as $\mathbf{c}(t) = (\cos^3 t, \sin^3 t)$, $t \in [0, 2\pi]$.

11. Choose $y = 2\cos t$, $z = 2\sin t$, so that $y^2 + z^2 = 4$. From the equation $x = yz$ it follows that $x = (2\cos t)(2\sin t) = 2\sin 2t$, and thus one possible parametrization is given by $\mathbf{c}(t) = (2\sin 2t, 2\cos t, 2\sin t)$, $t \in [0, 2\pi]$.

13. Choose $x = \sqrt{2}\sin t$, $z = \sqrt{2}\cos t$, so that $x^2 + z^2 = 2$. Then $y = \pm\sqrt{2}\sin t$ (so that $y^2 + z^2 = 2$) and a parametrization is

$$\mathbf{c}(t) = (\sqrt{2}\sin t, \pm\sqrt{2}\sin t, \sqrt{2}\cos t),$$

where $t \in [0, 2\pi]$.

15. Eliminating t from $x = t - 3$ and $y = t^2$ we get $y = (x + 3)^2$. Since $0 \le t \le 2$, it follows that $-3 \le x \le -1$, and therefore $\mathbf{c}(t)$ represents the part of the parabola $y = (x + 3)^2$ (that is, the graph of $y = x^2$ moved 3 units to the left) from $(-3, 0)$ to $(-1, 4)$; see Figure 3.1.15.

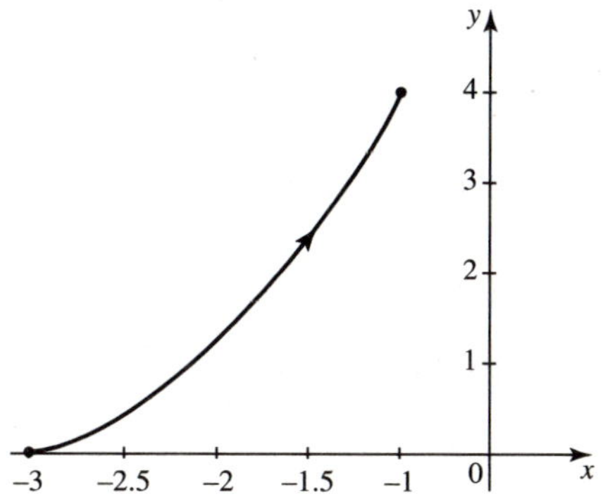

Figure 3.1.15

17. From $x = 2\cosh t$ and $y = 2\sinh t$ it follows that $x^2 - y^2 = 4$. Hence $\mathbf{c}$ is a part (or all, we have not figured it out yet) of the hyperbola with asymptotes $y = \pm x$ and x-intercepts at $(\pm 2, 0)$. Now $x = 2\cosh t \geq 0$ for all t, but $y = 2\sinh t$ can be any real number. It follows that $\mathbf{c}$ is the right branch of the hyperbola $x^2 - y^2 = 4$. It has no endpoints, and is oriented so that, as $t \to \infty$, both components of $\mathbf{c}(t)$ approach $+\infty$ (i.e., as t increases, $\mathbf{c}(t)$ gets closer and closer to its asymptote $y = x$ in the first quadrant); see Figure 3.1.17. The orientation can also be determined from the fact that for negative values of t, $\mathbf{c}(t)$ lies below the x-axis, and for $t > 0$, $\mathbf{c}(t)$ is above the x-axis.

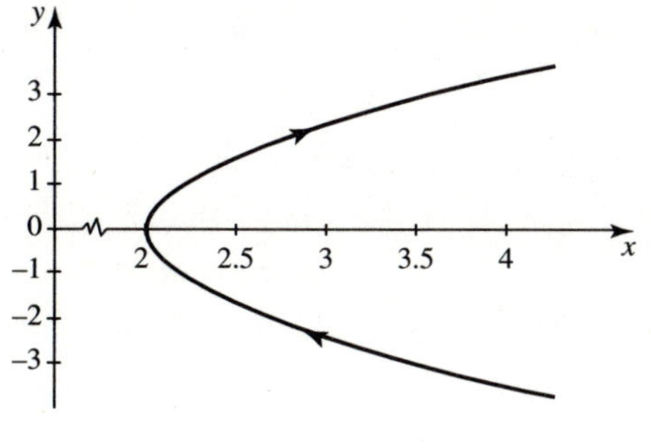

Figure 3.1.17

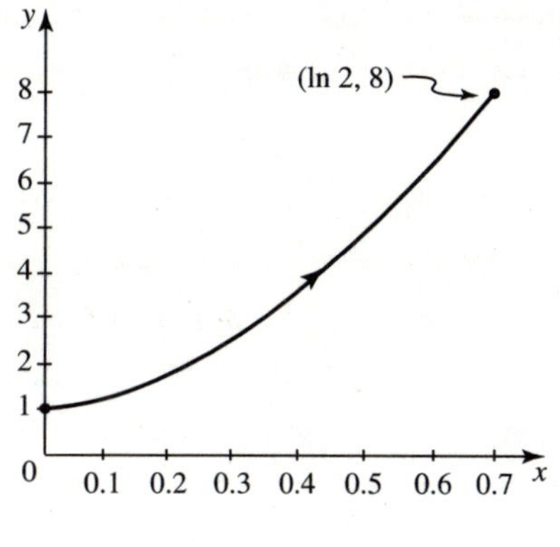

Figure 3.1.19

19. Combining $x = t$ and $y = e^{3t}$ we get $y = e^{3x}$. It follows that $\mathbf{c}(t)$ represents the part of the graph of the exponential function $y = e^{3x}$ between $(0, 1)$ and $(\ln 2, 8)$. Since $t \in [0, \ln 2]$, it follows that $x \in [0, \ln 2]$ and $\mathbf{c}(t)$ is oriented from $(0, 1)$ to $(\ln 2, 8)$; see Figure 3.1.19.

21. From $x = \cos t$, $y = \sin t$ it follows that $x^2 + y^2 = 1$. Since $z = 3$, it follows that $\mathbf{c}(t)$ represents a circle of radius 1 in the plane $z = 3$ with the center at $(0, 0, 3)$ on the z-axis. Its initial and terminal points are at $(1, 0, 3)$, and the orientation is counterclockwise, as seen by somebody standing on the plane $z = 3$.

23. From $x = \cos t$, $y = \sin t$, it follows that $x^2 + y^2 = 1$; i.e., $\mathbf{c}(t)$ lies on the cylinder of radius 1 whose axis of rotation is the z-axis. As t changes from 0 to π, the height (the z-coordinate) changes from 0 to π^3. So a point on $\mathbf{c}(t)$ has to "rotate" counterclockwise, and, at the same time, increase its height according to $z = t^3$. It follows that $\mathbf{c}(t)$ represents a circular helix (see Figure 3.1.23) with initial point $\mathbf{c}(0) = (1, 0, 0)$ and terminal point $\mathbf{c}(\pi) = (-1, 0, \pi^3)$.

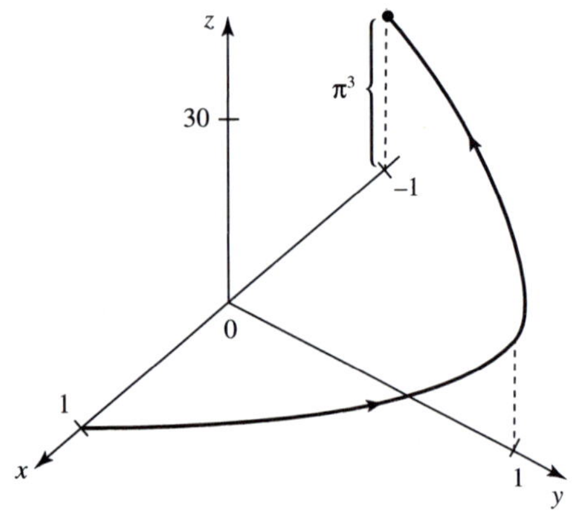

Figure 3.1.23

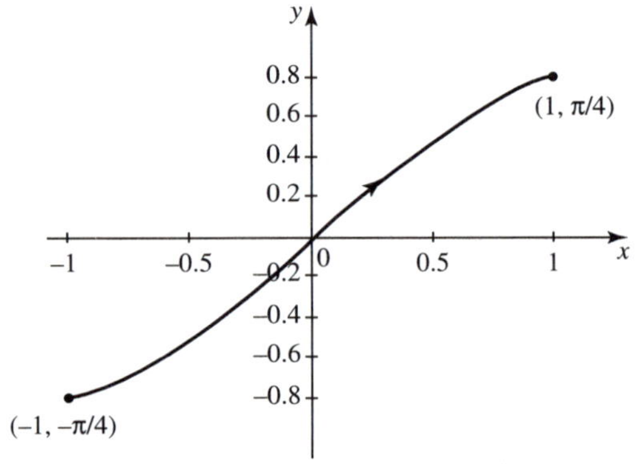

Figure 3.1.25

25. Eliminating t, we get $y = \arctan x$. So $\mathbf{c}(t)$ represents the part of the graph of the function $y = \arctan x$ between $(-1, \arctan(-1)) = (-1, -\pi/4)$ and $(1, \arctan(1)) = (1, \pi/4)$; see Figure 3.1.25.

27. We have to eliminate t to get an equation in x and y. The trick is to square and subtract: from $x^2 = \left(t + t^{-1}\right)^2 = t^2 + 2 + t^{-2}$ and $y^2 = \left(t - t^{-1}\right)^2 = t^2 - 2 + t^{-2}$ it follows that $x^2 - y^2 = 4$. This equation represents the hyperbola with asymptotes $y = \pm x$ and x-intercepts at $(\pm 2, 0)$. When $t = 1$, $\mathbf{c}(1) = (2, 0)$, and when $t = 2$, $\mathbf{c}(2) = (5/2, 3/2)$. It follows that $\mathbf{c}(t)$ is the part of the right branch of the hyperbola $x^2 - y^2 = 4$, with endpoints $(2, 0)$ and $(5/2, 3/2)$, oriented from $(2, 0)$ to $(5/2, 3/2)$.

29. A parametrization of the form $\mathbf{c}(t) = (a \sin t, a \cos t)$, $a > 0$, $t \in [0, 2\pi]$ represents the circle in the xy-plane of radius a, oriented clockwise, with initial and terminal points at $(0, a)$. The parameter t represents the angle measured clockwise from the positive y-axis. As t changes from 0 to 2π, the point $(0, a)$ is rotated, thus describing a circle.

This is not what is needed, but will help. Replace a by $e^{t/4}$ and think of t as described above. This time, as t changes, the "radius" keeps increasing from e^0 (when $t = 0$) to $e^{\pi/2}$ (when $t = 2\pi$). It follows that $\mathbf{c}(t)$ describes a clockwise spiral, starting at $(0, 1)$ and ending at $(0, e^{\pi/2} \approx 4.81048)$; see Figure 3.1.29.

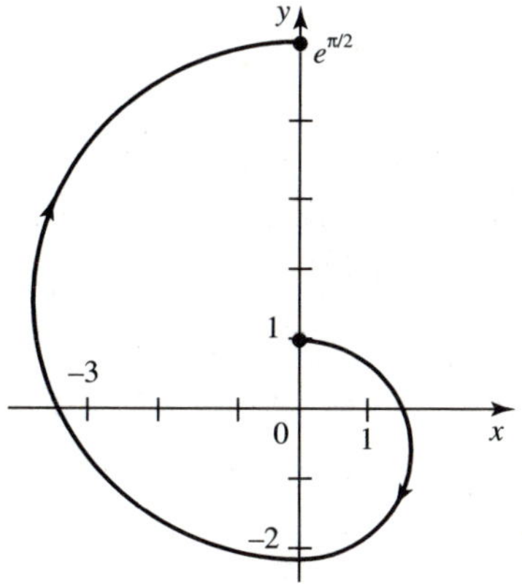

Figure 3.1.29

31. In all four cases, $x^2 + y^2 = 4$.

(a) The initial point of $\mathbf{c}_1$ is $\mathbf{c}_1(0) = (0,2)$; as t starts increasing from 0, both components of $\mathbf{c}_1$ become positive; i.e., $\mathbf{c}_1$ goes from $(0,2)$ into the first quadrant. In other words, $\mathbf{c}_1$ is oriented clockwise. Since sin and cos have period 2π, $\mathbf{c}_1$ winds around the circle once.

(b) The initial point of $\mathbf{c}_2$ is $\mathbf{c}_2(0) = (2,0)$ on the x-axis. Since the y-coordinate of $\mathbf{c}_2$ is positive as t starts increasing from 0, it follows that $\mathbf{c}_2$ is oriented counterclockwise; it winds around the circle once.

(c) The parametrization $\mathbf{c}_3$ gives a clockwise orientation to the circle $x^2 + y^2 = 4$. The initial and terminal points are at $(0,2)$. Since the period of $\sin 3t$ and $\cos 3t$ is $2\pi/3$, $\mathbf{c}_3$ winds around the circle three times.

(d) The initial point of $\mathbf{c}_4$ is the point $\mathbf{c}_4(0) = (-2,0)$ on the negative x-axis. As t starts increasing from 0, $y = 2\sin(t/2) > 0$; so $\mathbf{c}_4$ is oriented clockwise. The period of $\cos(t/2)$ and $\sin(t/2)$ is 4π, hence $\mathbf{c}_4$ winds around the circle once.

There are infinitely many possible parametrizations. For example,

$$\mathbf{c}_5(t) = (2\cos 5t, 2\sin 5t), \qquad t \in [0, 4\pi]$$

(counterclockwise, winds around the circle ten times), or, more "exotic",

$$\mathbf{c}_6(t) = (\sqrt{2}(\sin t + \cos t), \sqrt{2}(\sin t - \cos t)), \qquad t \in [0, 2\pi]$$

(counterclockwise, winds around the circle once).

33. Using the fact that $f(t) = t^{1/3}$ is not differentiable at $t = 0$ (recall that $f'(t) = t^{-2/3}/3$) we get a parametrization $\mathbf{c}(t) = (t^{1/3}, 2t^{1/3})$, $t \in \mathbb{R}$, of the line $y = 2x$ that is not differentiable at $t = 0$.

Remark: we could have used $\mathbf{c}_1(t) = (t^{3/5}, 2t^{3/5})$ (not differentiable at $t = 0$); or,

$\mathbf{c}_1(t) = ((t+1)^{2/3}, 2(t+1)^{2/3})$ (not differentiable at $t=-1$), etc.

35. The parametrization $\mathbf{c}(t)$ is continuous, but since $\mathbf{c}'(t) = (t^{-2/3}/3, 1)$ it is not differentiable at $t=0$. For the same reason it is not C^1 at $t=0$.

37. The parametrization $\mathbf{c}(t)$ is continuous, since both of its components are continuous. If $t \geq 0$, then $\mathbf{c}(t) = (t^2, t^6)$ and $\mathbf{c}'(t) = (2t, 6t^5)$; if $t < 0$, then $\mathbf{c}(t) = (-t^2, -t^6)$ and $\mathbf{c}'(t) = (-2t, -6t^5)$. It follows that $\mathbf{c}(t)$ is differentiable and C^1 if $t \neq 0$. Since $\mathbf{c}'(0) = (0,0)$ (i.e., $\mathbf{c}'(0)$ exists) it follows that $\mathbf{c}$ is differentiable at $t=0$ as well. Now $\lim_{t\to 0} \mathbf{c}'(t) = (0,0) = \mathbf{c}'(0)$ implies that $\mathbf{c}$ is C^1 at $t=0$.

3.2 Tangents, Velocity and Acceleration

1. The velocity is $\mathbf{v}(t) = \mathbf{c}'(t) = (3t^2, -t^{-2}, 0)$, and the speed is $\|\mathbf{c}'(t)\| = (9t^4 + t^{-4})^{1/2}$. The acceleration is $\mathbf{a}(t) = \mathbf{c}''(t) = (6t, 2t^{-3}, 0)$.

3. The velocity is $\mathbf{c}'(t) = (-\sin t/\cos t)\mathbf{i} + (\cos t/\sin t)\mathbf{j} = -\tan t\mathbf{i} + \cot t\mathbf{j}$ and the speed is $\|\mathbf{c}'(t)\| = \sqrt{\tan^2 t + \cot^2 t}$. The acceleration is $\mathbf{c}''(t) = -\sec^2 t\mathbf{i} - \csc^2 t\mathbf{j}$.

5. The velocity is computed by the product rule:

$$\begin{aligned}\mathbf{v}(t) = \mathbf{c}'(t) &= (2e^{2t}\sin(2t) + 2e^{2t}\cos(2t))\mathbf{i} + (2e^{2t}\cos(2t) - 2e^{2t}\sin(2t))\mathbf{j} \\ &= 2e^{2t}[(\sin(2t) + \cos(2t))\mathbf{i} + (\cos(2t) - \sin(2t))\mathbf{j}].\end{aligned}$$

It follows that the speed is

$$\begin{aligned}\|\mathbf{c}'(t)\| = 2e^{2t}(&\sin^2(2t) + \cos^2(2t) + 2\sin(2t)\cos(2t) + \sin^2(2t) \\ &+ \cos^2(2t) - 2\sin(2t)\cos(2t))^{1/2} = 2\sqrt{2}e^{2t}.\end{aligned}$$

Using the product rule again, we get

$$\begin{aligned}\mathbf{c}''(t) &= 4e^{2t}[(\sin(2t) + \cos(2t))\mathbf{i} + (\cos(2t) - \sin(2t))\mathbf{j}] \\ &\quad + 2e^{2t}[(2\cos(2t) - 2\sin(2t))\mathbf{i} + (-2\sin(2t) - 2\cos(2t))\mathbf{j}] \\ &= 4e^{2t}(2\cos(2t)\mathbf{i} - 2\sin(2t)\mathbf{j} \\ &= 8e^{2t}(\cos(2t)\mathbf{i} - \sin(2t)\mathbf{j}).\end{aligned}$$

7. The velocity is $\mathbf{v}(t) = \mathbf{c}'(t) = (t^{-1/2}/2, 1, 3t^{1/2}/2)$; the speed is

$$\|\mathbf{c}'(t)\| = \sqrt{t^{-1}/4 + 1 + 9t/4} = \tfrac{1}{2}\sqrt{t^{-1} + 4 + 9t};$$

the acceleration is $\mathbf{a}(t) = (-t^{-3/2}/4, 0, 3t^{-1/2}/4)$.

9. The velocity is

$$\mathbf{v}(t) = \int \mathbf{a}(t)\,dt = \int (-1, 1, 0)\,dt = (-t + C_1, t + C_2, C_3).$$

Since $\mathbf{v}(0) = (1,2,0)$ it follows that $(C_1, C_2, C_3) = (1,2,0)$; i.e., $C_1 = 1$, $C_2 = 2$ and $C_3 = 0$, and hence $\mathbf{v}(t) = (-t+1, t+2, 0)$. Now

$$\mathbf{c}(t) = \int \mathbf{v}(t)\,dt = \int (-t+1, t+2, 0)\,dt = \left(-\tfrac{1}{2}t^2 + t + D_1, \tfrac{1}{2}t^2 + 2t + D_2, D_3\right).$$

Since $\mathbf{c}(0) = (0,2,0)$ it follows that $(D_1, D_2, D_3) = (0,2,0)$; i.e., $D_1 = 0$, $D_2 = 2$ and $D_3 = 0$. Therefore, $\mathbf{c}(t) = (t - t^2/2, 2t + 2 + t^2/2, 0)$.

11. The velocity is $\mathbf{v}(t) = \int \mathbf{a}(t)\,dt = (t^2/2 + C_1, t + C_2, t + C_3)$. Since $\mathbf{v}(0) = (0,1,0)$ it follows that $(C_1, C_2, C_3) = (0,1,0)$; i.e., $C_1 = 0$, $C_2 = 1$ and $C_3 = 0$, so $\mathbf{v}(t) = (t^2/2, t+1, t)$. Integrating again, we get

$$\mathbf{c}(t) = \int \mathbf{v}(t)\,dt = \left(\tfrac{1}{6}t^3 + D_1, \tfrac{1}{2}t^2 + t + D_2, \tfrac{1}{2}t^2 + D_3\right).$$

Since $\mathbf{c}(0) = (-2,0,3)$ it follows that $(D_1, D_2, D_3) = (-2,0,3)$; i.e., $D_1 = -2$, $D_2 = 0$ and $D_3 = 3$. Therefore, $\mathbf{c}(t) = (t^3/6 - 2, t^2/2 + t, t^2/2 + 3)$.

13. The velocity is

$$\mathbf{v}(t) = \int \mathbf{a}(t)\,dt = \int (t\mathbf{i} + t^2\mathbf{j} + t\mathbf{k})\,dt = (\tfrac{1}{2}t^2 + C_1)\mathbf{i} + (\tfrac{1}{3}t^3 + C_2)\mathbf{j} + (\tfrac{1}{2}t^2 + C_3)\mathbf{k}.$$

Since $\mathbf{v}(0) = 2\mathbf{j} - 3\mathbf{k}$ it follows that $C_1\mathbf{i} + C_2\mathbf{j} + C_3\mathbf{k} = 2\mathbf{j} - 3\mathbf{k}$; i.e., $C_1 = 0$, $C_2 = 2$ and $C_3 = -3$, and hence $\mathbf{v}(t) = (t^2/2)\mathbf{i} + (2 + t^3/3)\mathbf{j} + (-3 + t^2/2)\mathbf{k}$. Now

$$\mathbf{c}(t) = \int \mathbf{v}(t)\,dt = \left(\tfrac{1}{6}t^3 + D_1\right)\mathbf{i} + \left(2t + \tfrac{1}{12}t^4 + D_2\right)\mathbf{j} + \left(-3t + \tfrac{1}{6}t^3 + D_3\right)\mathbf{k}.$$

Since $\mathbf{c}(0) = 4\mathbf{i} + 2\mathbf{j} - 6\mathbf{k}$, it follows that $D_1\mathbf{i} + D_2\mathbf{j} + D_3\mathbf{k} = 4\mathbf{i} + 2\mathbf{j} - 6\mathbf{k}$; i.e., $D_1 = 4$, $D_2 = 2$ and $D_3 = -6$. So, $\mathbf{c}(t) = (4 + t^3/6)\mathbf{i} + (2 + 2t + t^4/12)\mathbf{j} + (-6 - 3t + t^3/6)\mathbf{k}$.

15. The velocity of the particle is $\mathbf{c}'(t) = (-t^{-2}, 0, 2t)$. We have to find the maximum value of its speed $s(t) = \sqrt{t^{-4} + 4t^2}$ on $[1,4]$. Recall that the maximum of $s(t)$ on $[1,4]$ will occur either at the endpoints $t = 1$ and $t = 4$ or at the critical points inside $[1,4]$. Now $s'(t) = 0$ implies that $(t^{-4} + 4t^2)^{-1/2}(-4t^{-5} + 8t) = 0$ and hence $-4t^{-5} + 8t = 0$ and $t = (0.5)^{1/6}$ or $t = 0$. It follows that there are no critical points inside $[1,4]$. From $s(1) = \sqrt{5}$ and $s(4) = \sqrt{4^{-4} + 64} \approx 8.0002$ it follows that, on the interval $[1,4]$, the particle reaches its maximum speed of approximately 8.0002 when $t = 4$.

17. To find the highest position of the particle, we have to maximize the z-coordinate of the position vector. From $\mathbf{v}(t) = \int \mathbf{a}(t)dt = \int (3,0,1)dt = (3t + C_1, C_2, t + C_3)$ and $\mathbf{v}(0) = (1,3,2)$ it follows that $(C_1, C_2, C_3) = (1,3,2)$, i.e., $\mathbf{v}(t) = (3t+1, 3, t+2)$. Now

$$\mathbf{c}(t) = \int \mathbf{v}(t)\,dt = \int (3t+1, 3, t+2)\,dt = \left(\tfrac{3}{2}t^2 + t + D_1, 3t + D_2, \tfrac{1}{2}t^2 + 2t + D_3\right).$$

Since $\mathbf{c}(0) = (0,0,0)$, it follows that $D_1 = D_2 = D_3 = 0$ and so $\mathbf{c}(t) = (3t^2/2 + t, 3t, t^2/2 + 2t)$. The height $h(t)$ of the particle is given by $h(t) = t^2/2 + 2t$. Since $h'(t) = t + 2 > 0$ when $t \geq 0$, it follows that $h(t)$ is increasing for $t \geq 0$. Hence its maximum on $[0,12]$ occurs when $t = 12$.

19. Choose a coordinate system so that the y-axis represents the height and the x-axis the horizontal distance. Since the acceleration is $\mathbf{a}(t) = -g\mathbf{j}$, it follows that $\mathbf{v}(t) = \int \mathbf{a}(t)\,dt = C_1\mathbf{i} + (-gt + C_2)\mathbf{j}$. The initial condition on the velocity is $\|\mathbf{v}(0)\| = 700$ and the angle of $\mathbf{v}(0)$ with respect to the x-axis is $\theta = \pi/3$. Hence $\mathbf{v}(0) = \|\mathbf{v}(0)\|(\cos\theta\mathbf{i} + \sin\theta\mathbf{j}) = 350\mathbf{i} + 350\sqrt{3}\mathbf{j}$. From the formula for $\mathbf{v}(t)$, we get $\mathbf{v}(0) = C_1\mathbf{i} + C_2\mathbf{j}$; so $C_1 = 350$ and $C_2 = 350\sqrt{3}$ and $\mathbf{v}(t) = 350\mathbf{i} + (-gt + 350\sqrt{3})\mathbf{j}$.

Now $\mathbf{c}(t) = (350t + D_1)\mathbf{i} + (-gt^2/2 + 350\sqrt{3}\,t + D_2)\mathbf{j}$, and since $\mathbf{c}(0) = \mathbf{0}$, we get $\mathbf{c}(t) = 350t\mathbf{i} + (-gt^2/2 + 350\sqrt{3}t)\mathbf{j}$. The highest point corresponds to the maximum value of the y-coordinate $y(t) = -gt^2/2 + 350\sqrt{3}\,t$. Since $y'(t) = 350\sqrt{3} - gt = 0$, it follows that $t = 350\sqrt{3}/g$; assuming that $g \approx 9.80$ m/s^2, we get $t \approx 61.86$ s (and this gives a maximum, since $y''(t) = -g < 0$). The maximum height reached is $y(61.86) \approx 18,750$ m. The projectile will hit the ground when $y(t) = -gt^2/2 + 350\sqrt{3}t = 0$; i.e., when $t = 700\sqrt{3}/g \approx 123.72$ s (i.e., double the time needed for the projectile to reach its highest position). The range of the projectile is the value of the x-coordinate at the moment of impact, $x(123.72) = (350)(123.72) \approx 43,302$ m. Finally, the speed at time of impact is $\|\mathbf{v}(123.72)\| = \|350\mathbf{i} - 606.24\mathbf{j}\| = 700$ m/s.

21. We could use the implicit derivative to compute the slope of the tangent. Alternatively, rewrite the equation of the ellipse as $x^2/3 + y^2/(3/4) = 1$, and parametrize it using $\mathbf{c}(t) = (\sqrt{3}\cos t, \sqrt{3/4}\sin t) = (\sqrt{3}\cos t, (\sqrt{3}/2)\sin t)$, $t \in [0, 2\pi]$ (in this case, $x = \sqrt{3}$ is obtained by choosing $t = 0$). It follows that $\mathbf{c}'(0) = (-\sqrt{3}\sin t, (\sqrt{3}/2)\cos t)\Big|_{t=0} = (0, \sqrt{3}/2)$; consequently, the tangent at $\mathbf{c}(0) = (\sqrt{3}, 0)$ is vertical. Its parametric equation is (we can take any vector parallel to $\mathbf{c}'(0)$, so take $(0,1)$) $\ell(t) = (\sqrt{3}, 0) + t(0,1) = (\sqrt{3}, t)$, $t \in \mathbb{R}$.

23. The point $(0, 3, 2\pi)$ is obtained for $t = \pi/2$; so, to get the tangent, we need $\mathbf{c}'(\pi/2)$. From $\mathbf{c}'(t) = (-3\sin t, 3\cos t, 4)$ it follows that $\mathbf{c}'(\pi/2) = (-3, 0, 4)$ and hence

$$\ell(t) = (0, 3, 2\pi) + t(-3, 0, 4) = (-3t, 3, 2\pi + 4t),$$

where $t \in \mathbb{R}$, is a parametrization of the desired tangent line.

25. Since $\cosh 0 = 1$ and $\sinh 0 = 0$, it follows that the point of tangency is $\mathbf{c}(0)$. From $\mathbf{c}'(0) = (-\sinh t, \cosh t)\Big|_{t=0} = (0, 1)$, it follows that the tangent is $\ell(t) = (-1, 1) + t(0, 1) = (-1, 1 + t)$, $t \in \mathbb{R}$.

27. Using the product rule and the chain rule we get

$$\mathbf{c}' = (r'\cos\theta + r\theta'(-\sin\theta), r'\sin\theta + r\theta'\cos\theta) = r'(\cos\theta, \sin\theta) + r\theta'(-\sin\theta, \cos\theta).$$

Similarly,

$$\begin{aligned}\mathbf{c}'' &= r''(\cos\theta, \sin\theta) + r'(-\theta'\sin\theta, \theta'\cos\theta) + r'\theta'(-\sin\theta, \cos\theta)\\ &\qquad + r\theta''(-\sin\theta, \cos\theta) + r\theta'(-\theta'\cos\theta, -\theta'\sin\theta)\\ &= r''(\cos\theta, \sin\theta) + 2r'\theta'(-\sin\theta, \cos\theta) + r\theta''(-\sin\theta, \cos\theta) - r(\theta')^2(\cos\theta, \sin\theta)\\ &= (r'' - r(\theta')^2)(\cos\theta, \sin\theta) + (2r'\theta' + r\theta'')(-\sin\theta, \cos\theta).\end{aligned}$$

29. The velocity is

$$\mathbf{c}'(t) = (-e^{-t}\cos t - e^{-t}\sin t, -e^{-t}\sin t + e^{-t}\cos t) = -e^{-t}(\cos t + \sin t, \sin t - \cos t).$$

The velocity is horizontal when its y component $\sin t - \cos t$ is zero. Hence $\sin t = \cos t$; i.e., $\tan t = 1$ and $t = \pi/4$, $t = 5\pi/4$ and $t = 9\pi/4$. The velocity is vertical when the x component of $\mathbf{c}'(t) = 0$; i.e., when $\cos t + \sin t = 0$. In that case, $\tan t = -1$ and $t = 3\pi/4$, $t = 7\pi/4$ and $t = 11\pi/4$. It follows that the velocity is horizontal at $\mathbf{c}(\pi/4) = e^{-\pi/4}(\sqrt{2}/2, \sqrt{2}/2)$, $\mathbf{c}(5\pi/4) = -e^{-5\pi/4}(\sqrt{2}/2, \sqrt{2}/2)$ and $\mathbf{c}(9\pi/4) = e^{-9\pi/4}(\sqrt{2}/2, \sqrt{2}/2)$. It is vertical at the points $\mathbf{c}(3\pi/4) = e^{-3\pi/4}(-\sqrt{2}/2, \sqrt{2}/2)$, $\mathbf{c}(7\pi/4) = e^{-7\pi/4}(\sqrt{2}/2, -\sqrt{2}/2)$ and $\mathbf{c}(11\pi/4) = e^{-11\pi/4}(-\sqrt{2}/2, \sqrt{2}/2)$.

31. The composition $\mathbf{d}(t) = \mathbf{F}(\mathbf{c}(t))$ is the image of the curve $\mathbf{c}$ under $\mathbf{F}$; we need its tangent. By the chain rule, $\mathbf{d}'(t) = D\mathbf{F}(\mathbf{c}(t)) \cdot \mathbf{c}'(t)$ and hence

$$\begin{aligned}\mathbf{d}'(0) &= D\mathbf{F}(\mathbf{c}(0)) \cdot \mathbf{c}'(0) = D\mathbf{F}(0,0) \cdot \mathbf{c}'(0)\\ &= \begin{bmatrix} 2xy - 3x^2 & x^2 \\ ye^x & e^x \end{bmatrix}_{at\ (0,0)} \cdot \begin{bmatrix} \cos t \\ 2t - 1 \end{bmatrix}_{at\ t=0} = \begin{bmatrix} 0 & 0 \\ 0 & 1 \end{bmatrix} \cdot \begin{bmatrix} 1 \\ -1 \end{bmatrix} = \begin{bmatrix} 0 \\ -1 \end{bmatrix}.\end{aligned}$$

In words, $\mathbf{d}'(0) = -\mathbf{j}$ is the tangent to the image of $\mathbf{c}$ under $\mathbf{F}$ at $t = 0$.

33. The image $\mathbf{d}(t)$ of $\mathbf{c}(t)$ under $\mathbf{F}$ is given by the composition $\mathbf{d}(t) = \mathbf{F}(\mathbf{c}(t))$, and the corresponding tangents are related by $\mathbf{d}'(t) = D\mathbf{F}(\mathbf{c}(t)) \cdot \mathbf{c}'(t)$.

We have to compute $D\mathbf{F}$. Let $A = \begin{bmatrix} a_{11} & a_{12} \\ a_{21} & a_{22} \end{bmatrix}$ and $\mathbf{x} = \begin{bmatrix} x \\ y \end{bmatrix}$. Then

$$\mathbf{F}(\mathbf{x}) = \begin{bmatrix} a_{11} & a_{12} \\ a_{21} & a_{22} \end{bmatrix} \cdot \begin{bmatrix} x \\ y \end{bmatrix} = \begin{bmatrix} a_{11}x + a_{12}y \\ a_{21}x + a_{22}y \end{bmatrix}$$

and it follows that $D\mathbf{F} = A$. Hence

$$\mathbf{d}'(0) = D\mathbf{F}(\mathbf{c}(0)) \cdot \mathbf{c}'(0) = \begin{bmatrix} a_{11} & a_{12} \\ a_{21} & a_{22} \end{bmatrix}_{at\ (0,0)} \cdot \begin{bmatrix} c_1 \\ c_2 \end{bmatrix} = \begin{bmatrix} a_{11}c_1 + a_{12}c_2 \\ a_{21}c_1 + a_{22}c_2 \end{bmatrix},$$

i.e., $\mathbf{d}'(0) = (a_{11}c_1 + a_{12}c_2)\mathbf{i} + (a_{21}c_1 + a_{22}c_2)\mathbf{j}$.

3.3 Length of a Curve

1. The polygonal path p_5 is obtained by joining $\mathbf{c}(1)$, $\mathbf{c}(1.2)$, $\mathbf{c}(1.4)$, $\mathbf{c}(1.6)$, $\mathbf{c}(1.8)$ and $\mathbf{c}(2)$ with straight line segments; see Figure 3.3.1. From $\mathbf{c}'(t) = (1, -t^{-2})$ it follows that $\|\mathbf{c}'(t)\| = \sqrt{1 + t^{-4}}$ and hence, using (3.2),

$$\ell(p_5) \approx \|\mathbf{c}'(1)\|\Delta t_1 + \|\mathbf{c}'(1.2)\|\Delta t_2 + \|\mathbf{c}'(1.4)\|\Delta t_3 + \|\mathbf{c}'(1.6)\|\Delta t_4 + \|\mathbf{c}'(1.8)\|\Delta t_5.$$

Since $\Delta t_1 = \ldots = \Delta t_5 = 0.2$, we get

$$\begin{aligned}\ell(p_5) &\approx (1.4142)(0.2) + (1.2175)(0.2) + (1.1226)(0.2)\\ &\quad + (1.0736)(0.2) + (1.0465)(0.2) = 1.17488.\end{aligned}$$

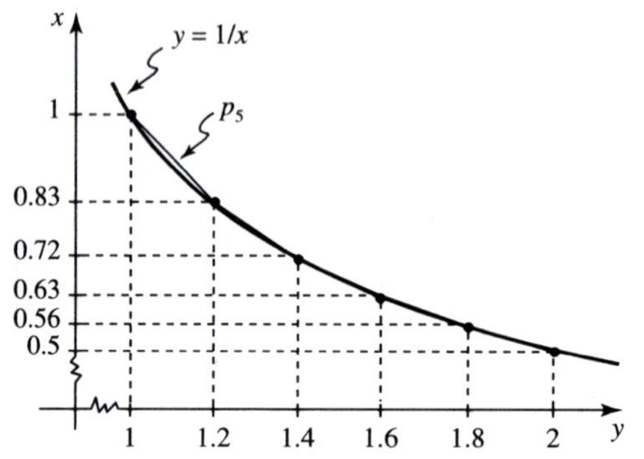

Figure 3.3.1

3. No integration is needed: $\mathbf{c}(t)$ is the semi-circle in the first and the fourth quadrants of radius 1 (it goes from $\mathbf{c}(0) = (0,1)$ to $\mathbf{c}(\pi/2) = (0,-1)$). So its length is π.

5. From

$$\mathbf{c}'(t) = (e^t \cos t - e^t \sin t)\mathbf{i} + (e^t \sin t + e^t \cos t)\mathbf{j} = e^t[(\cos t - \sin t)\mathbf{i} + (\sin t + \cos t)\mathbf{j}]$$

it follows that

$$\|\mathbf{c}'(t)\| = e^t \|(\cos t - \sin t)\mathbf{i} + (\sin t + \cos t)\mathbf{j}\| = e^t\sqrt{2}$$

and therefore the length of $\mathbf{c}$ is

$$\ell(\mathbf{c}) = \int_0^\pi e^t\sqrt{2}\,dt = \sqrt{2}\,e^t\Big|_0^\pi = \sqrt{2}(e^\pi - 1) \approx 31.31167.$$

7. From $\mathbf{c}'(t) = (1, 3(1+t)^{1/2}/2)$ it follows that

$$\|\mathbf{c}'(t)\| = \sqrt{1 + \tfrac{9}{4}(1+t)} = \tfrac{1}{2}\sqrt{9t+13},$$

and therefore the length of $\mathbf{c}$ is

$$\ell(\mathbf{c}) = \int_0^1 \tfrac{1}{2}\sqrt{9t+13}\;dt = \tfrac{1}{27}(9t+13)^{3/2}\Big|_0^1 = \tfrac{1}{27}(22^{3/2} - 13^{3/2}) \approx 2.08581.$$

9. From $\mathbf{c}'(t) = (1, \sinh t)$ it follows that

$$\|\mathbf{c}'(t)\| = \sqrt{1 + \sinh^2 t} = \sqrt{\cosh^2 t} = \cosh t.$$

The length of $\mathbf{c}$ is

$$\ell(\mathbf{c}) = \int_0^1 \cosh t\,dt = \sinh t\Big|_0^1 = \sinh 1 \approx 1.1752.$$

11. From $\mathbf{c}'(t) = (2-2t)\mathbf{i} + 4t^{1/2}\mathbf{j}$ it follows that

$$\|\mathbf{c}'(t)\| = \sqrt{4 + 4t^2 - 8t + 16t} = \sqrt{4t^2 + 8t + 4} = \sqrt{(2t+2)^2} = 2t + 2,$$

since $2t+2>0$ when $1 \le t \le 3$. The length of $\mathbf{c}$ is

$$\ell(\mathbf{c}) = \int_1^3 (2t+2)\,dt = \left.(t^2+2t)\right|_1^3 = 12.$$

13. Using the formula of Example 3.25 we get that the length of the curve $y = \sin x$, $x \in [0, 2\pi]$ (call it $\mathbf{c}_1$) is

$$\ell(\mathbf{c}_1) = \int_0^{2\pi} \sqrt{1+\cos^2 x}\,dx.$$

Similarly, the length of the curve $y = 2\sin x$, $x \in [0, 2\pi]$ (call it $\mathbf{c}_2$) is

$$\ell(\mathbf{c}_2) = \int_0^{2\pi} \sqrt{1+4\cos^2 x}\,dx.$$

We will show that

$$\sqrt{1+4\cos^2 x} \le \sqrt{2.5}\sqrt{1+\cos^2 x}$$

and therefore (by integrating both sides from 0 to 2π), $\ell(\mathbf{c}_2) \le \sqrt{2.5}\,\ell(\mathbf{c}_1)$. Since $\sqrt{2.5} < 2$, it follows that $\mathbf{c}_2$ is not twice as long as $\mathbf{c}_1$.

Consider the function $f(t) = (1+4t^2)/(1+t^2)$. From $f'(t) = 6t/(1+t^2)^2$, it follows that $f(t)$ is increasing for $t \ge 0$. Now let $t = \cos x$. Since $f(t)$ is increasing, its largest value will be obtained for the largest possible value of t, which is 1 (since $\cos x \le 1$). So

$$f(\cos x) = \frac{1+4\cos^2 x}{1+\cos^2 x} \le \tfrac{5}{2},$$

and (by computing the square root) $\sqrt{1+4\cos^2 x} \le \sqrt{5/2}\,\sqrt{1+\cos^2 x}$.

15. From $dr/d\theta = -a\sin\theta$ it follows that the length of the curve is (since $a > 0$)

$$\ell = \int_0^{\pi/4} \sqrt{a^2\cos^2\theta + a^2\sin^2\theta}\,d\theta = \int_0^{\pi/4} a\,d\theta = \frac{a\pi}{4}.$$

17. From $dr/d\theta = 6\theta$ it follows that

$$\ell = \int_1^2 \sqrt{9\theta^4 + 36\theta^2}\,d\theta = \int_1^2 \sqrt{9\theta^2(\theta^2+4)}\,d\theta = \int_1^2 3\theta\sqrt{\theta^2+4}\,d\theta,$$

since $\theta > 0$ (so that $\sqrt{\theta^2} = |\theta| = \theta$). We proceed by substitution $u = \theta^2 + 4$, thus getting

$$\ell = \tfrac{3}{2}\int_5^8 \sqrt{u}\,du = \left.u^{3/2}\right|_5^8 = 8^{3/2} - 5^{3/2} \approx 11.447.$$

19. Since $r^2 + (dr/d\theta)^2 = (1+\sin\theta)^2 + \cos^2\theta = 2 + 2\sin\theta$, it follows that

$$\ell = \int_{-\pi/2}^{\pi/2} \sqrt{2}\sqrt{1+\sin\theta}\,d\theta = \sqrt{2}\int_{-\pi/2}^{\pi/2} \sqrt{1+\sin\theta}\frac{\sqrt{1-\sin\theta}}{\sqrt{1-\sin\theta}}\,d\theta = \sqrt{2}\int_{-\pi/2}^{\pi/2} \frac{\cos\theta}{\sqrt{1-\sin\theta}}\,d\theta$$

(we used the fact that $\sqrt{1-\sin^2\theta} = \sqrt{\cos^2\theta} = |\cos\theta| = \cos\theta$, since $-\pi/2 \le \theta \le \pi/2$). Proceed by the substitution $u = 1 - \sin\theta$ to get

$$\ell = \sqrt{2}(-2)(1-\sin\theta)^{1/2}\Big|_{-\pi/2}^{\pi/2} = -2\sqrt{2}(-\sqrt{2}) = 4.$$

21. From $\mathbf{c}'(t) = (\sin 2t + 2t\cos 2t, \cos 2t - 2t\sin 2t, 2t^{1/2})$ we get

$$\begin{aligned}\|\mathbf{c}'(t)\| &= \sqrt{\sin^2 2t + 4t^2\cos^2 2t + \cos^2 2t + 4t^2\sin^2 2t + 4t}\\ &= \sqrt{1+4t+4t^2} = \sqrt{(2t+1)^2} = 2t+1,\end{aligned}$$

(since $t \geq 0$ implies $2t+1 \geq 0$ and so $\sqrt{(2t+1)^2} = |2t+1| = 2t+1$). Consequently, the arc-length function $s(t)$ is given by

$$s(t) = \int_0^t \|\mathbf{c}'(\tau)\|\, d\tau = \int_0^t (2\tau+1)\, d\tau = t^2 + t,$$

where $0 \leq t \leq 2\pi$.

23. From $\mathbf{c}'(t) = -5\sin t\mathbf{i} + 5\cos t\mathbf{j} + 12\mathbf{k}$ we get $\|\mathbf{c}'(t)\| = 13$ and the arc-length function of $\mathbf{c}$ is

$$s(t) = \int_0^t \|\mathbf{c}'(\tau)\|\, d\tau = \int_0^t 13\, d\tau = 13t,$$

where $t \in [0, \pi/4]$. The path $\mathbf{c}$, parametrized by the arc-length function s is (from $s = 13t$ it follows that $t = s/13$)

$$\mathbf{c}(s) = 5\cos(s/13)\,\mathbf{i} + 5\sin(s/13)\,\mathbf{j} + \tfrac{12}{13}s\mathbf{k}, \qquad \text{where} \qquad s \in [0, 13\pi/4].$$

25. From $\mathbf{c}(t) = e^t\cos t\,\mathbf{i} + e^t\sin t\,\mathbf{j}$ we get

$$\mathbf{c}'(t) = (e^t\cos t - e^t\sin t)\mathbf{i} + (e^t\sin t + e^t\cos t)\mathbf{j} = e^t[(\cos t - \sin t)\mathbf{i} + (\sin t + \cos t)\mathbf{j}];$$

it follows that

$$\|\mathbf{c}'(t)\| = e^t\|(\cos t - \sin t)\mathbf{i} + (\sin t + \cos t)\mathbf{j}\| = e^t\sqrt{2},$$

and therefore

$$s(t) = \int_0^t \|\mathbf{c}'(\tau)\|\, d\tau = \int_0^t \sqrt{2}\, e^\tau\, d\tau = \sqrt{2}(e^t - 1), \qquad \text{where} \qquad 0 \leq t \leq 1.$$

Solving $s = \sqrt{2}(e^t - 1)$ for t, we get $t = \ln(1 + s/\sqrt{2})$ and the parametrization by the arc-length function is (since $e^t = s/\sqrt{2} + 1$)

$$\mathbf{c}(s) = \left(\frac{s}{\sqrt{2}} + 1\right)\cos\left(\ln(1+s/\sqrt{2})\right)\mathbf{i} + \left(\frac{s}{\sqrt{2}} + 1\right)\sin\left(\ln(1+s/\sqrt{2})\right)\mathbf{j},$$

where $s \in [0, \sqrt{2}(e-1)]$.

27. Since $\mathbf{c}_2(t) = \mathbf{c}_1(t/2)$ and $\mathbf{c}_3(t) = \mathbf{c}_1(-t)$, it follows that $\mathbf{c}_1$, $\mathbf{c}_2$ and $\mathbf{c}_3$ are reparametrizations of each other, so they have the same image (i.e., represent the same curve). From $\mathbf{c}_1'(t) = (-2\sin 2t, 2\cos 2t, 1)$, $\mathbf{c}_2'(t) = (-\sin t, \cos t, 1/2)$ and $\mathbf{c}_3'(t) = (-2\sin 2t, -2\cos 2t, -1)$ we conculde that $\mathbf{c}_1'(t)$, $\mathbf{c}_2'(t)$, and $\mathbf{c}_3'(t)$ are non-zero for all t. Moreover, $\mathbf{c}_1$ maps distinct points to distinct points (for two different values of t the z-coordinates are different), so $\mathbf{c}_1$ is a smooth parametrization. For the same reasons $\mathbf{c}_2$ and $\mathbf{c}_3$ are smooth as well. Consequently, it does not matter what parametrization is used in the computation of the length. Using $\mathbf{c}_1$, we get

$$\ell = \int_0^\pi \|\mathbf{c}_1'(t)\|\, dt = \int_0^\pi \sqrt{5}\, dt = \pi\sqrt{5}.$$

29. From $\mathbf{c}'(t) = (-t^{-2} - 1, -t^{-2} + 1, 0)$, we get

$$\mathbf{T}(1) = \mathbf{c}'(1)/\|\mathbf{c}'(1)\| = (-2, 0, 0)/2 = (-1, 0, 0) = -\mathbf{i}.$$

31. The curve $\mathbf{c}(t)$ is not smooth, since $\mathbf{c}(\pi) = (0, 0) = \mathbf{c}(0)$; i.e., $\mathbf{c}$ intersects itself.

3.4 Acceleration and Curvature

1. From $\mathbf{c}(t) = (t^2, t, t^2)$ we get $\mathbf{c}'(t) = (2t, 1, 2t)$, $\|\mathbf{c}'(t)\| = \sqrt{8t^2+1}$ and the acceleration $\mathbf{a}(t) = (2, 0, 2)$. From $ds/dt = \|\mathbf{c}'(t)\| = \sqrt{8t^2+1}$ it follows that

$$\frac{d^2s}{dt^2} = \frac{1}{2\sqrt{8t^2+1}}16t = \frac{8t}{\sqrt{8t^2+1}};$$

so the tangential component is (recall that $\mathbf{T}(t) = \mathbf{c}'(t)/\|\mathbf{c}'(t)\|$)

$$\mathbf{a}_T(t) = \frac{d^2s}{dt^2}\mathbf{T}(t) = \frac{8t}{\sqrt{8t^2+1}}\frac{(2t,1,2t)}{\sqrt{8t^2+1}} = \frac{(16t^2, 8t, 16t^2)}{8t^2+1}.$$

The normal component is

$$\begin{aligned}\mathbf{a}_N = \mathbf{a} - \mathbf{a}_T &= (2,0,2) - \frac{(16t^2, 8t, 16t^2)}{8t^2+1}\\ &= \frac{(16t^2+2, 0, 16t^2+2)}{8t^2+1} - \frac{(16t^2, 8t, 16t^2)}{8t^2+1} = \frac{(2, -8t, 2)}{8t^2+1}.\end{aligned}$$

3. From $\mathbf{c}(t) = 5t\,\mathbf{i} + 12\sin t\,\mathbf{j} + 12\cos t\,\mathbf{k}$ we get $\mathbf{c}'(t) = 5\mathbf{i} + 12\cos t\,\mathbf{j} - 12\sin t\,\mathbf{k}$, $\|\mathbf{c}'(t)\| = \sqrt{5^2+12^2} = 13$ and the acceleration $\mathbf{a}(t) = -12\sin t\,\mathbf{j} - 12\cos t\,\mathbf{k}$. Since $ds/dt = \|\mathbf{c}'(t)\| = 13$ it follows that $d^2s/dt^2 = 0$ and the tangential component $\mathbf{a}_T = (d^2s/dt^2)\mathbf{T}(t)$ of $\mathbf{a}$ is zero. Therefore $\mathbf{a}_N = \mathbf{a} = -12\sin t\,\mathbf{j} - 12\cos t\,\mathbf{k}$.

5. From $\mathbf{c}'(t) = (1-\cos t)\mathbf{i} + \sin t\mathbf{j}$ it follows that $\|\mathbf{c}'(t)\| = \sqrt{2-2\cos t}$ and $\mathbf{a}(t) = \sin t\,\mathbf{i} + \cos t\,\mathbf{j}$. Next, $ds/dt = \|\mathbf{c}'(t)\| = \sqrt{2-2\cos t}$ and

$$\frac{d^2s}{dt^2} = \frac{2\sin t}{2\sqrt{2-2\cos t}} = \frac{\sin t}{\sqrt{2-2\cos t}}.$$

The unit tangent is

$$\mathbf{T}(t) = \frac{\mathbf{c}'(t)}{\|\mathbf{c}'(t)\|} = \frac{(1-\cos t)\mathbf{i} + \sin t\mathbf{j}}{\sqrt{2-2\cos t}}$$

and so

$$\mathbf{a}_T = \frac{d^2s}{dt^2}\mathbf{T}(t) = \frac{\sin t}{\sqrt{2-2\cos t}}\frac{(1-\cos t)\mathbf{i} + \sin t\mathbf{j}}{\sqrt{2-2\cos t}} = \frac{\sin t(1-\cos t)\mathbf{i} + \sin^2 t\mathbf{j}}{2-2\cos t}.$$

The normal component is

$$\begin{aligned}\mathbf{a}_N = \mathbf{a} - \mathbf{a}_T &= \sin t\mathbf{i} + \cos t\mathbf{j} - \frac{\sin t(1-\cos t)\mathbf{i} + \sin^2 t\mathbf{j}}{2-2\cos t}\\ &= \frac{\sin t(1-\cos t)\mathbf{i} + (2\cos t - 2\cos^2 t - \sin^2 t)\mathbf{j}}{2-2\cos t}.\end{aligned}$$

7. Drop t from the notation. From $\mathbf{c}' = (x', y')$ we get $\|\mathbf{c}'\| = \sqrt{(x')^2 + (y')^2}$ and $\mathbf{c}'' = (x'', y'')$. The unit tangent vector is

$$\mathbf{T} = \mathbf{c}'/\|\mathbf{c}'\| = (x'/\sqrt{(x')^2+(y')^2}, y'/\sqrt{(x')^2+(y')^2}).$$

Now

$$\begin{aligned}\frac{d}{dt}\left(\frac{x'}{\sqrt{(x')^2+(y')^2}}\right) &= \frac{x''\sqrt{(x')^2+(y')^2} - x'\frac{1}{2}((x')^2+(y')^2)^{-1/2}(2x'x''+2y'y'')}{(x')^2+(y')^2}\\ &= \frac{x''((x')^2+(y')^2) - (x')^2x'' - x'y'y''}{((x')^2+(y')^2)^{3/2}} = \frac{x''(y')^2 - x'y'y''}{((x')^2+(y')^2)^{3/2}}.\end{aligned}$$

By symmetry,

$$\frac{d}{dt}\left(\frac{y'}{\sqrt{(x')^2+(y')^2}}\right) = \frac{y''(x')^2 - x'y'x''}{((x')^2+(y')^2)^{3/2}},$$

and therefore

$$\begin{aligned}\mathbf{T}' &= \left(\frac{x''(y')^2 - x'y'y''}{((x')^2+(y')^2)^{3/2}}, \frac{y''(x')^2 - x'y'x''}{((x')^2+(y')^2)^{3/2}}\right)\\ &= ((x')^2+(y')^2)^{-3/2}\left(x''(y')^2 - x'y'y'', y''(x')^2 - x'y'x''\right).\end{aligned}$$

The norm of $\mathbf{T}'$ is

$$\begin{aligned}\|\mathbf{T}'\| &= ((x')^2+(y')^2)^{-3/2}\big((x'')^2(y')^4 + (x')^2(y')^2(y'')^2 - 2x'(y')^3x''y''\\ &\quad + (y'')^2(x')^4 + (x')^2(y')^2(x'')^2 - 2(x')^3y'x''y''\big)^{1/2}\\ &= ((x')^2+(y')^2)^{-3/2}\big((x'')^2(y')^2((x')^2+(y')^2) + (x')^2(y'')^2((x')^2+(y')^2)\\ &\quad - 2x'y'x''y''((x')^2+(y')^2)\big)^{1/2}\\ &= ((x')^2+(y')^2)^{-3/2}((x')^2+(y')^2)^{1/2}\big((x'')^2(y')^2 + (x')^2(y'')^2 - 2x'y'x''y''\big)^{1/2}\\ &= ((x')^2+(y')^2)^{-1}\big((x''y' - x'y'')^2\big)^{1/2} = ((x')^2+(y')^2)^{-1}|x''y' - x'y''|.\end{aligned}$$

Finally,

$$\kappa = \frac{\|\mathbf{T}'\|}{\|\mathbf{c}'\|} = \frac{((x')^2+(y')^2)^{-1}|x''y'-x'y''|}{\sqrt{(x')^2+(y')^2}} = \frac{|x''y'-x'y''|}{((x')^2+(y')^2)^{3/2}}.$$

9. We use the formula of Exercise 7. From $x(t) = 2-2t^3$ we get $x'(t) = -6t^2$ and $x''(t) = -12t$; similarly, $y(t) = t^3+1$ implies $y'(t) = 3t^2$ and $y''(t) = 6t$. Hence

$$\kappa(t) = \frac{|x'(t)y''(t) - x''(t)y'(t)|}{((x'(t))^2+(y'(t))^2)^{3/2}} = \frac{|(-6t^2)6t - (-12t)3t^2|}{(36t^4+9t^4)^{3/2}} = 0.$$

We can write $\mathbf{c}(t)$ as $\mathbf{c}(t) = (2,1) + t^3(-2,1)$, $t \in \mathbb{R}$. In words, $\mathbf{c}(t)$ is the line through $(2,1)$ in the direction of the vector $(-2,1)$. The curvature of a line is zero, and that confirms our calculation.

11. We will use the formula of Exercise 7. From $x(t) = t^2$ we get $x'(t) = 2t$ and $x''(t) = 2$; similarly, $y(t) = 3-t$ implies $y'(t) = -1$ and $y''(t) = 0$. Hence

$$\kappa(t) = \frac{|x'(t)y''(t) - x''(t)y'(t)|}{((x'(t))^2+(y'(t))^2)^{3/2}} = \frac{|2t(0) - 2(-1)|}{(4t^2+1)^{3/2}} = \frac{2}{(4t^2+1)^{3/2}}.$$

The curvature is largest when the denominator is smallest; i.e., when $t = 0$. It follows that the maximum curvature $\kappa(0) = 2$ occurs at $\mathbf{c}(0) = (0, 3)$. As $t \to \infty$, $\kappa(t) = 2/(4t^2 + 1)^{3/2} \to 0$.

13. From $\mathbf{c}(t) = (e^t \sin t, 0, e^t \cos t)$ we get

$$\mathbf{c}'(t) = (e^t \sin t + e^t \cos t, 0, e^t \cos t - e^t \sin t) = e^t(\sin t + \cos t, 0, \cos t - \sin t)$$

and hence

$$\|\mathbf{c}'(t)\| = e^t \|(\sin t + \cos t, 0, \cos t - \sin t)\| = e^t\sqrt{2}.$$

The unit tangent vector is

$$\mathbf{T}(t) = \frac{\mathbf{c}'(t)}{\|\mathbf{c}'(t)\|} = \frac{1}{\sqrt{2}}(\sin t + \cos t, 0, \cos t - \sin t).$$

Now

$$\mathbf{T}'(t) = \frac{1}{\sqrt{2}}(\cos t - \sin t, 0, -\sin t - \cos t),$$

and since

$$\|\mathbf{T}'(t)\| = \frac{1}{\sqrt{2}}\|(\cos t - \sin t, 0, -\sin t - \cos t)\| = \frac{1}{\sqrt{2}}\sqrt{2} = 1,$$

it follows that the principal unit normal is

$$\mathbf{N}(t) = \mathbf{T}'(t) = \frac{1}{\sqrt{2}}(\cos t - \sin t, 0, -\sin t - \cos t).$$

The curvature of $\mathbf{c}$ is

$$\kappa(t) = \frac{\|\mathbf{T}'(t)\|}{\|\mathbf{c}'(t)\|} = \frac{1}{\sqrt{2}e^t},$$

and

$$\begin{aligned}\mathbf{a}_N &= \|\mathbf{c}'(t)\|\mathbf{T}'(t) = \sqrt{2}e^t \frac{1}{\sqrt{2}}(\cos t - \sin t, 0, -\sin t - \cos t) \\ &= e^t(\cos t - \sin t, 0, -\sin t - \cos t)\end{aligned}$$

is the normal component of the acceleration.

15. From $\mathbf{c}(t) = (t, \cos t, 1 - \sin t)$ we get $\mathbf{c}'(t) = (1, -\sin t, -\cos t)$ and hence $\|\mathbf{c}'(t)\| = \sqrt{2}$. The unit tangent vector is

$$\mathbf{T}(t) = \frac{\mathbf{c}'(t)}{\|\mathbf{c}'(t)\|} = \frac{1}{\sqrt{2}}(1, -\sin t, -\cos t).$$

Now

$$\mathbf{T}'(t) = \frac{1}{\sqrt{2}}(0, -\cos t, \sin t),$$

and since $\|\mathbf{T}'(t)\| = 1/\sqrt{2}$, it follows that the principal unit normal is

$$\mathbf{N}(t) = \frac{\mathbf{T}'(t)}{\|\mathbf{T}'(t)\|} = (0, -\cos t, \sin t).$$

The curvature of $\mathbf{c}$ is

$$\kappa(t) = \frac{\|\mathbf{T}'(t)\|}{\|\mathbf{c}'(t)\|} = \frac{1/\sqrt{2}}{\sqrt{2}} = \frac{1}{2},$$

and

$$\mathbf{a}_N = \|\mathbf{c}'(t)\|\mathbf{T}'(t) = (0, -\cos t, \sin t)$$

is the normal component of the acceleration.

17. From $\mathbf{c}(t) = (e^{-t}\cos t, e^{-t}\sin t, e^{-t})$ we get $\mathbf{c}'(t) = e^{-t}(-\cos t - \sin t, -\sin t + \cos t, -1)$ and hence $\|\mathbf{c}'(t)\| = e^{-t}\sqrt{3}$. The unit tangent vector is

$$\mathbf{T}(t) = \frac{\mathbf{c}'(t)}{\|\mathbf{c}'(t)\|} = \frac{1}{\sqrt{3}}(-\cos t - \sin t, -\sin t + \cos t, -1).$$

Now

$$\mathbf{T}'(t) = \frac{1}{\sqrt{3}}(\sin t - \cos t, -\sin t - \cos t, 0),$$

and since $\|\mathbf{T}'(t)\| = (1/\sqrt{3})\sqrt{2} = \sqrt{2/3}$, it follows that the principal unit normal is

$$\mathbf{N}(t) = \frac{\mathbf{T}'(t)}{\|\mathbf{T}'(t)\|} = \frac{1}{\sqrt{2}}(\sin t - \cos t, -\sin t - \cos t, 0).$$

The curvature of $\mathbf{c}$ is

$$\kappa(t) = \frac{\|\mathbf{T}'(t)\|}{\|\mathbf{c}'(t)\|} = \frac{\sqrt{2}}{3}e^t,$$

and

$$\mathbf{a}_N = \|\mathbf{c}'(t)\|\mathbf{T}'(t) = e^{-t}(\sin t - \cos t, -\sin t - \cos t, 0)$$

is the normal component of the acceleration.

19. We will use the formula in Exercise 7 to compute the curvature. From $x(t) = t^3$ we get $x'(t) = 3t^2$ and $x''(t) = 6t$; similarly, $y(t) = t$ implies $y'(t) = 1$ and $y''(t) = 0$. Hence

$$\kappa(t) = \frac{|x'(t)y''(t) - x''(t)y'(t)|}{((x'(t))^2 + (y'(t))^2)^{3/2}} = \frac{|3t^2(0) - 6t(1)|}{(9t^4 + 1)^{3/2}} = \frac{6|t|}{(9t^4 + 1)^{3/2}}.$$

It follows that the radius of the osculating circle at $(8, 2)$ is $1/\kappa(2) = (145)^{3/2}/12 \approx 145.5026$. Since $\mathbf{c}(t)$ lies in the xy-plane, the osculating circle lies in it as well. Its center is on the line normal to $\mathbf{c}$ at $\mathbf{c}(2) = (8, 2)$. To get a direction normal to $\mathbf{c}$, take any vector perpendicular to $\mathbf{c}' = (3t^2, 1)$; for example, $(1, -3t^2)$. Hence a normal direction at $t = 2$ is

$$\mathbf{N} = (1, -3t^2)\Big|_{t=2} = (1, -12).$$

The equation of the normal line at $(8, 2)$ is

$$\ell(u) = (8, 2) + u(1, -12) = (8 + u, 2 - 12u),$$

where $u \in \mathbb{R}$. To locate the center of the osculating circle, we have to find the point on that line that is $1/\kappa(2) = (145)^{3/2}/12$ units away from $(8, 2)$ in the direction towards which $\overline{\mathbf{N}}$ points (that is, with $u > 0$). The distance between $(8, 2)$ and $(8 + u, 2 - 12u)$ is

$$\sqrt{u^2 + 144u^2} = \sqrt{145u^2},$$

so $\sqrt{145u^2} = (145)^{3/2}/12$ implies that $145u^2 = 145^3/144$ or $u = 145/12$ ($u = -145/12$ gives a point on the wrong side). Hence

$$\ell(145/12) = (8 + (145/12), 2 - 12(145/12)) = (241/12, -143)$$

is the center of the osculating circle. Its equation is $(x - 241/12)^2 + (y + 143)^2 = 145^3/144$.

21. The osculating plane goes through $\mathbf{c}(\pi/2) = (0, 2, \pi/2)$ and is spanned by the vectors $\mathbf{T}(\pi/2)$ and $\mathbf{N}(\pi/2)$. From $\mathbf{c}'(t) = (-2\sin t, 2\cos t, 1)$ it follows that $\|\mathbf{c}'(t)\| = \sqrt{5}$ and

$$\mathbf{T}(t) = \frac{\mathbf{c}'(t)}{\|\mathbf{c}'(t)\|} = \left(-\frac{2}{\sqrt{5}}\sin t, \frac{2}{\sqrt{5}}\cos t, \frac{1}{\sqrt{5}}\right).$$

The derivative of $\mathbf{T}$ is

$$\mathbf{T}'(t) = \left(-\frac{2}{\sqrt{5}}\cos t, -\frac{2}{\sqrt{5}}\sin t, 0\right),$$

its norm is $\|\mathbf{T}'(t)\| = 2/\sqrt{5}$ and so the principal unit normal is $\mathbf{N}(t) = \mathbf{T}'(t)/\|\mathbf{T}'(t)\| = (-\cos t, -\sin t, 0)$. To get a normal to the plane, we compute the cross product of the tangent vector $\mathbf{T}(t)$ and the vector $\mathbf{N}(t)$ at $t = \pi/2$:

$$\mathbf{T}(\pi/2) \times \mathbf{N}(\pi/2) = \begin{vmatrix} \mathbf{i} & \mathbf{j} & \mathbf{k} \\ -2/\sqrt{5} & 0 & 1/\sqrt{5} \\ 0 & -1 & 0 \end{vmatrix} = \frac{1}{\sqrt{5}}\mathbf{i} + \frac{2}{\sqrt{5}}\mathbf{k}.$$

The equation of the osculating plane is

$$\frac{1}{\sqrt{5}}(x - 0) + 0(y - 2) + \frac{2}{\sqrt{5}}(z - \frac{\pi}{2}) = 0,$$

i.e., $x + 2z - \pi = 0$.

23. Using the formula of Exercise 22 we get ($y = x^2$, so $y' = 2x$ and $y'' = 2$)

$$\kappa(x_0) = \left.\frac{|2|}{(1 + 4x^2)^{3/2}}\right|_{x=x_0} = \frac{2}{(1 + 4x_0^2)^{3/2}}.$$

25. Using the formula of Exercise 22 we get ($y = \ln x$, so $y' = 1/x$ and $y'' = -1/x^2$)

$$\kappa(x) = \frac{|-1/x^2|}{(1 + (1/x)^2)^{3/2}} = \frac{1/x^2}{(x^2 + 1)^{3/2}/x^3} = \frac{x}{(x^2 + 1)^{3/2}}.$$

From

$$\kappa'(x) = \frac{(x^2 + 1)^{3/2} - x\frac{3}{2}(x^2 + 1)^{1/2}2x}{(x^2 + 1)^3} = \frac{1 - 2x^2}{(x^2 + 1)^{5/2}} = 0$$

it follows that $x = 1/\sqrt{2}$ ($x = -1/\sqrt{2}$ makes no sense, since $\ln x$ is not defined for negative x). If $0 < x < 1/\sqrt{2}$, then $\kappa'(x) > 0$ and if $x > 1/\sqrt{2}$, then $\kappa'(x) < 0$. Therefore $x = 1/\sqrt{2}$ is the point where maximum curvature occurs. Its value is

$$\kappa(1/\sqrt{2}) = \frac{1/\sqrt{2}}{(3/2)^{3/2}} = \frac{2}{3\sqrt{3}} \approx 0.3849.$$

27. The osculating plane goes through $\mathbf{c}(1) = (1, 0, 2)$ and is spanned by the vectors $\mathbf{T}(1)$ and $\mathbf{N}(1)$. From $\mathbf{c}'(t) = (1, -2t, 4t)$ it follows that $\|\mathbf{c}'(t)\| = \sqrt{1 + 20t^2}$ and

$$\mathbf{T}(t) = \frac{\mathbf{c}'(t)}{\|\mathbf{c}'(t)\|} = \frac{(1, -2t, 4t)}{\sqrt{1 + 20t^2}}.$$

The derivative of $\mathbf{T}$ is

$$\mathbf{T}'(t) = \frac{(0,-2,4)(1+20t^2)^{1/2} - (1,-2t,4t)\frac{1}{2}(1+20t^2)^{-1/2}40t}{1+20t^2} = \frac{(-20t,-2,4)}{(1+20t^2)^{3/2}}.$$

We do not really need the unit tangent and the principal unit normal vectors — we only need their directions. So instead of $\mathbf{T}(1)$ take $\mathbf{c}'(1) = (1,-2,4)$ and instead of $\mathbf{N}(1)$ or $\mathbf{T}'(1)$ take the numerator of $\mathbf{T}'(1)$, which is $(-20,-2,4)$. From

$$\begin{vmatrix} \mathbf{i} & \mathbf{j} & \mathbf{k} \\ 1 & -2 & 4 \\ -20 & -2 & 4 \end{vmatrix} = -84\mathbf{j} - 42\mathbf{k} = -42(2\mathbf{j}+\mathbf{k})$$

it follows that we can take a normal vector to be $(0,2,1)$; the equation of the osculating plane is $0(x-1)+2(y-0)+1(z-2)=0$; i.e., $2y+z-2=0$.

29. Since $y' = \cos x$ and $y'' = -\sin x$, it follows from Exercise 22 that $\kappa(x) = |-\sin x|/(1+\cos^2 x)^{3/2}$ and so $\kappa(\pi/2) = 1$. It follows that the osculating circle at $(\pi/2, 1)$ has radius $1/\kappa(\pi/2) = 1$. The graph of $y = \sin x$ has a horizontal tangent at $x = \pi/2$, and so its normal is the vertical line $x = \pi/2$. So the osculating circle has a center at $(\pi/2, 0)$ (i.e., along the normal, 1 unit away from $(\pi/2,1)$) and is of radius 1; its equation is $(x-\pi/2)^2 + y^2 = 1$.

3.5 Introduction to Differential Geometry of Curves

1. By the product rule and (3.6)

$$\begin{aligned}\frac{d}{ds}(\mathbf{T}(s)\cdot\mathbf{T}(s)) &= \frac{d\mathbf{T}(s)}{ds}\cdot\mathbf{T}(s) + \mathbf{T}(s)\cdot\frac{d\mathbf{T}(s)}{ds} \\ &= 2\mathbf{T}(s)\cdot\frac{d\mathbf{T}(s)}{ds} = 2\mathbf{T}(s)\cdot\kappa(s)\mathbf{N}(s) = 2\kappa(s)\mathbf{T}(s)\cdot\mathbf{N}(s) = 0,\end{aligned}$$

since $\mathbf{T}(s)$ and $\mathbf{N}(s)$ are orthogonal. Next,

$$\frac{d\mathbf{c}(t)}{dt}\cdot\mathbf{T}(t) = \mathbf{c}'(t)\cdot\frac{\mathbf{c}'(t)}{\|\mathbf{c}'(t)\|} = \|\mathbf{c}'(t)\|,$$

since $\mathbf{c}'(t)\cdot\mathbf{c}'(t) = \|\mathbf{c}'(t)\|^2$. Using (3.7), we get

$$\frac{d\mathbf{N}(s)}{ds}\cdot\mathbf{B}(s) = \big(-\kappa(s)\mathbf{T}(s) + \tau(s)\mathbf{B}(s)\big)\cdot\mathbf{B}(s) = \tau(s),$$

since $\mathbf{T}(s)\cdot\mathbf{B}(s) = 0$ and $\mathbf{B}(s)\cdot\mathbf{B}(s) = \|\mathbf{B}(s)\|^2 = 1$.

3. The dot product of $\mathbf{c}''$ and $\mathbf{B}$ is equal to $\mathbf{c}''\cdot\mathbf{B} = \mathbf{c}''\cdot(\mathbf{T}\times\mathbf{N})$; i.e., it is equal to the scalar triple product of $\mathbf{c}''$, $\mathbf{T}$ and $\mathbf{N}$. Since $\mathbf{T} = \mathbf{c}'/\|\mathbf{c}'\|$, it follows that

$$\mathbf{N} = \frac{1}{\|\mathbf{T}'\|}\mathbf{T}' = \frac{1}{\|\mathbf{T}'\|}\frac{\|\mathbf{c}'\|\mathbf{c}'' - \|\mathbf{c}'\|'\mathbf{c}'}{\|\mathbf{c}'\|^2} = \frac{1}{\|\mathbf{c}'\|\,\|\mathbf{T}'\|}\mathbf{c}'' - \frac{\|\mathbf{c}'\|'}{\|\mathbf{c}'\|^2\,\|\mathbf{T}'\|}\mathbf{c}'.$$

In words, $\mathbf{N}$ is a linear combination of $\mathbf{c}'$ and $\mathbf{c}''$. So the three vectors $\mathbf{c}''$, $\mathbf{T}$ and $\mathbf{N}$ lie in the plane spanned by $\mathbf{c}'$ and $\mathbf{c}''$. Consequently, their scalar triple product is zero, and so $\mathbf{c}''$ is perpendicular to $\mathbf{B}$.

5. From $\mathbf{c}(t) = e^t(\cos t\mathbf{i} + \sin t\mathbf{k})$ we get (using the product rule)

$$\mathbf{c}'(t) = e^t(\cos t\mathbf{i} + \sin t\mathbf{k}) + e^t(-\sin t\mathbf{i} + \cos t\mathbf{k}) = e^t((\cos t - \sin t)\mathbf{i} + (\sin t + \cos t)\mathbf{k})$$

and $\|\mathbf{c}'(t)\| = e^t\sqrt{2}$. The unit tangent vector is

$$\mathbf{T}(t) = \frac{(\cos t - \sin t)\mathbf{i} + (\sin t + \cos t)\mathbf{k}}{\sqrt{2}}.$$

From

$$\mathbf{T}'(t) = \frac{(-\sin t - \cos t)\mathbf{i} + (\cos t - \sin t)\mathbf{k}}{\sqrt{2}}$$

and $\|\mathbf{T}'(t)\| = 1$ it follows that the principal unit normal is

$$\mathbf{N}(t) = \mathbf{T}'(t) = \frac{(-\sin t - \cos t)\mathbf{i} + (\cos t - \sin t)\mathbf{k}}{\sqrt{2}}.$$

The binormal vector is

$$\mathbf{B}(t) = \mathbf{T}(t) \times \mathbf{N}(t) = \frac{1}{\sqrt{2}}\frac{1}{\sqrt{2}}\begin{vmatrix} \mathbf{i} & \mathbf{j} & \mathbf{k} \\ \cos t - \sin t & 0 & \sin t + \cos t \\ -\sin t - \cos t & 0 & \cos t - \sin t \end{vmatrix}$$

$$= -\tfrac{1}{2}((\cos t - \sin t)^2 + (\sin t + \cos t)^2)\mathbf{j} = -\mathbf{j}.$$

At $t = \pi/2$, $\mathbf{T}(\pi/2) = (-\mathbf{i} + \mathbf{k})/\sqrt{2}$, $\mathbf{N}(\pi/2) = (-\mathbf{i} - \mathbf{k})/\sqrt{2}$, $\mathbf{B}(\pi/2) = -\mathbf{j}$.

7. From $\mathbf{c}(t) = (\sin t, \sin t, \sqrt{2}\cos t)$ we get $\mathbf{c}'(t) = (\cos t, \cos t, -\sqrt{2}\sin t)$ and $\|\mathbf{c}'(t)\| = \sqrt{2}$. The unit tangent vector is

$$\mathbf{T}(t) = \frac{\mathbf{c}'(t)}{\|\mathbf{c}'(t)\|} = \left(\frac{\cos t}{\sqrt{2}}, \frac{\cos t}{\sqrt{2}}, -\sin t\right).$$

From

$$\mathbf{T}'(t) = \left(\frac{-\sin t}{\sqrt{2}}, \frac{-\sin t}{\sqrt{2}}, -\cos t\right)$$

and $\|\mathbf{T}'(t)\| = 1$ it follows that the principal unit normal is

$$\mathbf{N}(t) = \mathbf{T}'(t) = \left(\frac{-\sin t}{\sqrt{2}}, \frac{-\sin t}{\sqrt{2}}, -\cos t\right).$$

The binormal vector is

$$\mathbf{B}(t) = \mathbf{T}(t) \times \mathbf{N}(t) = \begin{vmatrix} \mathbf{i} & \mathbf{j} & \mathbf{k} \\ \cos t/\sqrt{2} & \cos t/\sqrt{2} & -\sin t \\ -\sin t/\sqrt{2} & -\sin t/\sqrt{2} & -\cos t \end{vmatrix} = \frac{-\mathbf{i} + \mathbf{j}}{\sqrt{2}} = \left(-\frac{1}{\sqrt{2}}, \frac{1}{\sqrt{2}}, 0\right).$$

9. Differentiating $\mathbf{N} \cdot \mathbf{N} = \|\mathbf{N}\|^2 = 1$ we get $(d\mathbf{N}/ds) \cdot \mathbf{N} + \mathbf{N} \cdot (d\mathbf{N}/ds) = 0$, and hence $(d\mathbf{N}/ds) \cdot \mathbf{N} = 0$. From

$$\left(\frac{d\mathbf{N}}{ds} + \kappa\mathbf{T}\right) \cdot \mathbf{N} = \frac{d\mathbf{N}}{ds} \cdot \mathbf{N} + \kappa\mathbf{T} \cdot \mathbf{N} = 0$$

we conclude that $d\mathbf{N}/ds + \kappa\mathbf{T}$ is perpendicular to $\mathbf{N}$.

Differentiating $\mathbf{T} \cdot \mathbf{N} = 0$ we get $(d\mathbf{T}/ds) \cdot \mathbf{N} + \mathbf{T} \cdot (d\mathbf{N}/ds) = 0$, and therefore

$$\left(\frac{d\mathbf{N}}{ds} + \kappa\mathbf{T}\right) \cdot \mathbf{T} = \frac{d\mathbf{N}}{ds} \cdot \mathbf{T} + \kappa\|\mathbf{T}\|^2 = -\frac{d\mathbf{T}}{ds} \cdot \mathbf{N} + \kappa,$$

since $\|\mathbf{T}\| = 1$. By the chain rule,

$$\frac{d\mathbf{T}}{ds} = \frac{d\mathbf{T}}{dt}\frac{dt}{ds} = \frac{\mathbf{T}'(t)}{\|\mathbf{c}'(t)\|},$$

and

$$\frac{d\mathbf{T}}{ds} \cdot \mathbf{N} = \frac{\mathbf{T}'(t)}{\|\mathbf{c}'(t)\|} \cdot \frac{\mathbf{T}'(t)}{\|\mathbf{T}'(t)\|} = \frac{\|\mathbf{T}'(t)\|}{\|\mathbf{c}'(t)\|} = \kappa.$$

It follows that $(d\mathbf{N}/ds + \kappa\mathbf{T}) \cdot \mathbf{T} = 0$. Since $d\mathbf{N}/ds + \kappa\mathbf{T}$ is perpendicular to both $\mathbf{N}$ and $\mathbf{T}$, it must be parallel to $\mathbf{B}$ (i.e., it is a multiple of $\mathbf{B}$). So we can define τ to be that multiple; that is, τ is the real number such that $d\mathbf{N}/ds + \kappa\mathbf{T} = \tau\mathbf{B}$.

Comment: in the above computation, we used the fact that

$$\mathbf{N}(s) = \frac{\mathbf{T}'(s)}{\|\mathbf{T}'(s)\|} = \frac{\mathbf{T}'(t)/\|\mathbf{c}'(t)\|}{\|\mathbf{T}'(t)\|/\|\mathbf{c}'(t)\|} = \frac{\mathbf{T}'(t)}{\|\mathbf{T}'(t)\|} = \mathbf{N}(t).$$

11. From $\mathbf{T}(t) = \mathbf{c}'(t)/\|\mathbf{c}'(t)\|$ we get $\mathbf{c}'(t) = \|\mathbf{c}'(t)\|\,\mathbf{T}(t) = (ds/dt)\mathbf{T}(t)$; differentiate $\mathbf{c}'(t) = (ds/dt)\mathbf{T}(t)$ with respect to t:

$$\mathbf{c}''(t) = \frac{d^2s}{dt^2}\mathbf{T} + \frac{ds}{dt}\frac{d\mathbf{T}}{dt} = \frac{d^2s}{dt^2}\mathbf{T} + \frac{ds}{dt}\frac{d\mathbf{T}}{ds}\frac{ds}{dt} = \frac{d^2s}{dt^2}\mathbf{T} + \left(\frac{ds}{dt}\right)^2\frac{d\mathbf{T}}{ds}.$$

(We used the chain rule in computing $d\mathbf{T}/dt$.) By (3.6), $d\mathbf{T}(s)/ds = \kappa(s)\mathbf{N}(s)$. Since $\kappa(s) = \kappa(t)$ (see the text immediately following the definition of the curvature in Section 3.4) and $\mathbf{N}(s) = \mathbf{N}(t)$ (see the comment at the end of Exercise 9) it follows that

$$\frac{d\mathbf{T}(s)}{ds} = \kappa(s)\mathbf{N}(s) = \kappa(t)\mathbf{N}(t)$$

and therefore

$$\mathbf{c}''(t) = \frac{d^2s}{dt^2}\mathbf{T}(t) + \left(\frac{ds}{dt}\right)^2\kappa(t)\mathbf{N}(t).$$

13. From Exercise 11 we get

$$\mathbf{c}'(t) \times \mathbf{c}''(t) = \frac{ds}{dt}\mathbf{T}(t) \times \left(\frac{d^2s}{dt^2}\mathbf{T}(t) + \kappa(t)\left(\frac{ds}{dt}\right)^2\mathbf{N}(t)\right).$$

From the distributivity of the cross product and the fact that $\mathbf{T}(t) \times \mathbf{T}(t) = \mathbf{0}$, it follows that

$$\mathbf{c}'(t) \times \mathbf{c}''(t) = \frac{ds}{dt}\kappa(t)\left(\frac{ds}{dt}\right)^2\mathbf{T}(t) \times \mathbf{N}(t) = \kappa(t)\left(\frac{ds}{dt}\right)^3\mathbf{B}(t),$$

by the definition of $\mathbf{B}(t)$.

15. From $\mathbf{c}(t) = a\cos t\mathbf{i} + a\sin t\mathbf{j} + bt\mathbf{k}$ we get $\mathbf{c}'(t) = -a\sin t\mathbf{i} + a\cos t\mathbf{j} + b\mathbf{k}$, $\mathbf{c}''(t) = -a\cos t\mathbf{i} - a\sin t\mathbf{j}$ and $\mathbf{c}'''(t) = a\sin t\mathbf{i} - a\cos t\mathbf{j}$ and hence

$$\mathbf{c}'(t) \times \mathbf{c}''(t) = \begin{vmatrix} \mathbf{i} & \mathbf{j} & \mathbf{k} \\ -a\sin t & a\cos t & b \\ -a\cos t & -a\sin t & 0 \end{vmatrix} = ab\sin t\mathbf{i} - ab\cos t\mathbf{j} + a^2\mathbf{k}.$$

So

$$\|\mathbf{c}'(t) \times \mathbf{c}''(t)\| = \sqrt{a^2b^2\sin^2 t + a^2b^2\cos^2 t + a^4} = \sqrt{a^2(a^2+b^2)} = a\sqrt{a^2+b^2},$$

since $a > 0$. Finally,

$$(\mathbf{c}'(t) \times \mathbf{c}''(t)) \cdot \mathbf{c}'''(t) = (ab \sin t\mathbf{i} - ab \cos t\mathbf{j} + a^2\mathbf{k}) \cdot (a \sin t\mathbf{i} - a \cos t\mathbf{j}) = a^2 b$$

and

$$\tau(t) = \frac{(\mathbf{c}'(t) \times \mathbf{c}''(t)) \cdot \mathbf{c}'''(t)}{\|\mathbf{c}'(t) \times \mathbf{c}''(t)\|^2} = \frac{a^2 b}{a^2(a^2 + b^2)} = \frac{b}{(a^2 + b^2)}.$$

17. The curve $\mathbf{c}(t)$ is a plane curve, so its torsion is zero. We use the formula of Exercise 28 in Section 3.4 (we could also use the formula of Exercise 7 in the same section). From $\mathbf{c}(t) = at^2\mathbf{i} + 2at\mathbf{j}$ we get $\mathbf{c}'(t) = 2at\mathbf{i} + 2a\mathbf{j}$ and $\mathbf{c}''(t) = 2a\mathbf{i}$ and hence

$$\mathbf{c}'(t) \times \mathbf{c}''(t) = \begin{vmatrix} \mathbf{i} & \mathbf{j} & \mathbf{k} \\ 2at & 2a & 0 \\ 2a & 0 & 0 \end{vmatrix} = -4a^2\mathbf{k}.$$

Therefore

$$\kappa(t) = \frac{\|\mathbf{c}'(t) \times \mathbf{c}''(t)\|}{\|\mathbf{c}'(t)\|^3} = \frac{4a^2}{(4a^2t^2 + 4a^2)^{3/2}} = \frac{1}{2a(t^2 + 1)^{3/2}}.$$

The curvature is largest when the denominator is smallest; i.e., when $t = 0$. So $\mathbf{c}(t)$ has maximum curvature at $\mathbf{c}(0) = (0, 0)$.

19. From $\mathbf{c}(t) = (t + \cos t, t - \sin t, t)$ we get $\mathbf{c}'(t) = (1 - \sin t, 1 - \cos t, 1)$, $\mathbf{c}''(t) = (-\cos t, \sin t, 0)$ and $\mathbf{c}'''(t) = (\sin t, \cos t, 0)$. So

$$\mathbf{c}'(t) \times \mathbf{c}''(t) = \begin{vmatrix} \mathbf{i} & \mathbf{j} & \mathbf{k} \\ 1 - \sin t & 1 - \cos t & 1 \\ -\cos t & \sin t & 0 \end{vmatrix} = -\sin t\mathbf{i} - \cos t\mathbf{j} + (\sin t + \cos t - 1)\mathbf{k}.$$

Since

$$\begin{aligned} \|\mathbf{c}'(t) \times \mathbf{c}''(t)\| &= \sqrt{\sin^2 t + \cos^2 t + \sin^2 t + \cos^2 t + 1 + 2\sin t\cos t - 2\sin t - 2\cos t} \\ &= \sqrt{3 + 2(\sin t \cos t - \sin t - \cos t)} \end{aligned}$$

and

$$\begin{aligned} \|\mathbf{c}'(t)\| &= \sqrt{1 + \sin^2 t - 2\sin t + 1 + \cos^2 t - 2\cos t + 1} \\ &= \sqrt{4 - 2\sin t - 2\cos t}, \end{aligned}$$

it follows from the formula of Exercise 28 in Section 3.4 that

$$\kappa(t) = \frac{\|\mathbf{c}'(t) \times \mathbf{c}''(t)\|}{\|\mathbf{c}'(t)\|^3} = \frac{(3 + 2\sin t\cos t - 2\sin t - 2\cos t)^{1/2}}{(4 - 2\sin t - 2\cos t)^{3/2}}.$$

From

$$(\mathbf{c}'(t) \times \mathbf{c}''(t)) \cdot \mathbf{c}'''(t) = (-\sin t\,\mathbf{i} - \cos t\,\mathbf{j} + (\sin t + \cos t - 1)\mathbf{k}) \cdot (\sin t\,\mathbf{i} + \cos t\,\mathbf{j}) = -1,$$

we get (see Exercise 14 in this section)

$$\tau(t) = \frac{(\mathbf{c}'(t) \times \mathbf{c}''(t)) \cdot \mathbf{c}'''(t)}{\|\mathbf{c}'(t) \times \mathbf{c}''(t)\|^2} = \frac{-1}{3 + 2\sin t\cos t - 2\sin t - 2\cos t}.$$

3.6 Flow Lines

1. Drop t to simplify notation and use $'$ to denote the derivative with respect to t. Let $\mathbf{c} = (x, y)$. Then $\mathbf{F}(\mathbf{c}) = \mathbf{c}'$ implies that $(x, 2y) = (x'; y')$, and hence $x' = x$ and $y' = 2y$. The solutions are (both equations represent a model of an exponential growth) $x = C_1 e^t$ and $y = C_2 e^{2t}$, where C_1 and C_2 are constants. Eliminating t we get $y = C_2(e^t)^2 = C_2 C_1^{-2} x^2$, hence the flow lines are parabolas $y = Cx^2$, $C =$ constant

Similar computation shows that the flow lines of $\mathbf{F}_1(x, y) = (3x, 6y)$ are given by $x = C_1 e^{3t}$ and $y = C_2 e^{6t}$. Eliminating t, we get a family of parabolas $y = Cx^2$.

The flow lines of $\mathbf{F}_2(x, y) = (-2x, -4y)$ are given by $x = C_1 e^{-2t}$ and $y = C_2 e^{-4t}$; i.e., are (again) parabolas $y = Cx^2$.

However, there is a difference. Since $\|\mathbf{c}'\| = \|\mathbf{F}(\mathbf{c})\|$, it follows that the speed of flow lines of $\mathbf{F}$ is $\sqrt{x^2 + 4y^2}$. Similarly, the speed of flow lines of $\mathbf{F}_1$ is $\sqrt{9x^2 + 36y^2} = 3\sqrt{x^2 + 4y^2}$ and the speed of flow lines of $\mathbf{F}_2$ is $\sqrt{4x^2 + 16y^2} = 2\sqrt{x^2 + 4y^2}$. So the flow lines of $\mathbf{F}$ are the slowest whereas those of $\mathbf{F}_1$ have the largest speed. Flow lines of $\mathbf{F}_1$ "flow" in the same direction as those of $\mathbf{F}$, but the flow lines of $\mathbf{F}_2$ "flow" in the opposite direction.

3. $\mathbf{F}(x, y) = 2\|\mathbf{r}\|^{-1}\mathbf{r}$ is a radial vector field and therefore its flow lines are straight lines emerging from the origin. In particular, the flow line going through the point $(3, 2)$ has the equation $\mathbf{c}(t) = a(3t, 2t)$, $a \neq 0$. From $\mathbf{F}(\mathbf{c}(t)) = \mathbf{c}'(t)$ it follows that

$$\frac{2(3at, 2at)}{\sqrt{9a^2t^2 + 4a^2t^2}} = (3a, 2a) \qquad \text{and} \qquad \frac{2at(3, 2)}{\sqrt{13}|a|\,|t|} = a(3, 2).$$

Now $t/|t| = \pm 1$ implies that $|a| = 2/\sqrt{13}$; i.e., $a = \pm 2/\sqrt{13}$ and so $\mathbf{c}(t) = \pm(2/\sqrt{13})(3t, 2t) = \pm\left(6t/\sqrt{13}, 4t/\sqrt{13}\right)$, $t \geq 0$. Since the flow lines have the same direction as the vector field we have to choose the $+$ sign.

5. See Figure 3.6.5

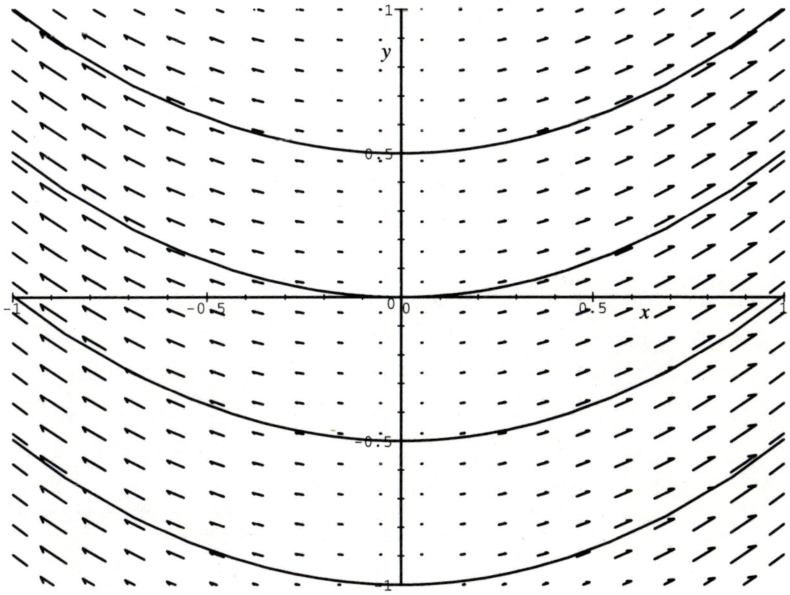

Figure 3.6.5

7. See Figure 3.6.7

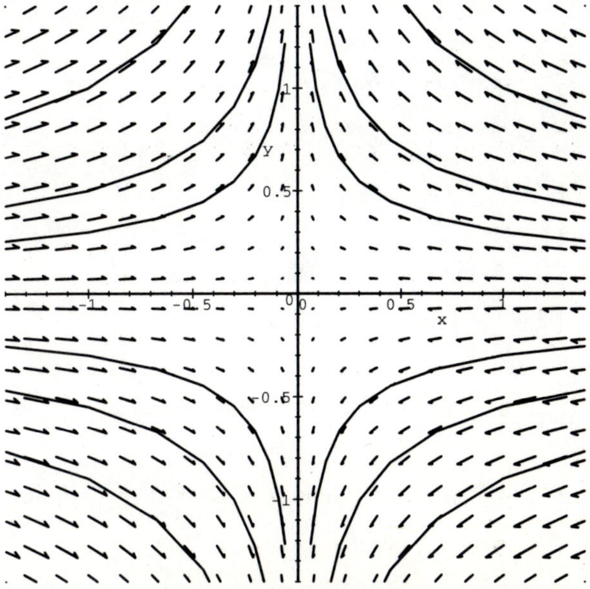

Figure 3.6.7

9. Let $\mathbf{F}(x,y,z) = (F_1(x,y,z), F_2(x,y,z), F_3(x,y,z))$. We have to find F_1, F_2 and F_3 so that (write $\mathbf{F}(\mathbf{c}(t)) = \mathbf{c}'(t)$ in components)

$$F_1(t^2, 2t, t) = 2t, \qquad F_2(t^2, 2t, t) = 2 \qquad \text{and} \qquad F_3(t^2, 2t, t) = 1.$$

Take, for example, $F_1(x,y,z) = y$ (then, clearly, $F_1(t^2, 2t, t) = 2t$). Let (keep in mind that we need 2 for the value of F_2) $F_2(x,y,z) = y/z$ and (we need 1 for the value of F_3) $F_3(x,y,z) = 2z/y$. Hence $\mathbf{F}(x,y,z) = (y, y/z, 2z/y)$ is a vector field whose flow line is $\mathbf{c}(t) = (t^2, 2t, t)$.

There are infinitely many answers. For example, $\mathbf{F}(x,y,z) = (2x/z, 2x/z^2, x/z^2)$ or $\mathbf{F}(x,y,z) = (2z, yz/x^2, z^2/x)$ have the same curve $\mathbf{c}(t) = (t^2, 2t, t)$ as their flow line.

11. Notice that the values of both t and $a+b-t$ belong to the interval $[a,b]$; hence $\mathbf{c}$ and $\boldsymbol{\gamma}$ represent the same curve. From $\boldsymbol{\gamma}(a) = \mathbf{c}(b)$ and $\boldsymbol{\gamma}(b) = \mathbf{c}(a)$ it follows that they are of opposite orientations. Assuming that $\mathbf{c}(t)$ is a flow line of $\mathbf{F}$, we get $\mathbf{G}(\boldsymbol{\gamma}(t)) = (-\mathbf{F})(\boldsymbol{\gamma}(t)) = -\mathbf{F}(\mathbf{c}(a+b-t)) = -\mathbf{c}'(a+b-t)$. From $\boldsymbol{\gamma}(t) = \mathbf{c}(a+b-t)$ it follows that $\boldsymbol{\gamma}'(t) = -\mathbf{c}'(a+b-t)$; hence $\mathbf{G}(\boldsymbol{\gamma}(t)) = (-\mathbf{F})(\boldsymbol{\gamma}(t)) = \boldsymbol{\gamma}'(t)$, and so $\boldsymbol{\gamma}$ is a flow line of $-\mathbf{F}$.

This argument implies that the flow lines of $-\mathbf{F}$ are obtained from those of $\mathbf{F}$ by reversing the orientation.

3.7 Chapter Review

Solutions to Computer Projects

1. Parametrize the ellipse as $\mathbf{c}(t) = (\cos t, 2\sin t)$, $t \in [0, 2\pi]$. Then $x' = -\sin t$, $x'' = -\cos t$, $y' = 2\cos t$, $y'' = -2\sin t$ and so (using Exercise 7 in Section 3.4)

$$\kappa(t) = \frac{|2\sin^2 t + 2\cos^2 t|}{(\sin^2 t + 4\cos^2 t)^{3/2}} = \frac{2}{(\sin^2 t + 4\cos^2 t)^{3/2}}.$$

Since $\mathbf{c}(t)$ is a planar curve, its torsion is zero.

(b) See Figure 3.7.1 (a). From

$$\begin{aligned}\kappa'(t) &= 2\left(-\tfrac{3}{2}\right)\left(\sin^2 t + 4\cos^2 t\right)^{-5/2}(2\sin t\cos t - 8\cos t\sin t) \\ &= \frac{18\sin t\cos t}{\left(\sin^2 t + 4\cos^2 t\right)^{5/2}} = \frac{9\sin 2t}{\left(\sin^2 t + 4\cos^2 t\right)^{5/2}} = 0\end{aligned}$$

it follows that $\sin 2t = 0$ and $t = 0$, $\pi/2$, π, $3\pi/2$ and 2π. Thus $\kappa(t)$ is largest (equal to 2) at $\mathbf{c}(\pi/2) = (0,2)$ and at $\mathbf{c}(3\pi/2) = (0,-2)$; it is smallest (equal to 1/4) at $\mathbf{c}(0) = \mathbf{c}(2\pi) = (1,0)$ and at $\mathbf{c}(\pi) = (-1,0)$.

(c) The point $(1,0)$ corresponds to $t = 0$. Since $\kappa(0) = 2/4^{3/2}$, it follows that the osculating circle at $(1,0)$ has radius $4^{3/2}/2 = 4$. From

$$\mathbf{c}'(0) = (-\sin t, 2\cos t)\Big|_{t=0} = (0,2) = 2\mathbf{j}$$

we conclude that the normal direction at $(1, 0)$ is the direction of the x-axis. Hence the center of the osculating circle is on the x-axis, at $x = -3$; its equation is $(x+3)^2 + y^2 = 16$; see Figure 3.7.1 (b).

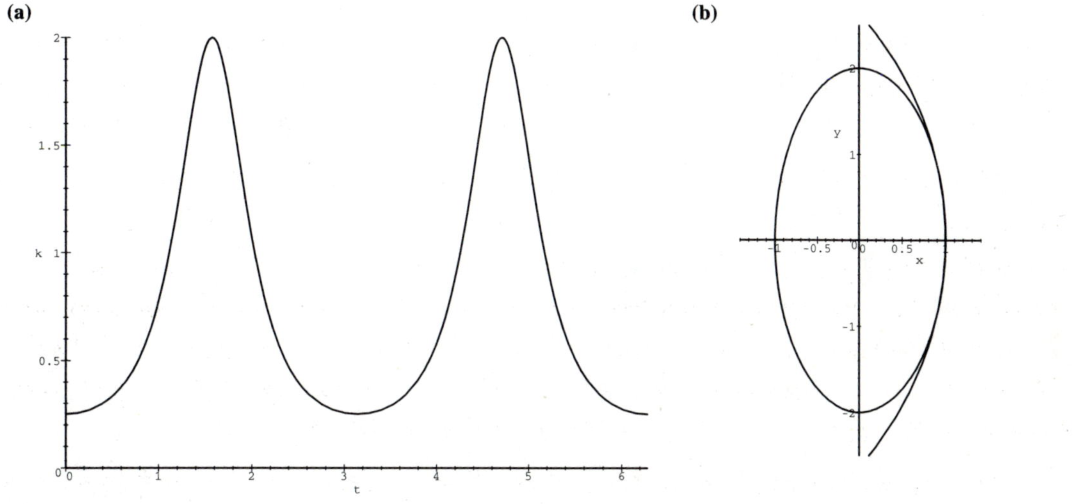

Figure 3.7.1

2. Clearly, $\mathbf{c}_1(0) = (0,0)$, $\mathbf{c}_1(1) = (1,1)$; $\mathbf{c}_2(0) = (0,0)$, $\mathbf{c}_2(\pi) = (1,1)$; $\mathbf{c}_3(3\pi/2) = (0,0)$, $\mathbf{c}_3(2\pi) = (1,1)$; and $\mathbf{c}_4(0) = (0,1)$, $\mathbf{c}_4(1) = (1,1)$.

Next, we compute the lengths $\ell(\mathbf{c}_i)$ of $\mathbf{c}_i$ and the corresponding times T_i, $i = 1,2,3,4$.

The curve $\mathbf{c}_1$ is a straight line segment joining $(0,0)$ and $(1,1)$; so $\ell(\mathbf{c}_1) = \sqrt{2} \approx 1.41421$. The time T_1 for $\mathbf{c}_1$ is

$$T_1 = \frac{1}{\sqrt{2g}} \int_0^1 \sqrt{\frac{2}{t}}\, dt = \frac{1}{\sqrt{g}} \int_0^1 t^{-1/2}\, dt = 2\,\frac{1}{\sqrt{g}}.$$

From $\mathbf{c}_2'(t) = ((1-\cos t)/\pi, \sin t/2)$ we get (using a computer)

$$\ell(\mathbf{c}_2) = \int_0^1 \left(\frac{(1-\cos t)^2}{\pi^2} + \frac{\sin^2 t}{4} \right)^{1/2} dt \approx 1.53775;$$

the time is

$$T_2 = \frac{1}{\sqrt{2g}} \int_0^1 \left(\frac{(1-\cos t)^2}{\pi^2} + \frac{\sin^2 t}{4} \right)^{1/2} \Bigg/ \left(\frac{(1-\cos t)}{2} \right)^{1/2} dt \approx 1.84030\,\frac{1}{\sqrt{g}}.$$

The curve $\mathbf{c}_3$ is a quarter-circle of radius 1, so $\ell(\mathbf{c}_3) = \pi/2 \approx 1.57080$. The corresponding time T_3 is

$$T_3 = \frac{1}{\sqrt{2g}} \int_0^{\pi/2} \sqrt{\frac{1}{\cos t}}\, dt \approx 1.85407\,\frac{1}{\sqrt{g}}.$$

From $\mathbf{c}_4'(t) = (1, (\pi/2)\cos(\pi t/2))$ we get (using a computer)

$$\ell(\mathbf{c}_4) = \int_0^1 \left(1 + \frac{\pi^2}{4}\cos^2(\pi t/2)\right)^{1/2} dt \approx 1.46370$$

and

$$T_4 = \frac{1}{\sqrt{2g}} \int_0^1 \left(\frac{1 + (\pi^2/4)\cos^2(\pi t/2)}{\sin(\pi t/2)}\right)^{1/2} dt \approx 1.88734\,\frac{1}{\sqrt{g}}.$$

Contrary to (maybe) our expectations, the line is not the "fastest" curve (although it is the shortest); see Figure 3.7.2. The "fastest" curve is $\mathbf{c}_2$.

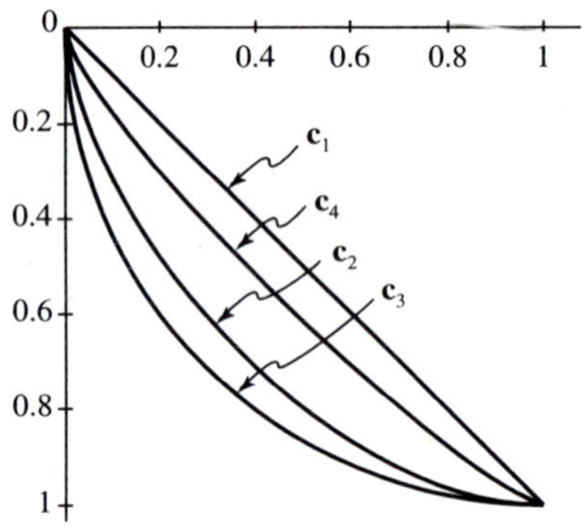

Figure 3.7.2

3. (a) Let $\mathbf{c}(t) = (a_1t^3 + b_1t^2 + c_1t + d_1, a_2t^3 + b_2t^2 + c_2t + d_2)$. Assume that, at time $t = 0$, the airplane is located at P, and at $t = 1$ it is located at Q; consequently, from $\mathbf{c}(0) = (0, -a)$ it follows that $d_1 = 0$ and $d_2 = -a$. Similarly, from $\mathbf{c}(1) = (a, 0)$ it follows that

$$a_1 + b_1 + c_1 = a$$

and

$$a_2 + b_2 + c_2 - a = 0.$$

The velocity of $\mathbf{c}$ is

$$\mathbf{c}'(t) = (3a_1t^2 + 2b_1t + c_1, 3a_2t^2 + 2b_2t + c_2).$$

At P, the velocity has to be horizontal, in the direction of the positive x-axis. So $\mathbf{c}'(0) = (c_1, c_2)$, implies that $c_1 > 0$ and $c_2 = 0$. At Q, the velocity has to be vertical, in the direction of the positive y-axis. So $\mathbf{c}'(1) = (3a_1 + 2b_1 + c_1, 3a_2 + 2b_2 + c_2)$ implies that $3a_1 + 2b_1 + c_1 = 0$ and $3a_2 + 2b_2 > 0$ (since $c_2 = 0$). We have recovered some coefficients; at the moment,

$$\mathbf{c}(t) = (a_1t^3 + b_1t^2 + c_1t, a_2t^3 + b_2t^2 - a).$$

Now let us compute the curvature (using the formula in Exercise 7 in Section 3.4; with the help of a computer, or by hand). From $x = a_1t^3 + b_1t^2 + c_1t$ we get $x' = 3a_1t^2 + 2b_1t + c_1$ and

$x'' = 6a_1t + 2b_1$; from $y = a_2t^3 + b_2t^2 - a$ we get $y' = 3a_2t^2 + 2b_2t$ and $y'' = 6a_2t + 2b_2$. So

$$\kappa(t) = \frac{|(3a_1t^2 + 2b_1t + c_1)(6a_2t + 2b_2) - (6a_1t + 2b_1)(3a_2t^2 + 2b_2t)|}{((3a_1t^2 + 2b_1t + c_1)^2 + (3a_2t^2 + 2b_2t)^2)^{3/2}}.$$

The requirement that $\kappa(0) = 0$ implies that $0 = 2c_1b_2/\sqrt{c_1^2} = 2b_2$, since $c_1 > 0$. So $b_2 = 0$.

From $a_2 + b_2 + c_2 - a = 0$ it follows that $a_2 = a$, and so the y-component of $\mathbf{c}(t)$ is $at^3 - a$. The requirement that $\kappa(1) = 1/a$ implies that

$$\frac{1}{a} = \frac{|(3a_1 + 2b_1 + c_1)(6a) - (6a_1 + 2b_1)(3a)|}{((3a_1 + 2b_1 + c_1)^2 + (3a)^2)^{3/2}} = \frac{3a|6a_1 + 2b_1|}{27a^3}$$

and thus $|6a_1 + 2b_1| = 9a$. In order to find the remaining coefficients, we have to solve the system

$$\begin{aligned} a_1 + b_1 + c_1 &= a \\ 3a_1 + 2b_1 + c_1 &= 0 \\ 6a_1 + 2b_1 &= \pm 9a. \end{aligned}$$

In fact, we have to solve two systems of equations, due to a choice in the third equation. However, one of the systems will give a solution for c_1 that will not satisfy the requirement $c_1 > 0$. So there will be only one solution; using a computer (or doing it by hand) we get $a_1 = 11/2$, $b_1 = -12$ and $c_1 = 15/2$, and so

$$\mathbf{c}(t) = \left(\tfrac{11}{2}at^3 - 12at^2 + \tfrac{15}{2}at, at^3 - a\right).$$

Here comes a disappointment: see Figure 3.7.3. Although the curvature does change continuously from 0 to $1/a$, it does not assume only the values between them, but grows much larger! Certainly, no airplane would choose a route that we suggest. Obviously, our requirements did not suffice — additional conditions have to be imposed (like curvature assuming only the values between 0 and $1/a$, probably some assumptions on the speed, etc.), thus increasing the number of unknown coefficients. We could also try other functions for components of $\mathbf{c}(t)$ — there is no reason to expect that polynomials should be the best choice. As a matter of fact, finding a really optimal route is quite difficult.

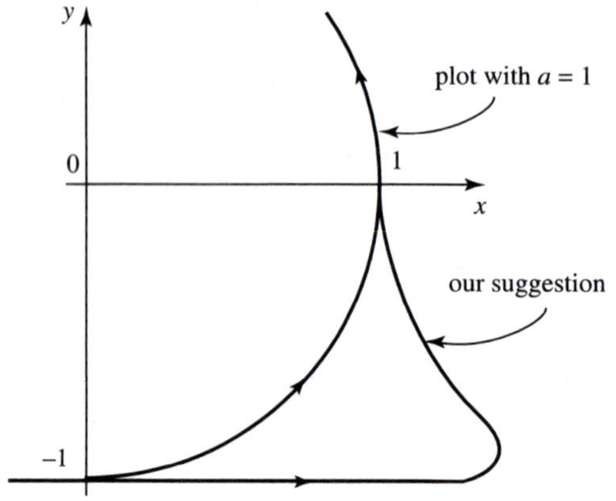

Figure 3.7.3

Solutions to Further Explorations

1. Since $\mathbf{T}$, $\mathbf{N}$ and $\mathbf{B}$ are mutually orthogonal vectors, we can write $\mathbf{W} = a\mathbf{T} + b\mathbf{N} + c\mathbf{B}$, where a, b and c are scalars. It follows that

$$\mathbf{W} \times \mathbf{T} = a\mathbf{T} \times \mathbf{T} + b\mathbf{N} \times \mathbf{T} + c\mathbf{B} \times \mathbf{T} = -b\mathbf{B} + c\mathbf{N},$$

since $\mathbf{N} \times \mathbf{T} = -\mathbf{T} \times \mathbf{N} = -\mathbf{B}$ and $\mathbf{B} \times \mathbf{T} = \mathbf{N}$ (the TNB frame is a right coordinate system, so $\mathbf{T} \times \mathbf{N} = \mathbf{B}$, $\mathbf{N} \times \mathbf{B} = \mathbf{T}$ and $\mathbf{B} \times \mathbf{T} = \mathbf{N}$). Substituting $\mathbf{W} \times \mathbf{T} = -b\mathbf{B} + c\mathbf{N}$ and $d\mathbf{T}/ds = \kappa\mathbf{N}$ into $d\mathbf{T}/ds = \mathbf{W} \times \mathbf{T}$, we get $-b\mathbf{B} + c\mathbf{N} = \kappa\mathbf{N}$ and so $b = 0$ and $c = \kappa$. From

$$\mathbf{W} \times \mathbf{N} = (a\mathbf{T} + b\mathbf{N} + c\mathbf{B}) \times \mathbf{N} = a\mathbf{B} - c\mathbf{T}$$

and $d\mathbf{N}/ds = -\kappa\mathbf{T} + \tau\mathbf{B}$ it follows that $-\kappa\mathbf{T} + \tau\mathbf{B} = a\mathbf{B} - c\mathbf{T}$ and so $a = \tau$ and $c = \kappa$. Therefore, $\mathbf{W} = \tau\mathbf{T} + \kappa\mathbf{B}$. We found $\mathbf{W}$ — but still have to check that it satisfies the third equation. From the Serret-Frenet fromula (3.8) we get

$$\mathbf{W} \times \mathbf{B} = (\tau\mathbf{T} + \kappa\mathbf{B}) \times \mathbf{B} = \tau\mathbf{T} \times \mathbf{B} = -\tau\mathbf{N} = \frac{d\mathbf{B}(s)}{ds}.$$

3. Recall that $\kappa(s) = \|d\mathbf{T}(s)/ds\|$, where s is the arc-length parameter of $\mathbf{c}$. Consider the tangent vectors $\mathbf{T}(s)$ at $\mathbf{c}(s)$ and $\mathbf{T}(s + \Delta s)$ at $\mathbf{c}(s + \Delta s)$ and let $\theta(s)$ be the angle between $\mathbf{T}(s)$ and the positive x-axis and $\theta(s + \Delta s)$ the angle between $\mathbf{T}(s + \Delta s)$ and the positive x-axis. Now

$$\kappa(s) = \left\| \frac{d\mathbf{T}(s)}{ds} \right\| = \lim_{\Delta s \to 0} \frac{\|\mathbf{T}(s + \Delta s) - \mathbf{T}(s)\|}{|\Delta s|}.$$

Since $\mathbf{T}(s)$ and $\mathbf{T}(s + \Delta s)$ are unit vectors, $\|\mathbf{T}(s + \Delta s) - \mathbf{T}(s)\|$ is the length of the base of an isosceles triangle with sides of length 1 and angle $|\theta(s + \Delta s) - \theta(s)|$ opposite to the base; see Figure 3.7.3. Hence $\|\mathbf{T}(s + \Delta s) - \mathbf{T}(s)\| = 2\sin(|\theta(s + \Delta s) - \theta(s)|/2)$ and therefore

$$\left\| \frac{d\mathbf{T}(s)}{ds} \right\| = \lim_{\Delta s \to 0} \frac{\left|2\sin\left((\theta(s + \Delta s) - \theta(s))/2\right)\right|}{|\Delta s|}.$$

Using the linear approximation for $\sin x$ at zero, we get $\sin x \approx x$ (for x close to 0). Hence $2\sin((\theta(s+\Delta s)-\theta(s))/2) \approx 2(\theta(s+\Delta s)-\theta(s))/2 = \theta(s+\Delta s)-\theta(s)$ and

$$\left\|\frac{d\mathbf{T}(s)}{ds}\right\| = \lim_{\Delta s\to 0}\frac{|\theta(s+\Delta s)-\theta(s)|}{|\Delta s|} = \lim_{\Delta s\to 0}\left|\frac{\theta(s+\Delta s)-\theta(s)}{\Delta s}\right| = \left|\frac{d\theta}{ds}\right|.$$

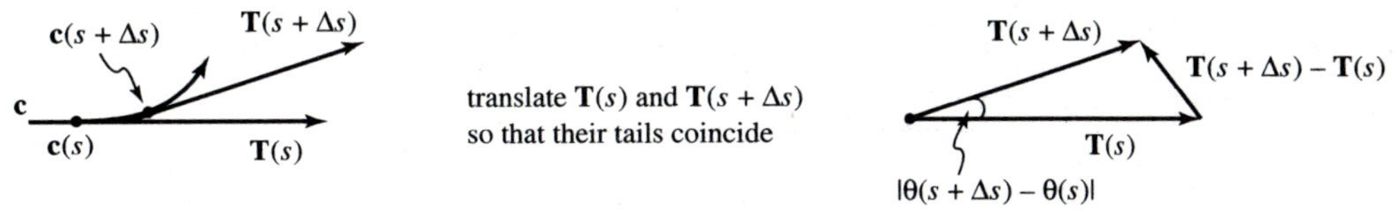

Figure 3.7.3

5. Let $\mathbf{c}(t) = (r(t), \theta(t))$ be the trajectory of the particle. We will show that

$$\frac{d}{dt}\left(r^2\frac{d\theta}{dt}\right) = 2r\frac{dr}{dt}\frac{d\theta}{dt} + r^2\frac{d^2\theta}{dt^2} = r(2r'\theta' + r\theta'') = 0.$$

From Newton's Second Law $\mathbf{F} = m\mathbf{a}$ and Exercise 27 in Section 3.2 it follows that

$$(f(r)\cos\theta, f(r)\sin\theta) = m\left[(r'' - r(\theta')^2)(\cos\theta, \sin\theta) + (2r'\theta' + r\theta'')(-\sin\theta, \cos\theta)\right]$$

and

$$(f(r)\cos\theta, f(r)\sin\theta) = \big(m(r'' - r(\theta')^2)\cos\theta - (2r'\theta' + r\theta'')\sin\theta,$$
$$m(r'' - r(\theta')^2)\sin\theta + (2r'\theta' + r\theta'')\cos\theta\big).$$

Comparing either the x-coordinates or the y-coordinates we get $2r'\theta' + r\theta'' = 0$, and so $(d/dt)(r^2\theta') = 0$; i.e., $r^2\theta'$ is constant.

7. Using the formula of Exercise 22 in Section 3.4 we get

$$\kappa(x) = \frac{4e^{2x}}{(1+4e^{4x})^{3/2}}.$$

We need to find the critical points of κ. From

$$\kappa'(x) = \frac{8e^{2x}(1+4e^{4x}) - 96e^{6x}}{(1+4e^{4x})^{5/2}} = \frac{8e^{2x} - 64e^{6x}}{(1+4e^{4x})^{5/2}} = \frac{8e^{2x}(1-8e^{4x})}{(1+4e^{4x})^{5/2}} = 0$$

it follows that $8e^{2x} - 64e^{6x} = 0$, and $e^{4x} = 1/8$ and $x = \ln(1/8)/4 = -(\ln 8)/4 \approx -0.51986$. Since $\kappa'(x) < 0$ if $x > -(\ln 8)/4$ and $\kappa'(x) > 0$ if $x < -(\ln 8)/4$, it follows that the largest curvature occurs at the point where $x = -(\ln 8)/4$; the y-coordinate is $y = e^{-(\ln 8)/2} = 8^{-1/2} = 1/\sqrt{8}$.

9. From $\mathbf{c}(t) = (e^t\cos t, e^t\sin t, e^t)$ we get

$$\mathbf{v}(t) = (e^t\cos t - e^t\sin t, e^t\sin t + e^t\cos t, e^t) = e^t(\cos t - \sin t, \sin t + \cos t, 1);$$

so the speed of the particle, $\|\mathbf{v}(t)\| = e^t\sqrt{3}$, is increasing (and is hence non-constant). The acceleration is

$$\begin{aligned}\mathbf{a}(t) = \mathbf{v}'(t) &= e^t(\cos t - \sin t, \sin t + \cos t, 1) + e^t(-\sin t - \cos t, \cos t - \sin t, 0)\\ &= e^t(-2\sin t, 2\cos t, 1).\end{aligned}$$

The magnitude of $\mathbf{a}(t)$ is non-constant, since $\|\mathbf{a}(t)\| = e^t\sqrt{5}$.

As a matter of fact, there was no need to compute the acceleration. The amount of increase in speed per unit time increases (since the speed $\|\mathbf{v}(t)\|$ depends exponentially on t), and so acceleration cannot be constant.

4. SCALAR AND VECTOR FIELDS

4.1 Gradients and Directional Derivatives

1. The unit vector in the direction of $\mathbf{v}$ is $\mathbf{u} = \mathbf{v}/\|\mathbf{v}\| = 2\mathbf{i}/\sqrt{5} - \mathbf{j}/\sqrt{5}$. The gradient of f is computed by the product rule:

$$\nabla f = (ye^{xy}(\cos x + \sin y) + e^{xy}(-\sin x))\mathbf{i} + (xe^{xy}(\cos x + \sin y) + e^{xy}\cos y)\mathbf{j},$$

and therefore $\nabla f(\pi/2, 0) = -\mathbf{i} + \mathbf{j}$. It follows that

$$D_{\mathbf{u}} f(\pi/2, 0) = \nabla f(\pi/2, 0) \cdot \mathbf{u} = -3/\sqrt{5}.$$

3. The unit vector in the direction of $\mathbf{v}$ is $\mathbf{u} = \mathbf{v}/\|\mathbf{v}\| = 3\mathbf{i}/5 + 4\mathbf{j}/5$. The gradient of f is

$$\nabla f = \ln y^2 \mathbf{i} + (2 + 2x/y)\mathbf{j},$$

and therefore $\nabla f(1, 2) = \ln 4\mathbf{i} + 3\mathbf{j}$. It follows that

$$D_{\mathbf{u}} f(1, 2) = \nabla f(1, 2) \cdot \mathbf{u} = (\ln 4\mathbf{i} + 3\mathbf{j}) \cdot (\tfrac{3}{5}\mathbf{i} + \tfrac{4}{5}\mathbf{j}) = \tfrac{1}{5}(3\ln 4 + 12).$$

5. The polar form of a vector is $\mathbf{u} = \|\mathbf{u}\|(\cos\theta\mathbf{i} + \sin\theta\mathbf{j})$. In our case, $\|\mathbf{u}\| = 1$ and $\theta = \pi/2$, so $\mathbf{u} = \mathbf{j}$. From $\nabla f(\mathbf{p}) = ye^{xy}\mathbf{i} + xe^{xy}\mathbf{j}\Big|_{(0,1)} = \mathbf{i}$ it follows that

$$D_{\mathbf{u}} f(0, 1) = \nabla f(0, 1) \cdot \mathbf{u} = \mathbf{i} \cdot \mathbf{j} = 0.$$

7. The gradient of f at $(2, 3)$ is

$$\nabla f(2, 3) = (3x^2y + 4xy^2 - y^3, x^3 + 4x^2y - 3xy^2)\Big|_{(2,3)} = (81, 2).$$

The rate of change of f in the direction of $\mathbf{v} = (1, -1)$ is (let $\mathbf{u} = \mathbf{v}/\|\mathbf{v}\| = (1/\sqrt{2}, -1/\sqrt{2})$)

$$D_{\mathbf{u}} f(2, 3) = \nabla f(2, 3) \cdot \mathbf{u} = (81, 2) \cdot (1/\sqrt{2}, -1/\sqrt{2}) = 79/\sqrt{2}.$$

9. The gradient of f at $(0, \pi)$ is

$$\nabla f(0, \pi) = (e^x \cos y, -e^x \sin y)\Big|_{(0,\pi)} = (-1, 0).$$

The rate of change of f in the direction of $\mathbf{v} = (-1, -\pi)$ is (we need a unit vector, so let $\mathbf{u} = \mathbf{v}/\|\mathbf{v}\| = (-1/\sqrt{\pi^2+1}, -\pi/\sqrt{\pi^2+1})$)

$$D_{\mathbf{u}} f(0, \pi) = \nabla f(0, \pi) \cdot \mathbf{u} = (-1, 0) \cdot (-1/\sqrt{\pi^2+1}, -\pi/\sqrt{\pi^2+1}) = 1/\sqrt{\pi^2+1}.$$

11. The gradient of f at $(0, -1, 2)$ is

$$\begin{aligned}\nabla f(0, -1, 2) &= (-2xe^{-x^2-y^2-z^2}\mathbf{i} - 2ye^{-x^2-y^2-z^2}\mathbf{j} - 2ze^{-x^2-y^2-z^2}\mathbf{k})\Big|_{(0,-1,2)} \\ &= 2e^{-5}\mathbf{j} - 4e^{-5}\mathbf{k}.\end{aligned}$$

The rate of change of f in the direction of $\mathbf{v} = \mathbf{i} + \mathbf{j} + \mathbf{k}$ is (we need a unit vector, so let $\mathbf{u} = \mathbf{v}/\|\mathbf{v}\| = (\mathbf{i} + \mathbf{j} + \mathbf{k})/\sqrt{3}$)

$$D_{\mathbf{u}} f(0, -1, 2) = \nabla f(0, -1, 2) \cdot \mathbf{u} = (2e^{-5}\mathbf{j} - 4e^{-5}\mathbf{k}) \cdot (\mathbf{i} + \mathbf{j} + \mathbf{k})/\sqrt{3} = -2e^{-5}/\sqrt{3}.$$

13. The maximum rate of change of f at $(\pi/4, \pi/4)$ occurs in the direction of

$$\nabla f(\pi/4, \pi/4) = (\sec x \tan x \tan y, \sec x \sec^2 y)\Big|_{(\pi/4,\pi/4)} = (\sqrt{2}, 2\sqrt{2}).$$

The maximum rate of change is $\|\nabla f(\pi/4, \pi/4)\| = \sqrt{10}$.

15. The maximum rate of change of f at $(1, 2, -1)$ occurs in the direction of

$$\nabla f(1, 2, -1) = (y^{-1} - zx^{-2}, -xy^{-2} + z^{-1}, -yz^{-2} + x^{-1})\Big|_{(1,2,-1)} = (3/2, -5/4, -1).$$

The maximum rate of change is $\|\nabla f(1, 2, -1)\| = \sqrt{77}/4$.

17. Near $\mathbf{p} = (3, -2)$, $f(x, y) = |xy| = -xy$, since $x > 0$ and $y < 0$. Therefore

$$\nabla f(3, -2) = (-y, -x)\Big|_{(3,-2)} = (2, -3).$$

Maximum rate of change equals $\|\nabla f(3, -2)\| = \sqrt{13}$, and it occurs in the direction of the vector $2\mathbf{i} - 3\mathbf{j}$.

19. (a) Since $\mathbf{u} = (\mathbf{i}+\mathbf{j})/\|\mathbf{i}+\mathbf{j}\| = (\mathbf{i}+\mathbf{j})/\sqrt{2}$, it follows that the rate of change of P at $(0, 1)$ in the direction of the vector $\mathbf{i}+\mathbf{j}$ is

$$\begin{aligned} D_{\mathbf{u}}P(0,1) &= \nabla P(0,1) \cdot \mathbf{u} \\ &= (-200xe^{-x^2-2y^2}, -400ye^{-x^2-2y^2})\Big|_{(0,1)} \cdot (1/\sqrt{2}, 1/\sqrt{2}) \\ &= (0, -400e^{-2}) \cdot (1/\sqrt{2}, 1/\sqrt{2}) = -400e^{-2}/\sqrt{2}. \end{aligned}$$

(b) The most rapid increase of P at $\mathbf{p}$ occurs in the direction $\nabla P(0, 1) = (0, -400e^{-2})$. The most rapid decrease of P at $\mathbf{p}$ occurs in the direction $-\nabla P(0, 1) = (0, 400e^{-2})$.

(c) Maximum rate of increase at $\mathbf{p}$ is $\|\nabla P(0, 1)\| = 400e^{-2}$.

(d) At $\mathbf{p}$, the pressure does not change in the directions perpendicular to the gradient $\nabla P(0, 1) = (0, -400e^{-2})$. These directions are $\mathbf{u} = \pm\mathbf{i}$.

21. Let $\mathbf{u} = (a, b)$ be a unit vector (so that $a^2 + b^2 = 1$). From

$$\nabla f(1, 2) = (2y, 2x)\Big|_{(1,2)} = (4, 2)$$

it follows that $D_{\mathbf{u}}f(1, 2) = \nabla f(1, 2) \cdot \mathbf{u} = 4a + 2b$. The requirement that $D_{\mathbf{u}}f(1, 2) = 4$ implies that $4a + 2b = 4$, i.e., $b = 2 - 2a$. Substitute the expression for b into $a^2 + b^2 = 1$, thus getting $a^2 + (2 - 2a)^2 = 1$; i.e., $5a^2 - 8a + 3 = 0$. Solving for a, we get $a = 3/5$ (and then $b = 4/5$) and $a = 1$ (and $b = 0$). Therefore, the directional derivative of f at $(1, 2)$ in the direction of the vector $(3/5, 4/5)$ or in the direction of the vector $(1, 0)$ is equal to 4.

23. It is given that $D_{\mathbf{u}_1}f(0, 2) = 4$, where $\mathbf{u}_1 = (\mathbf{i} + 2\mathbf{j})/\sqrt{5}$ and $D_{\mathbf{u}_2}f(0, 2) = 12$, where $\mathbf{u}_2 = (2\mathbf{i} - \mathbf{j})/\sqrt{5}$. Let $\nabla f(0, 2) = (a, b)$. Then $D_{\mathbf{u}_1}f(0, 2) = 4$ implies that $\nabla f(0, 2) \cdot \mathbf{u}_1 = 4$; i.e., $a/\sqrt{5} + 2b/\sqrt{5} = 4$ or $a + 2b = 4\sqrt{5}$. Similarly, from $D_{\mathbf{u}_2}f(0, 2) = 12$ it follows that $2a - b = 12\sqrt{5}$. Solving for a and b, we get $a = 28/\sqrt{5}$ and $b = -4/\sqrt{5}$; i.e., $\nabla f(0, 2) =$

$(28/\sqrt{5}, -4/\sqrt{5})$. The directional derivative at $(0,2)$ in the direction of $3\mathbf{i}+3\mathbf{j}$ is (we need a unit vector, hence $\mathbf{u} = (\mathbf{i}+\mathbf{j})/\sqrt{2}$)

$$D_{\mathbf{u}}f(0,2) = \nabla f(0,2)\cdot\mathbf{u} = (28/\sqrt{5}, -4/\sqrt{5})\cdot(1/\sqrt{2}, 1/\sqrt{2}) = 24/\sqrt{10}.$$

25. Assume (for convenience) that $f, g\colon \mathbb{R}^2 \to \mathbb{R}$ are differentiable functions of two variables (the proof works for any number of variables). Then

$$\begin{aligned}\nabla(af \pm bg) &= ((af \pm bg)_x, (af \pm bg)_y)\\ &= (af_x \pm bg_x, af_y \pm bg_y) = (af_x, af_y) \pm (bg_x, bg_y) = a\nabla f \pm b\nabla g.\end{aligned}$$

27. Assume, for simplicity, that $f, g\colon \mathbb{R}^2 \to \mathbb{R}$ (a proof in $\mathbb{R}^n$ is analogous). Using the definition of the gradient and the quotient rule we get

$$\begin{aligned}\nabla\left(\frac{f}{g}\right) &= \left(\left(\frac{f}{g}\right)_x, \left(\frac{f}{g}\right)_y\right)\\ &= \left(\frac{f_x g - f g_x}{g^2}, \frac{f_y g - f g_y}{g^2}\right) = \frac{g(f_x, f_y)}{g^2} - \frac{f(g_x, g_y)}{g^2} = \frac{g\nabla f - f\nabla g}{g^2}.\end{aligned}$$

29. The angle between two surfaces at their common point is the angle between their tangent planes at that point — and that angle is equal to the angle between the normal vectors to the two surfaces (at that point).

The normal vector to $x^3y^3 - 3yz = 8$ at $(1,2,0)$ is given by (think of the surface as a level surface of $f_1(x,y,z) = x^3y^3 - 3yz$ of value 8)

$$\nabla f_1(1,2,0) = (3x^2y^3\mathbf{i} + (3x^3y^2 - 3z)\mathbf{j} - 3y\mathbf{k})\Big|_{(1,2,0)} = 24\mathbf{i} + 12\mathbf{j} - 6\mathbf{k}.$$

Since we need the direction only, we will take $\mathbf{n}_1 = 4\mathbf{i} + 2\mathbf{j} - \mathbf{k}$ as a normal, to simplify computations. Similarly, the surface $x + 3z^2 = y^2 - 3$ can be viewed as a level surface of $f_2 = x + 3z^2 - y^2$ of value -3. It follows that

$$\nabla f_2(1,2,0) = (\mathbf{i} - 2y\mathbf{j} + 6z\mathbf{k})\Big|_{(1,2,0)} = \mathbf{i} - 4\mathbf{j};$$

i.e., $\mathbf{n}_2 = \mathbf{i} - 4\mathbf{j}$ is a normal vector to the surface $x + 3z^2 = y^2 - 3$ at the point $(1,2,0)$. Now

$$\cos\theta = \frac{\mathbf{n}_1\cdot\mathbf{n}_2}{\|\mathbf{n}_1\|\,\|\mathbf{n}_2\|} = \frac{(4\mathbf{i}+2\mathbf{j}-\mathbf{k})\cdot(\mathbf{i}-4\mathbf{j})}{\sqrt{21}\,\sqrt{5}} = \frac{-4}{\sqrt{105}}.$$

Hence $\theta = \arccos(-4/\sqrt{105}) \approx 1.9718 > \pi/2$ rad. So the acute angle is $\pi - \theta \approx 1.1698$ rad.

31. Let $f(x,y,z) = \sin(xy) - 2\cos(yz)$, and think of the given surface as a level surface $f(x,y,z) = 0$. A normal vector to that surface at $(\pi/2, 1, \pi/3)$ is

$$\nabla f(\pi/2,1,\pi/3) = (y\cos(xy), x\cos(xy) + 2z\sin(yz), 2y\sin(yz))\Big|_{(\pi/2,1,\pi/3)} = (0, \pi\sqrt{3}/3, \sqrt{3}).$$

The tangent plane has the equation

$$0(x - \tfrac{\pi}{2}) + \tfrac{\pi\sqrt{3}}{3}(y-1) + \sqrt{3}(z - \tfrac{\pi}{3}) = 0;$$

i.e., $\pi y + 3z - 2\pi = 0$.

33. Rewrite the given equation as $f(x,y,z) = z^2 - (4x-y)/(x+y+1) = 0$ and think of it as a level surface of f of value 0. The vector

$$\nabla f(1,0,\sqrt{2}) = \left(-\frac{5y+4}{(x+y+1)^2}, \frac{5x+1}{(x+y+1)^2}, 2z\right)\Bigg|_{(1,0,\sqrt{2})} = (-1, 3/2, 2\sqrt{2}).$$

is perpendicular to the level surface $f(x,y,z) = 0$, and is therefore normal to the tangent plane. So

$$-1(x-1) + \tfrac{3}{2}(y-0) + 2\sqrt{2}(z-\sqrt{2}) = 0,$$

or $2x - 3y - 4\sqrt{2}z + 6 = 0$ is the equation of the tangent plane at $(1,0,\sqrt{2})$.

35. A normal direction to the given surface at the point $(2,4,\sqrt{3})$ is $\nabla f(2,4,\sqrt{3})$, where $f(x,y,z) = (x-2)^2+(y-3)^2+z^2-4$ (and the surface is viewed as a level surface $f(x,y,z) = 0$). From

$$\nabla f(2,4,\sqrt{3}) = (2(x-2), 2(y-3), 2z)\Big|_{(2,4,\sqrt{3})} = (0,2,2\sqrt{3})$$

we get a parametric equation of the line normal to the surface at $(2,4,\sqrt{3})$:

$$\ell_N(t) = (2,4,\sqrt{3}) + t(0,2,2\sqrt{3}) = (2, 4+2t, \sqrt{3}+2t\sqrt{3}),$$

where $t \in \mathbb{R}$. An equation of the tangent plane at $(2,4,\sqrt{3})$ is

$$0(x-2) + 2(y-4) + 2\sqrt{3}(z-\sqrt{3}) = 0;$$

i.e., $y + \sqrt{3}z - 7 = 0$.

Here is another way of finding a normal line: the given surface is the sphere centered at $(2,3,0)$ of radius 2. The normal to the sphere is the line going through its center; so, to find its equation all we have to do is find a parametric equation of the line going through the point $(2,3,0)$ and the (given) point $(2,4,\sqrt{3})$.

37. A normal direction to the surface at $(0,e,1)$ is $\nabla f(0,e,1)$, where $f(x,y,z) = \ln(x^2 + y^2) - 2$, and the surface is viewed as a level surface $f(x,y,z) = 0$. From

$$\nabla f(0,e,1) = \left(\frac{2x}{x^2+y^2}, \frac{2y}{x^2+y^2}, 0\right)\Bigg|_{(0,e,1)} = (0, 2/e, 0)$$

we get a parametric equation of the line normal to the surface at $(0,e,1)$ (to make things simpler, take $(0,1,0)$ as a normal)

$$\ell_N(t) = (0,e,1) + t(0,1,0) = (0, e+t, 1),$$

where $t \in \mathbb{R}$. An equation of the tangent plane at $(0,e,1)$ is

$$0(x-0) + 1(y-1) + 0(z-1) = 0;$$

i.e., $y = 1$.

39. View the given curve as the level curve of the function $f(x, y) = 3x - 2y - 4$ of value 0. A normal direction at $(0, -2)$ is given by $\nabla f(0, -2) = (3, -2)\Big|_{(0,-2)} = (3, -2)$. Consequently, the normal line has the equation

$$\ell_N(t) = (0, -2) + t(3, -2) = (3t, -2 - 2t),$$

$t \in \mathbb{R}$. To get a tangent direction, take any vector perpendicular to $(3, -2)$; i.e., take any vector whose dot product with $(3, -2)$ is zero — for example, $(2, 3)$. It follows that the tangent line at $(0, -2)$ has the equation

$$\ell_T(t) = (0, -2) + t(2, 3) = (2t, -2 + 3t),$$

where $t \in \mathbb{R}$.

41. View the given curve as the level curve of the function $f(x, y) = e^x \sin y$ of value 2. A normal direction at $(\ln 2, \pi/2)$ is given by

$$\nabla f(\ln 2, \pi/2) = (e^x \sin y, e^x \cos y)\Big|_{(\ln 2, \pi/2)} = (2, 0).$$

Consequently, the normal line has the equation

$$\ell_N(t) = (\ln 2, \pi/2) + t(2, 0) = (2t + \ln 2, \pi/2),$$

$t \in \mathbb{R}$. To get a tangent direction, take any vector perpendicular to $(2, 0)$; for example, $(0, 1)$. So the tangent line at $(\ln 2, \pi/2)$ has the equation

$$\ell_T(t) = (\ln 2, \pi/2) + t(0, 1) = (\ln 2, t + \pi/2),$$

where $t \in \mathbb{R}$.

43. Parallel planes have parallel normal vectors. The paraboloid $z = x^2 + y^2 - 5$ can be thought of as the level surface of $f(x, y, z) = x^2 + y^2 - z$ of value 5. Its normal (and hence a normal to its tangent plane) is given by $\nabla f(x, y, z) = (2x, 2y, -1)$. We are asked to find all points on the paraboloid where $\nabla f(x, y, z) = (2x, 2y, -1)$ is parallel to $(1, 3, -1)$. From $(2x, 2y, -1) = a(1, 3, -1)$ it follows that $2x = a$, $2y = 3a$ and $-1 = -a$. Therefore, $a = 1$, $x = 1/2$ and $y = 3/2$. So $z = 1/4 + 9/4 - 5 = -10/4$ and it follows that $(1/2, 3/2, -5/2)$ is the only point with the required property.

45. Let $f(x, y, z) = \cos(xy) - e^z$; the given surface is the level surface $f(x, y, z) = -1$. The gradient of f is $\nabla f(x, y, z) = (-y \sin(xy), -x \sin(xy), -e^z)$. At $(1, \pi/2, 0)$, the vector $\nabla f(1, \pi/2, 0) = (-\pi/2, -1, -1)$ defines a direction normal to the surface. The unit normals are

$$\pm \frac{\nabla f(1, \pi/2, 0)}{\|\nabla f(1, \pi/2, 0)\|} = \pm \frac{(-\pi/2, -1, -1)}{\sqrt{(\pi^2 + 8)/4}} = \pm \frac{(\pi, 2, 2)}{\sqrt{\pi^2 + 8}}.$$

47. The equation $(x - m)^2 + (y - n)^2 + (z - p)^2 = r^2$ represents the sphere of radius r centered at (m, n, p). Let $f(x, y, z) = (x - m)^2 + (y - n)^2 + (z - p)^2$ and (x_0, y_0, z_0) be a point on the sphere. Then $\nabla f(x_0, y_0, z_0) = (2(x_0 - m), 2(y_0 - n), 2(z_0 - p))$ is normal to the sphere and

$$\ell(t) = (x_0, y_0, z_0) + 2t(x_0 - m, y_0 - n, z_0 - p),$$

$t \in \mathbb{R}$, is the equation of the normal line through (x_0, y_0, z_0). Since $\ell(-1/2) = (m, n, p)$, the normal line goes through the center of the sphere.

49. Let $f(x, y) = xy - k$; then $\nabla f(x, y) = (y, x)$ is a vector field perpendicular to the family of curves $xy = k$. Similarly, defining $g(x, y) = x^2 - y^2 - m$, we obtain that $\nabla g(x, y) = (2x, -2y)$ is a vector field perpendicular to the family of curves $x^2 - y^2 = m$. Now let (x, y) be a point of intersection of two curves, one from each family. Since $\nabla f(x, y) \cdot \nabla g(x, y) = 0$, it follows that $\nabla f(x, y)$ is perpendicular to $\nabla g(x, y)$; therefore, the two curves are orthogonal to each other.

51. We use the fact that the gradient is perpendicular to a level surface. The normal to $x^2 + y^2 + z^2 = 16$ is $\mathbf{n}_1 = (2x, 2y, 2z)$ and the normal to $(x - 5)^2 + y^2 + z^2 = 9$ is $\mathbf{n}_2 = (2(x - 5), 2y, 2z)$. Hence $\mathbf{n}_1 \cdot \mathbf{n}_2 = 4x^2 - 20x + 4y^2 + 4z^2$. The two spheres intersect at points (x, y, z) whose coordinates satisfy $x^2+y^2+z^2 = 16$ and $(x-5)^2+y^2+z^2 = 9$. Subtracting the second equation from the first equation, we get $10x - 25 = 7$; i.e., $x = 32/10 = 3.2$. Then $y^2 + z^2 = 16 - (3.2)^2 = 5.76$ and

$$\mathbf{n}_1 \cdot \mathbf{n}_2 = 4x^2 - 20x + 4y^2 + 4z^2 = 4(3.2)^2 - 20(3.2) + 4(5.76) = 0.$$

53. From $\nabla V = -\mathbf{F} = -e^{xy}(1 + xy)\mathbf{i} - x^2 e^{xy}\mathbf{j}$ it follows that $V_x = -e^{xy}(1 + xy)$ and $V_y = -x^2 e^{xy}$. Integrating the second equation with respect to y, we get $V = -xe^{xy} + C(x)$, where $C(x)$ is a real number or a function of x only. Substitute V into the equation for V_x to get

$$-e^{xy} - xe^{xy}y + C'(x) = -e^{xy}(1 + xy)$$

and $C'(x) = 0$. Therefore, $C(x) = C$ (C is a constant) and $V(x, y) = -xe^{xy} + C$.

55. From $\nabla V = -\mathbf{F} = (2xy^2+3x^2z, 2x^2y-z^3, x^3-3yz^2)$ it follows that $\partial V/\partial x = 2xy^2+3x^2z$, $\partial V/\partial y = 2x^2y - z^3$ and $\partial V/\partial z = x^3 - 3yz^2$. Integrating the first equation with respect to x we get

$$V(x, y, z) = x^2y^2 + x^3z + C(y, z),$$

where the "constant" $C(y, z)$ of integration might depend on y and z, since these two variables were viewed as constants. We managed to partially recover V. To compute $C(y, z)$ substitute the expression for $V(x, y, z)$ into $\partial V/\partial y$, thus getting

$$2x^2y + \frac{\partial C(y, z)}{\partial y} = 2x^2y - z^3,$$

which implies that $\partial C(y, z)/\partial y = -z^3$ and $C(y, z) = -yz^3 + C(z)$, by integration with respect to y (the variable z was kept fixed, so the integration "constant" might still depend on z). Hence

$$V(x, y, z) = x^2y^2 + x^3z - yz^3 + C(z).$$

Finally, substituting this expression into the equation for $\partial V/\partial z$, we get

$$x^3 - 3yz^2 + C'(z) = x^3 - 3yz^2,$$

so that $C(z) = C$ after integrating with respect to z (C is a real number, not a function any longer). It follows that any function of the form

$$V(x, y, z) = x^2y^2 + x^3z - yz^3 + C$$

(where C is a real number) is a potential function for the given vector field.

57. From $\nabla V = -\mathbf{F} = -(xy^2 + 3x^2y)\mathbf{i} - (x^3 + yx^2)\mathbf{j}$ it follows that

$$V_x = -xy^2 - 3x^2y \qquad \text{and} \qquad V_y = -x^3 - yx^2.$$

Integrating the first equation with respect to x, we get $V = -x^2y^2/2 - x^3y + C(y)$, where $C(y)$ is a real number or a function of y only. Substitute V into the equation for V_y, thus getting $-x^2y - x^3 + C'(y) = -x^3 - x^2y$ and $C'(y) = 0$. Therefore, $C(y) = C$ (C is a constant) and so $V(x, y) = -x^2y^2/2 - x^3y + C$.

4.2 Divergence and Curl of a Vector Field

1. Since *grad* acts on scalar functions and *grad* f is a vector, the expression *grad* (*grad* f) has no meaning.

3. Since *curl* $\mathbf{F}$ is a vector field, *curl* (*curl* $\mathbf{F}$) is defined and is a vector field. $\mathbf{G}$ is also a vector field and therefore their difference *curl* (*curl* $\mathbf{F}$) $-$ $\mathbf{G}$ is a vector field.

5. *curl* $(\mathbf{F} - \mathbf{G})$ is a vector field; *div* $\mathbf{F}$ is a scalar function; so *grad* (*div* $\mathbf{F}$) is defined, and is a vector field. The expression in question is a vector field, as a cross product of two vector fields.

7. *grad* f is a vector field, and so *curl* (*grad* f) is defined and is a vector field. Since the divergence acts on vector fields, *div* (*curl* (*grad* f)) is defined, and is a scalar function.

9. By definition,

$$\mathit{div}\,\mathbf{F} = \frac{\partial}{\partial x}(y^2z) + \frac{\partial}{\partial y}(-xz) + \frac{\partial}{\partial z}(xyz) = 0 + 0 + xy = xy$$

and

$$\mathit{curl}\,\mathbf{F} = \begin{vmatrix} \mathbf{i} & \mathbf{j} & \mathbf{k} \\ \partial/\partial x & \partial/\partial y & \partial/\partial z \\ y^2z & -xz & xyz \end{vmatrix} = (xz + x)\,\mathbf{i} - \left(yz - y^2\right)\mathbf{j} + (-z - 2yz)\,\mathbf{k}.$$

11. By definition,

$$\mathit{div}\,\mathbf{F} = \frac{\partial}{\partial x}\left(-ze^{-y}\right) + \frac{\partial}{\partial y}(xe^y) + \frac{\partial}{\partial z}(0) = xe^y$$

and

$$\mathit{curl}\,\mathbf{F} = \begin{vmatrix} \mathbf{i} & \mathbf{j} & \mathbf{k} \\ \partial/\partial x & \partial/\partial y & \partial/\partial z \\ -ze^{-y} & xe^y & 0 \end{vmatrix} = -e^{-y}\mathbf{j} + \left(e^y - ze^{-y}\right)\mathbf{k}.$$

13. By definition,

$$div\,\mathbf{F} = \frac{\partial}{\partial x}(3(x^2+y^2+z^2)) + \frac{\partial}{\partial y}(x^2+y^2+z^2) - \frac{\partial}{\partial z}(x^2+y^2+z^2) = 6x+2y-2z$$

and

$$\begin{aligned} curl\,\mathbf{F} &= \begin{vmatrix} \mathbf{i} & \mathbf{j} & \mathbf{k} \\ \partial/\partial x & \partial/\partial y & \partial/\partial z \\ 3(x^2+y^2+z^2) & x^2+y^2+z^2 & -(x^2+y^2+z^2) \end{vmatrix} \\ &= (-2y-2z)\mathbf{i} - (-2x-6z))\,\mathbf{j} + (2x-6y)\mathbf{k} \\ &= -2(y+z)\mathbf{i} + 2(x+3z)\mathbf{j} + 2(x-3y)\mathbf{k}. \end{aligned}$$

15. From

$$curl\,\mathbf{F} = \begin{vmatrix} \mathbf{i} & \mathbf{j} & \mathbf{k} \\ \partial/\partial x & \partial/\partial y & \partial/\partial z \\ \cos y & \sin x & \tan z \end{vmatrix} = (\cos x + \sin y)\mathbf{k} \neq \mathbf{0}$$

it follows that $\mathbf{F}$ is not conservative.

17. The fact that

$$curl\,\mathbf{F} = \begin{vmatrix} \mathbf{i} & \mathbf{j} & \mathbf{k} \\ \partial/\partial x & \partial/\partial y & \partial/\partial z \\ 3x^2y & x^3+y^3 & 0 \end{vmatrix} = (3x^2-3x^2)\mathbf{k} = \mathbf{0}$$

implies that $\mathbf{F}$ is conservative.

From $\mathbf{F} = -\nabla V$ it follows that $V_x = -3x^2y$ and $V_y = -x^3 - y^3$. Integrating the first equation with respect to x, we get $V = -x^3y + C(y)$; substituting this into the expression for V_y, we get $-x^3 + C'(y) = -x^3 - y^3$ and therefore $C'(y) = -y^3$ and $C(y) = -y^4/4 + C$. It follows that all functions of the form $V(x,y) = -x^3y - y^4/4 + C$ (where C is a constant) are potential functions for $\mathbf{F}$.

19. From

$$curl\,\mathbf{F} = \begin{vmatrix} \mathbf{i} & \mathbf{j} & \mathbf{k} \\ \partial/\partial x & \partial/\partial y & \partial/\partial z \\ -y & -x & -3 \end{vmatrix} = \mathbf{0}$$

it follows that $\mathbf{F}$ is conservative.

The potential function V is determined from $\nabla V = -\mathbf{F} = y\mathbf{i} + x\mathbf{j} + 3\mathbf{k}$; i.e., $V_x = y$, $V_y = x$ and $V_z = 3$. Integrating the first equation with respect to x, we get $V = xy + C(y,z)$; substituting this into the expression for V_y produces a differential equation for $C(y,z)$: $V_y = x + \partial C(y,z)/\partial y = x$, and so $\partial C(y,z)/\partial y = 0$ and $C(y,z) = C(z)$. Consequently, $V = xy + C(z)$. Finally, $V_z = 3$ implies $C'(z) = 3$; i.e., $C(z) = 3z + C$. It follows that $V(x,y,z) = xy + 3z + C$ represents all potential functions of $\mathbf{F}$.

21. By the chain rule,

$$grad\,f = \left(\frac{-2x}{(x^2+y^2+z^2)^2}, \frac{-2y}{(x^2+y^2+z^2)^2}, \frac{-2z}{(x^2+y^2+z^2)^2}\right).$$

The **i** component of *curl* (*grad* f) is equal to

$$\begin{aligned}\frac{\partial}{\partial y}\left(\frac{-2z}{(x^2+y^2+z^2)^2}\right) - \frac{\partial}{\partial z}\left(\frac{-2y}{(x^2+y^2+z^2)^2}\right)\\ &= -2z(-2)(x^2+y^2+z^2)^{-3}\,2y - (-2y)(-2)(x^2+y^2+z^2)^{-3}\,2z\\ &= 8yz(x^2+y^2+z^2)^{-3} - 8yz(x^2+y^2+z^2)^{-3} = 0.\end{aligned}$$

The remaining two components are checked analogously.

23. Since $div\,(curl\,\mathbf{F}) = x^2+y^2+z^2 \neq 0$, it follows that there can be no C^2 vector field $\mathbf{F}$ such that $curl\,\mathbf{F} = xy^2\mathbf{i} + yz^2\mathbf{j} + zx^2\mathbf{k}$ (since, for C^2 vector fields, $div\,(curl\,\mathbf{F}) = 0$).

25. By definition,

$$\begin{aligned} curl\,\mathbf{F} &= \begin{vmatrix} \mathbf{i} & \mathbf{j} & \mathbf{k} \\ \partial/\partial x & \partial/\partial y & \partial/\partial z \\ f(x) & g(y) & h(z) \end{vmatrix}\\ &= \left(\frac{\partial h(z)}{\partial y} - \frac{\partial g(y)}{\partial z}\right)\mathbf{i} - \left(\frac{\partial h(z)}{\partial x} - \frac{\partial f(x)}{\partial z}\right)\mathbf{j} + \left(\frac{\partial g(y)}{\partial x} - \frac{\partial f(x)}{\partial y}\right)\mathbf{k}\\ &= \mathbf{0}.\end{aligned}$$

27. From

$$(x-iy)^3 = x^3 - 3x^2iy + 3xi^2y^2 - i^3y^3 = (x^3-3xy^2) + i(-3x^2y+y^3)$$

it follows that the vector field in question is $\mathbf{F} = (x^3-3xy^2, -3x^2y+y^3, 0)$. From $div\,\mathbf{F} = 3x^2 - 3y^2 - 3x^2 + 3y^2 = 0$, we conclude that $\mathbf{F}$ is incompressible. The fact that

$$curl\,\mathbf{F} = \begin{vmatrix} \mathbf{i} & \mathbf{j} & \mathbf{k} \\ \partial/\partial x & \partial/\partial y & \partial/\partial z \\ x^3-3xy^2 & -3x^2y+y^3 & 0 \end{vmatrix} = (-6xy+6xy)\mathbf{k} = \mathbf{0}$$

implies that $\mathbf{F}$ is irrotational.

29. For example, $\mathbf{F} = \mathbf{F}_1 + \mathbf{F}_2$, where $\mathbf{F}_1 = z\mathbf{i} + 2\mathbf{j}$ and $\mathbf{F}_2 = -z\mathbf{k}$. Clearly, $div\,\mathbf{F}_1 = 0$ and $curl\,\mathbf{F}_2 = \mathbf{0}$. The decomposition is not unique (actually there are infinitely many possibilities). Here are two more decompositions: $\mathbf{F} = \overline{\mathbf{F}}_1 + \overline{\mathbf{F}}_2$, where $\overline{\mathbf{F}}_1 = z\mathbf{i} + \mathbf{j}$ is incompressible and $\overline{\mathbf{F}}_2 = \mathbf{j} - z\mathbf{k}$ is irrotational; $\mathbf{F} = \overline{\overline{\mathbf{F}}}_1 + \overline{\overline{\mathbf{F}}}_2$, where $\overline{\overline{\mathbf{F}}}_1 = z\mathbf{i}$ is incompressible and $\overline{\overline{\mathbf{F}}}_2 = 2\mathbf{j} - z\mathbf{k}$ is irrotational.

31. There are infinitely many decompositions. For example, $\mathbf{F} = \mathbf{F}_1 + \mathbf{F}_2$, where $\mathbf{F}_1 = -4\mathbf{i}$ is incompressible and $\mathbf{F}_2 = e^{2y}\mathbf{j} + \arctan z\mathbf{k}$ is irrotational. Another possibility is $\mathbf{F} = \overline{\mathbf{F}}_1 + \overline{\mathbf{F}}_2$, where $\overline{\mathbf{F}}_1 = \mathbf{i} + \mathbf{k}$ is incompressible and $\overline{\mathbf{F}}_2 = -5\mathbf{i} + e^{2y}\mathbf{j} + (\arctan z - 1)\mathbf{k}$ is irrotational.

33. Let $\mathbf{F} = grad\,f$. Then $curl\,\mathbf{F} = curl\,(grad\,f) = \mathbf{0}$, and so $\mathbf{F}$ is irrotational. Since $div\,\mathbf{F} = div\,(grad\,f) = \Delta f = 0$ (by assumption), $\mathbf{F}$ is incompressible as well.

35. By definition,

$$div\,\mathbf{F} = 4x + 8y^2z + 3x^3 - 3x - 8y^2z + 2x^3 \neq 0,$$

so $\mathbf{F}$ is not incompressible. However,

$$\begin{aligned} div\,(xyz^2\mathbf{F}) &= div\,\left((2x^3yz^2 + 8x^2y^3z^3)\mathbf{i} + (3x^4y^2z^2 - 3x^2y^2z^2)\mathbf{j} - (4xy^3z^4 + 2x^4yz^3)\mathbf{k}\right) \\ &= 6x^2yz^2 + 16xy^3z^3 + 6x^4yz^2 - 6x^2yz^2 - 16xy^3z^3 - 6x^4yz^2 = 0; \end{aligned}$$

consequently, $\mathbf{G} = xyz^2\mathbf{F}$ is incompressible.

37. Since $f(\|\mathbf{r}\|)\mathbf{r} = f(\sqrt{x^2+y^2+z^2})(x\mathbf{i} + y\mathbf{j} + z\mathbf{k})$, it follows that

$$curl\,\mathbf{F} = \begin{vmatrix} \mathbf{i} & \mathbf{j} & \mathbf{k} \\ \partial/\partial x & \partial/\partial y & \partial/\partial z \\ xf(\sqrt{x^2+y^2+z^2}) & yf(\sqrt{x^2+y^2+z^2}) & zf(\sqrt{x^2+y^2+z^2}) \end{vmatrix}.$$

The $\mathbf{i}$ component of $curl\,\mathbf{F}$ is computed by the chain rule to be

$$\begin{aligned} &\frac{\partial}{\partial y}\left(zf(\sqrt{x^2+y^2+z^2})\right) - \frac{\partial}{\partial z}\left(yf(\sqrt{x^2+y^2+z^2})\right) \\ &\qquad = f'(\sqrt{x^2+y^2+z^2})\,\tfrac{1}{2}(x^2+y^2+z^2)^{-1/2}\,2yz \\ &\qquad - f'(\sqrt{x^2+y^2+z^2})\,\tfrac{1}{2}(x^2+y^2+z^2)^{-1/2}\,2zy = 0. \end{aligned}$$

The remaining two components are computed in the same way.

39. Let $\mathbf{F} = (F_1, F_2, F_3)$. Then $f\mathbf{F} = (fF_1, fF_2, fF_3)$ and (using the product rule)

$$\begin{aligned} curl\,(f\mathbf{F}) &= \begin{vmatrix} \mathbf{i} & \mathbf{j} & \mathbf{k} \\ \partial/\partial x & \partial/\partial y & \partial/\partial z \\ fF_1 & fF_2 & fF_3 \end{vmatrix} \\ &= \left(\frac{\partial f}{\partial y}F_3 + f\frac{\partial F_3}{\partial y} - \frac{\partial f}{\partial z}F_2 - f\frac{\partial F_2}{\partial z}\right)\mathbf{i} - \left(\frac{\partial f}{\partial x}F_3 + f\frac{\partial F_3}{\partial x} - \frac{\partial f}{\partial z}F_1 - f\frac{\partial F_1}{\partial z}\right)\mathbf{j} \\ &+ \left(\frac{\partial f}{\partial x}F_2 + f\frac{\partial F_2}{\partial x} - \frac{\partial f}{\partial y}F_1 - f\frac{\partial F_1}{\partial y}\right)\mathbf{k}. \end{aligned}$$

On the other hand,

$$\begin{aligned} f\,curl\,\mathbf{F} + (grad\,f)\times\mathbf{F} &= f\begin{vmatrix} \mathbf{i} & \mathbf{j} & \mathbf{k} \\ \partial/\partial x & \partial/\partial y & \partial/\partial z \\ F_1 & F_2 & F_3 \end{vmatrix} + \begin{vmatrix} \mathbf{i} & \mathbf{j} & \mathbf{k} \\ \partial f/\partial x & \partial f/\partial y & \partial f/\partial z \\ F_1 & F_2 & F_3 \end{vmatrix} \\ &= f\left(\frac{\partial F_3}{\partial y} - \frac{\partial F_2}{\partial z}\right)\mathbf{i} - f\left(\frac{\partial F_3}{\partial x} - \frac{\partial F_1}{\partial z}\right)\mathbf{j} + f\left(\frac{\partial F_2}{\partial x} - \frac{\partial F_1}{\partial y}\right)\mathbf{k} \\ &+ \left(\frac{\partial f}{\partial y}F_3 - F_2\frac{\partial f}{\partial z}\right)\mathbf{i} - \left(\frac{\partial f}{\partial x}F_3 - F_1\frac{\partial f}{\partial z}\right)\mathbf{j} + \left(\frac{\partial f}{\partial x}F_2 - F_1\frac{\partial f}{\partial y}\right)\mathbf{k}. \end{aligned}$$

The assumption needed is that both f and $\mathbf{F}$ are differentiable.

41. $div\,\mathbf{r} = div\,(x\mathbf{i} + y\mathbf{j} + z\mathbf{k}) = 3$.

43. Here is an alternative to a straightforward computation: using Exercise 38 with $f = \|\mathbf{r}\|$ and $\mathbf{F} = \mathbf{r}$ we get

$$div\,(\|\mathbf{r}\|\mathbf{r}) = \|\mathbf{r}\|\,div\,\mathbf{r} + \mathbf{r}\cdot grad\,\|\mathbf{r}\|.$$

Now proceed using Exercises 41 and 42: $div\,(\|\mathbf{r}\|\mathbf{r}) = 3\|\mathbf{r}\| + \mathbf{r}\cdot\mathbf{r}/\|\mathbf{r}\| = 4\|\mathbf{r}\|$, since $\mathbf{r}\cdot\mathbf{r} = \|\mathbf{r}\|^2$.

45. Assuming that f and g are of class C^2 we get

$$
\begin{aligned}
div\,(grad\, f \times grad\, g) &= div \begin{vmatrix} \mathbf{i} & \mathbf{j} & \mathbf{k} \\ f_x & f_y & f_z \\ g_x & g_y & g_z \end{vmatrix} \\
&= div\,((f_y g_z - f_z g_y)\mathbf{i} - (f_x g_z - g_x f_z)\mathbf{j} + (f_x g_y - g_x f_y)\mathbf{k}) \\
&= \frac{\partial}{\partial x}(f_y g_z - f_z g_y) + \frac{\partial}{\partial y}(-(f_x g_z - g_x f_z)) + \frac{\partial}{\partial z}(f_x g_y - g_x f_y) \\
&= f_{yx} g_z + f_y g_{zx} - f_{zx} g_y - f_z g_{yx} - f_{xy} g_z - f_x g_{zy} + g_{xy} f_z + g_x f_{zy} \\
&\quad + f_{xz} g_y + f_x g_{yz} - g_{xz} f_y - g_x f_{yz} = 0.
\end{aligned}
$$

Notice that we used the equality of mixed second partials. This is where the assumption that f and g are of class C^2 is needed.

47. Using Exercise 44, we get

$$div\,(\mathbf{F} \times \mathbf{r}) = \mathbf{r} \cdot curl\,\mathbf{F} - \mathbf{F} \cdot curl\,\mathbf{r} = -\mathbf{F} \cdot curl\,\mathbf{r},$$

since, by assumption, $curl\,\mathbf{F} = \mathbf{0}$. By Exercise 40, $curl\,\mathbf{r} = \mathbf{0}$ and therefore $div\,(\mathbf{F} \times \mathbf{r}) = 0$.

49. $\mathbf{r}$ is a radial vector field (its flow lines are lines through the origin). Since there are no rotations within the flow, $curl\,\mathbf{r} = \mathbf{0}$.

51. Here is an alternative to a straightforward computation: By Exercise 38,

$$grad\,(div\,\|\mathbf{r}\|^{-1}\mathbf{r}) = grad\,\left(\|\mathbf{r}\|^{-1} div\,\mathbf{r} + \mathbf{r} \cdot grad\,\|\mathbf{r}\|^{-1}\right)$$

(proceed using Exercise 41 in this section and Exercise 52 in Section 4.1)

$$
\begin{aligned}
&= grad\,\left(3\|\mathbf{r}\|^{-1} + \mathbf{r}(-\|\mathbf{r}\|^{-3}) \cdot \mathbf{r}\right) \\
&= grad\,(2\|\mathbf{r}\|^{-1}) = -2\|\mathbf{r}\|^{-3}\mathbf{r}.
\end{aligned}
$$

4.3 Some Identities of Vector Analysis

1. The left side is equal to

$$
\begin{aligned}
grad\,(\mathbf{F} \cdot \mathbf{G}) &= grad\,(F_1 G_1 + F_2 G_2 + F_3 G_3) \\
&= \left(\frac{\partial F_1}{\partial x} G_1 + F_1 \frac{\partial G_1}{\partial x} + \frac{\partial F_2}{\partial x} G_2 + F_2 \frac{\partial G_2}{\partial x} + \frac{\partial F_3}{\partial x} G_3 + F_3 \frac{\partial G_3}{\partial x}\right)\mathbf{i} \\
&\quad + A\mathbf{j} + B\mathbf{k},
\end{aligned}
$$

where A and B contain six terms each; A is obtained from the coefficient of $\mathbf{i}$ by replacing all partial derivatives with respect to x with partial derivatives with respect to y, and B is obtained from the coefficient of $\mathbf{i}$ by replacing all partial derivatives with respect to x with partial derivatives with respect to z. The first term on the right side is

$$\begin{aligned}(\mathbf{F}\cdot\nabla)\mathbf{G} &= (F_1,F_2,F_3)\cdot\left(\frac{\partial G_1}{\partial x},\frac{\partial G_1}{\partial y},\frac{\partial G_1}{\partial z}\right)\mathbf{i}\\ &\quad+(F_1,F_2,F_3)\cdot\left(\frac{\partial G_2}{\partial x},\frac{\partial G_2}{\partial y},\frac{\partial G_2}{\partial z}\right)\mathbf{j}+(F_1,F_2,F_3)\cdot\left(\frac{\partial G_3}{\partial x},\frac{\partial G_3}{\partial y},\frac{\partial G_3}{\partial z}\right)\mathbf{k}\\ &=\left(F_1\frac{\partial G_1}{\partial x}+F_2\frac{\partial G_1}{\partial y}+F_3\frac{\partial G_1}{\partial z}\right)\mathbf{i}+\left(F_1\frac{\partial G_2}{\partial x}+F_2\frac{\partial G_2}{\partial y}+F_3\frac{\partial G_2}{\partial z}\right)\mathbf{j}\\ &\quad+\left(F_1\frac{\partial G_3}{\partial x}+F_2\frac{\partial G_3}{\partial y}+F_3\frac{\partial G_3}{\partial z}\right)\mathbf{k}.\end{aligned}$$

Similarly (interchange $\mathbf{F}$ and $\mathbf{G}$),

$$\begin{aligned}(\mathbf{G}\cdot\nabla)\mathbf{F} &=\left(G_1\frac{\partial F_1}{\partial x}+G_2\frac{\partial F_1}{\partial y}+G_3\frac{\partial F_1}{\partial z}\right)\mathbf{i}+\left(G_1\frac{\partial F_2}{\partial x}+G_2\frac{\partial F_2}{\partial y}+G_3\frac{\partial F_2}{\partial z}\right)\mathbf{j}\\ &\quad+\left(G_1\frac{\partial F_3}{\partial x}+G_2\frac{\partial F_3}{\partial y}+G_3\frac{\partial F_3}{\partial z}\right)\mathbf{k}.\end{aligned}$$

Next,

$$\mathbf{F}\times \mathit{curl}\,\mathbf{G}=\mathbf{F}\times\begin{vmatrix}\mathbf{i} & \mathbf{j} & \mathbf{k}\\ \partial/\partial x & \partial/\partial y & \partial/\partial z\\ G_1 & G_2 & G_3\end{vmatrix}$$

$$\begin{aligned}&=\begin{vmatrix}\mathbf{i} & \mathbf{j} & \mathbf{k}\\ F_1 & F_2 & F_3\\ (\partial G_3/\partial y)-(\partial G_2/\partial z) & -(\partial G_3/\partial x)+(\partial G_1/\partial z) & (\partial G_2/\partial x)-(\partial G_1/\partial y)\end{vmatrix}\\ &=\left(F_2\frac{\partial G_2}{\partial x}-F_2\frac{\partial G_1}{\partial y}+F_3\frac{\partial G_3}{\partial x}-F_3\frac{\partial G_1}{\partial z}\right)\mathbf{i}\\ &\quad+\left(-F_1\frac{\partial G_2}{\partial x}+F_1\frac{\partial G_1}{\partial y}+F_3\frac{\partial G_3}{\partial y}-F_3\frac{\partial G_2}{\partial z}\right)\mathbf{j}\\ &\quad+\left(-F_1\frac{\partial G_3}{\partial x}+F_1\frac{\partial G_1}{\partial z}-F_2\frac{\partial G_3}{\partial y}+F_2\frac{\partial G_2}{\partial z}\right)\mathbf{k}.\end{aligned}$$

Interchanging $\mathbf{F}$ and $\mathbf{G}$ we get

$$\begin{aligned}\mathbf{G}\times \mathit{curl}\,\mathbf{F} &=\left(G_2\frac{\partial F_2}{\partial x}-G_2\frac{\partial F_1}{\partial y}+G_3\frac{\partial F_3}{\partial x}-G_3\frac{\partial F_1}{\partial z}\right)\mathbf{i}\\ &\quad+\left(-G_1\frac{\partial F_2}{\partial x}+G_1\frac{\partial F_1}{\partial y}+G_3\frac{\partial F_3}{\partial y}-G_3\frac{\partial F_2}{\partial z}\right)\mathbf{j}\\ &\quad+\left(-G_1\frac{\partial F_3}{\partial x}+G_1\frac{\partial F_1}{\partial z}-G_2\frac{\partial F_3}{\partial y}+G_2\frac{\partial F_2}{\partial z}\right)\mathbf{k}.\end{aligned}$$

The $\mathbf{i}$ component of the vector on the right side is equal to

$$\begin{aligned}&F_1\frac{\partial G_1}{\partial x}+F_2\frac{\partial G_1}{\partial y}+F_3\frac{\partial G_1}{\partial z}+G_1\frac{\partial F_1}{\partial x}+G_2\frac{\partial F_1}{\partial y}+G_3\frac{\partial F_1}{\partial z}\\ &\quad+F_2\frac{\partial G_2}{\partial x}-F_2\frac{\partial G_1}{\partial y}+F_3\frac{\partial G_3}{\partial x}-F_3\frac{\partial G_1}{\partial z}+G_2\frac{\partial F_2}{\partial x}-G_2\frac{\partial F_1}{\partial y}+G_3\frac{\partial F_3}{\partial x}-G_3\frac{\partial F_1}{\partial z}\\ &=F_1\frac{\partial G_1}{\partial x}+G_1\frac{\partial F_1}{\partial x}+F_2\frac{\partial G_2}{\partial x}+F_3\frac{\partial G_3}{\partial x}+G_2\frac{\partial F_2}{\partial x}+G_3\frac{\partial F_3}{\partial x},\end{aligned}$$

and that is the **i** component of $grad(\mathbf{F}\cdot\mathbf{G})$. The remaining two components are checked analogously.

3. By the product rule,

$$curl\,(f\mathbf{F}) = \begin{vmatrix} \mathbf{i} & \mathbf{j} & \mathbf{k} \\ \partial/\partial x & \partial/\partial y & \partial/\partial z \\ fF_1 & fF_2 & fF_3 \end{vmatrix}$$

$$= \left(\frac{\partial(fF_3)}{\partial y} - \frac{\partial(fF_2)}{\partial z}\right)\mathbf{i} + \left(\frac{\partial(fF_1)}{\partial z} - \frac{\partial(fF_3)}{\partial x}\right)\mathbf{j} + \left(\frac{\partial(fF_2)}{\partial x} - \frac{\partial(fF_1)}{\partial y}\right)\mathbf{k}$$

$$= \left(F_3\frac{\partial f}{\partial y} + f\frac{\partial F_3}{\partial y} - F_2\frac{\partial f}{\partial z} - f\frac{\partial F_2}{\partial z}\right)\mathbf{i}$$

$$+ \left(F_1\frac{\partial f}{\partial z} + f\frac{\partial F_1}{\partial z} - F_3\frac{\partial f}{\partial x} - f\frac{\partial F_3}{\partial x}\right)\mathbf{j}$$

$$+ \left(F_2\frac{\partial f}{\partial x} + f\frac{\partial F_2}{\partial x} - F_1\frac{\partial f}{\partial y} - f\frac{\partial F_1}{\partial y}\right)\mathbf{k}.$$

The right side is computed to be

$$f\,curl\,\mathbf{F} + grad\,f \times \mathbf{F} = f\begin{vmatrix} \mathbf{i} & \mathbf{j} & \mathbf{k} \\ \partial/\partial x & \partial/\partial y & \partial/\partial z \\ F_1 & F_2 & F_3 \end{vmatrix} + \begin{vmatrix} \mathbf{i} & \mathbf{j} & \mathbf{k} \\ \partial f/\partial x & \partial f/\partial y & \partial f/\partial z \\ F_1 & F_2 & F_3 \end{vmatrix}$$

$$= f\left(\left(\frac{\partial F_3}{\partial y} - \frac{\partial F_2}{\partial z}\right)\mathbf{i} + \left(\frac{\partial F_1}{\partial z} - \frac{\partial F_3}{\partial x}\right)\mathbf{j} + \left(\frac{\partial F_2}{\partial x} - \frac{\partial F_1}{\partial y}\right)\mathbf{k}\right)$$

$$+ \left(F_3\frac{\partial f}{\partial y} - F_2\frac{\partial f}{\partial z}\right)\mathbf{i} + \left(F_1\frac{\partial f}{\partial z} - F_3\frac{\partial f}{\partial x}\right)\mathbf{j} + \left(F_2\frac{\partial f}{\partial x} - F_1\frac{\partial f}{\partial y}\right)\mathbf{k}$$

$$= \left(f\frac{\partial F_3}{\partial y} - f\frac{\partial F_2}{\partial z} + F_3\frac{\partial f}{\partial y} - F_2\frac{\partial f}{\partial z}\right)\mathbf{i}$$

$$+ \left(f\frac{\partial F_1}{\partial z} - f\frac{\partial F_3}{\partial x} + F_1\frac{\partial f}{\partial z} - F_3\frac{\partial f}{\partial x}\right)\mathbf{j}$$

$$+ \left(f\frac{\partial F_2}{\partial x} - f\frac{\partial F_1}{\partial y} + F_2\frac{\partial f}{\partial x} - F_1\frac{\partial f}{\partial y}\right)\mathbf{k}.$$

5. For simplicity, assume that $f, g\colon \mathbb{R}^2 \to \mathbb{R}$ (the general case of functions of m variables, $m \geq 2$, is dealt with analogously). The left side is computed to be

$$\Delta(fg) = div\,(grad\,(fg)) = div\,((f_x g + f g_x)\mathbf{i} + (f_y g + f g_y)\mathbf{j})$$

$$= f_{xx}g + 2f_x g_x + f g_{xx} + f_{yy}g + 2f_y g_y + f g_{yy}.$$

The first two terms on the right side are

$$g\Delta f + f\Delta g = g(f_{xx} + f_{yy}) + f(g_{xx} + g_{yy});$$

the remaining term is equal to

$$2\,grad\,f \cdot grad\,g = 2\,(f_x\mathbf{i} + f_y\mathbf{j})\cdot(g_x\mathbf{i} + g_y\mathbf{j}) = 2(f_x g_x + f_y g_y).$$

It follows that the two sides are equal.

Here is an alternative proof: using product rules for the gradient and the divergence, we get

$$\begin{aligned}\Delta(fg) &= div\,(grad\,(fg)) = div\,(g\,grad\,f + f\,grad\,g)\\ &= g\,div\,(grad\,f) + grad\,g \cdot grad\,f + f\,div\,(grad\,g) + grad\,f \cdot grad\,g\\ &= g\Delta f + f\Delta g + 2\,grad\,f \cdot grad\,g.\end{aligned}$$

7. One way to prove this identity is by a straightforward computation, using appropriate definitions and the product rule. Here is a shorter alternative: start with a product rule for the divergence

$$div(g\,grad\,f \times f\,grad\,g) = (f\,grad\,g)\cdot curl\,(g\,grad\,f) - (g\,grad\,f)\cdot curl\,(f\,grad\,g),$$

apply a product rule for *curl*

$$\begin{aligned}&= (f\,grad\,g)\cdot[g\,curl\,(grad\,f) + (grad\,g)\times(grad\,f)]\\ &\quad - (g\,grad\,f)\cdot[f\,curl\,(grad\,g) + (grad\,f)\times(grad\,g)]\end{aligned}$$

and use the fact that the curl of the gradient is a zero vector

$$= f\,(grad\,g)\cdot((grad\,g)\times(grad\,f)) - g\,(grad\,f)\cdot((grad\,f)\times(grad\,g)) = 0.$$

In the last step we used the fact that the mixed product (of three vectors) with two equal factors is zero.

9. Using the product rule $div\,(f\mathbf{F}) = f div\,\mathbf{F} + \mathbf{F}\cdot grad\,f$ with $\mathbf{F} = grad\,g$, we get

$$div\,(f\,grad\,g) = f\,div\,(grad\,g) + (grad\,g)\cdot(grad\,f).$$

Now interchange f and g to get

$$div\,(g\,grad\,f) = g\,div\,(grad\,f) + (grad\,f)\cdot(grad\,g).$$

Subtracting the two equations and using the definition of the Laplace operator, we get

$$\begin{aligned}&div(f\,grad\,g - g\,grad\,f)\\ &\quad = f\,div\,(grad\,g) + (grad\,g)\cdot(grad\,f) - [g\,div\,(grad\,f) + (grad\,f)\cdot(grad\,g)]\\ &\quad = f\Delta g - g\Delta f.\end{aligned}$$

11. Using the product rule $div\,(f\mathbf{F}) = f div\,\mathbf{F} + \mathbf{F}\cdot grad\,f$ with $f = \|\mathbf{r}\|^{-3}$ and $\mathbf{F} = \mathbf{r}$, we get

$$div\,\|\mathbf{r}\|^{-3}\mathbf{r} = \|\mathbf{r}\|^{-3} div\,\mathbf{r} + \mathbf{r}\cdot grad\,(\|\mathbf{r}\|^{-3}) = 3\|\mathbf{r}\|^{-3} + \mathbf{r}\cdot grad\,(\|\mathbf{r}\|^{-3}),$$

since $div\,\mathbf{r} = 3$. From

$$\begin{aligned}grad\,(\|\mathbf{r}\|^{-3}) &= grad\,(x^2+y^2+z^2)^{-3/2}\\ &= \left(\frac{-3x}{(x^2+y^2+z^2)^{5/2}}, \frac{-3y}{(x^2+y^2+z^2)^{5/2}}, \frac{-3z}{(x^2+y^2+z^2)^{5/2}}\right)\\ &= -3\|\mathbf{r}\|^{-5}\mathbf{r}\end{aligned}$$

it follows that

$$div\,\|\mathbf{r}\|^{-3}\mathbf{r} = 3\|\mathbf{r}\|^{-3} + \mathbf{r}(-3)\|\mathbf{r}\|^{-5}\mathbf{r} = 3\|\mathbf{r}\|^{-3} - 3\|\mathbf{r}\|^{-3} = 0.$$

13. Using the chain rule,

$$\begin{aligned} grad\, f &= \left(\frac{\partial f}{\partial x}, \frac{\partial f}{\partial y}\right) \\ &= \left(D_1 g\frac{\partial u}{\partial x} + D_2 g\frac{\partial v}{\partial x} + D_3 g, D_1 g\frac{\partial u}{\partial y} + D_2 g\frac{\partial v}{\partial y} + D_4 g\right), \end{aligned}$$

where $D_i g$, $i = 1, 2, 3, 4$, are partial derivatives of g with respect to its variables; $D_i g$ are evaluated at $(u(x,y), v(x,y), x, y)$.

15. Since $\|\mathbf{r}\| = \sqrt{x^2+y^2+z^2}$, it follows that

$$\begin{aligned} grad\,(\ln\|\mathbf{r}\|) &= \tfrac{1}{2} grad\,(\ln(x^2+y^2+z^2)) \\ &= \tfrac{1}{2}\left(\frac{2x}{x^2+y^2+z^2}, \frac{2y}{x^2+y^2+z^2}, \frac{2z}{x^2+y^2+z^2}\right) = \frac{1}{x^2+y^2+z^2}(x,y,z) = \frac{\mathbf{r}}{\|\mathbf{r}\|^2}. \end{aligned}$$

17. Using the result of Exercise 15, we get

$$\Delta(\ln\|\mathbf{r}\|) = div\,(grad\,(\ln\|\mathbf{r}\|)) = div\,(\|\mathbf{r}\|^{-2}\mathbf{r}).$$

We proceed by a product rule for the divergence:

$$div\,(\|\mathbf{r}\|^{-2}\mathbf{r}) = \|\mathbf{r}\|^{-2} div\,\mathbf{r} + \mathbf{r}\cdot grad\,\|\mathbf{r}\|^{-2} = 3\|\mathbf{r}\|^{-2} + \mathbf{r}\cdot grad\,(\|\mathbf{r}\|^{-2}),$$

since $div\,\mathbf{r} = 3$. From

$$\begin{aligned} grad\,(\|\mathbf{r}\|^{-2}) &= grad\left(\frac{1}{x^2+y^2+z^2}\right) \\ &= \left(-\frac{2x}{(x^2+y^2+z^2)^2}, -\frac{2y}{(x^2+y^2+z^2)^2}, -\frac{2z}{(x^2+y^2+z^2)^2}\right) \\ &= -\frac{2}{(x^2+y^2+z^2)^2}(x,y,z) = -2\|\mathbf{r}\|^{-4}\mathbf{r}, \end{aligned}$$

it follows that

$$div\,(\|\mathbf{r}\|^{-2}\mathbf{r}) = 3\|\mathbf{r}\|^{-2} - 2\|\mathbf{r}\|^{-4}\mathbf{r}\cdot\mathbf{r} = \|\mathbf{r}\|^{-2}.$$

19. Since $\mathbf{F}(r,\theta,z) = r^2\mathbf{e}_r + rz\mathbf{e}_\theta + \sin\theta\mathbf{e}_z$ it follows that $F_r = r^2$, $F_\theta = rz$ and $F_z = \sin\theta$. Using the formulas derived in this section, we get

$$div\,\mathbf{F} = \frac{1}{r}\frac{\partial(r^3)}{\partial r} + \frac{1}{r}\frac{\partial(rz)}{\partial\theta} + \frac{\partial(\sin\theta)}{\partial z} = 3r$$

and

$$\begin{aligned} curl\,\mathbf{F} &= \frac{1}{r}\begin{vmatrix} \mathbf{e}_r & r\mathbf{e}_\theta & \mathbf{e}_z \\ \partial/\partial r & \partial/\partial\theta & \partial/\partial z \\ r^2 & r^2 z & \sin\theta \end{vmatrix} \\ &= \tfrac{1}{r}[(\cos\theta - r^2)\mathbf{e}_r - (0)r\mathbf{e}_\theta + 2rz\mathbf{e}_z] = \tfrac{1}{r}(\cos\theta - r^2)\mathbf{e}_r + 2z\mathbf{e}_z. \end{aligned}$$

21. The components of $\mathbf{F}$ are $F_\rho = \theta$, $F_\theta = \cos\theta\cos\phi$ and $F_\phi = -\rho$. Therefore,

$$\begin{aligned} div\,\mathbf{F} &= \frac{1}{\rho^2}\frac{\partial}{\partial\rho}(\rho^2\theta) + \frac{1}{\rho\sin\phi}\frac{\partial}{\partial\theta}(\cos\theta\cos\phi) + \frac{1}{\rho\sin\phi}\frac{\partial}{\partial\phi}(-\rho\sin\phi) \\ &= \frac{2\theta}{\rho} - \frac{\sin\theta\cot\phi}{\rho} - \cot\phi \end{aligned}$$

and

$$\begin{aligned} curl\,\mathbf{F} &= \frac{1}{\rho^2\sin\phi}\begin{vmatrix} \mathbf{e}_\rho & \rho\mathbf{e}_\phi & \rho\sin\phi\mathbf{e}_\theta \\ \frac{\partial}{\partial\rho} & \frac{\partial}{\partial\phi} & \frac{\partial}{\partial\theta} \\ \theta & -\rho^2 & \rho\sin\phi\cos\phi\cos\theta \end{vmatrix} \\ &= \frac{1}{\rho^2\sin\phi}\left[\rho\cos\theta\cos(2\phi)\mathbf{e}_\rho - \rho(\tfrac{1}{2}\cos\theta\sin(2\phi) - 1)\mathbf{e}_\phi + \rho\sin\phi(-2\rho)\mathbf{e}_\theta\right] \\ &= \frac{\cos\theta\cos(2\phi)}{\rho\sin\phi}\mathbf{e}_\rho - \frac{\frac{1}{2}\cos\theta\sin(2\phi) - 1}{\rho\sin\phi}\mathbf{e}_\phi - 2\mathbf{e}_\theta. \end{aligned}$$

23. We need unit vectors and scale factors first. From $\mathbf{r} = \cosh u\cos v\mathbf{i} + \sinh u\sin v\mathbf{j} + z\mathbf{k}$ it follows that

$$\frac{\partial\mathbf{r}}{\partial u} = \sinh u\cos v\mathbf{i} + \cosh u\sin v\mathbf{j}$$

and

$$\begin{aligned} h_u^2 = \left\|\frac{\partial\mathbf{r}}{\partial u}\right\|^2 &= \sinh^2 u\cos^2 v + \cosh^2 u\sin^2 v \\ &= \sinh^2 u\cos^2 v + \cosh^2 u(1 - \cos^2 v) \\ &= \cos^2 v(\sinh^2 u - \cosh^2 u) + \cosh^2 u = \cosh^2 u - \cos^2 v. \end{aligned}$$

From $\cosh^2 u - \sinh^2 u = 1$ and $\sin^2 v + \cos^2 v = 1$ it follows that $\cosh^2 u - \sinh^2 u = \sin^2 v + \cos^2 v$; i.e., $\cosh^2 u - \cos^2 v = \sinh^2 u + \sin^2 v$, and so

$$h_u = \sqrt{\cosh^2 u - \cos^2 v} = \sqrt{\sinh^2 u + \sin^2 v}.$$

It follows that

$$\mathbf{e}_u = \frac{\partial\mathbf{r}}{\partial u}\Big/\left\|\frac{\partial\mathbf{r}}{\partial u}\right\| = \frac{\sinh u\cos v\mathbf{i} + \cosh u\sin v\mathbf{j}}{\sqrt{\cosh^2 u - \cos^2 v}}.$$

Similarly,

$$\frac{\partial\mathbf{r}}{\partial v} = -\cosh u\sin v\mathbf{i} + \sinh u\cos v\mathbf{j}$$

and

$$h_v^2 = \left\|\frac{\partial\mathbf{r}}{\partial v}\right\|^2 = \cosh^2 u - \cos^2 v = \sinh^2 u + \sin^2 v;$$

So

$$\mathbf{e}_v = \frac{-\cosh u\sin v\mathbf{i} + \sinh u\cos v\mathbf{j}}{\sqrt{\cosh^2 u - \cos^2 v}},$$

$h_z = 1$ and, finally, $\mathbf{e}_z = \mathbf{k}$.

Vectors $\mathbf{e}_u$, $\mathbf{e}_v$ and $\mathbf{e}_z$ are orthonormal (by construction, they are unit vectors; the fact that they are orthogonal follows from $\mathbf{e}_u \cdot \mathbf{e}_v = \mathbf{e}_u \cdot \mathbf{e}_z = \mathbf{e}_v \cdot \mathbf{e}_z = 0$). So the vector field $\mathbf{F} = uv\mathbf{i} - \mathbf{j}$ can be written as $\mathbf{F} = F_u\mathbf{e}_u + F_v\mathbf{e}_v + F_z\mathbf{e}_z$, where

$$F_u = \mathbf{F} \cdot \mathbf{e}_u = (uv\mathbf{i} - \mathbf{j}) \cdot \frac{\sinh u \cos v\mathbf{i} + \cosh u \sin v\mathbf{j}}{\sqrt{\cosh^2 u - \cos^2 v}} = \frac{uv \sinh u \cos v - \cosh u \sin v}{\sqrt{\cosh^2 u - \cos^2 v}},$$

$$F_v = \mathbf{F} \cdot \mathbf{e}_v = (uv\mathbf{i} - \mathbf{j}) \cdot \frac{-\cosh u \sin v\mathbf{i} + \sinh u \cos v\mathbf{j}}{\sqrt{\cosh^2 u - \cos^2 v}} = \frac{-uv \cosh u \sin v - \sinh u \cos v}{\sqrt{\cosh^2 u - \cos^2 v}},$$

and $F_z = \mathbf{F} \cdot \mathbf{e}_z = (uv\mathbf{i} - \mathbf{j}) \cdot \mathbf{k} = 0$. From the formula for the divergence in curvilinear coordinate systems we get

$$\begin{aligned} div\,\mathbf{F} &= \frac{1}{h_u h_v h_z}\left(\frac{\partial}{\partial u}(h_v h_z F_u) + \frac{\partial}{\partial v}(h_u h_z F_v) + \frac{\partial}{\partial z}(h_u h_v F_z)\right) \\ &= \frac{1}{\cosh^2 u - \cos^2 v}\left(\frac{\partial}{\partial u}(uv \sinh u \cos v - \cosh u \sin v)\right. \\ &\quad \left. + \frac{\partial}{\partial v}(-uv \cosh u \sin v - \sinh u \cos v) + \frac{\partial}{\partial z}(0)\right) \\ &= \frac{1}{\cosh^2 u - \cos^2 v}(v \sinh u \cos v - u \cosh u \sin v). \end{aligned}$$

25. The scale factors were computed in Exercise 5 in Section 2.8 to be $h_u = h_v = a/(\cosh v - \cos u)$, $h_z = 1$. Therefore, if $\mathbf{F}$ is a vector field with components $\mathbf{F} = F_u\mathbf{e}_u + F_v\mathbf{e}_v + F_z\mathbf{e}_z$, then

$$curl\,\mathbf{F} = \frac{1}{A^2}\begin{vmatrix} A\mathbf{e}_u & A\mathbf{e}_v & \mathbf{e}_z \\ \partial/\partial u & \partial/\partial v & \partial/\partial z \\ AF_u & AF_v & F_z \end{vmatrix},$$

where $A = a/(\cosh v - \cos u)$.

27. We have to express $\Delta\phi$ in elliptic cylindrical coordinates. Recall that the scale factors are (take $a = 1$) $h_u = h_v = \sqrt{\sinh^2 u + \sin^2 v} = \sqrt{\cosh^2 u - \cos^2 v}$ (see the comment in Exercise 23 appearing after the computation of h_u) and $h_z = 1$. Hence

$$\begin{aligned} \Delta\phi &= \frac{1}{h_u h_v h_z}\left(\frac{\partial}{\partial u}\left(\frac{h_v h_z}{h_u}\frac{\partial\phi}{\partial u}\right) + \frac{\partial}{\partial v}\left(\frac{h_u h_z}{h_v}\frac{\partial\phi}{\partial v}\right) + \frac{\partial}{\partial z}\left(\frac{h_u h_v}{h_z}\frac{\partial\phi}{\partial z}\right)\right) \\ &= \frac{1}{\sinh^2 u + \sin^2 v}\left(\frac{\partial^2\phi}{\partial u^2} + \frac{\partial^2\phi}{\partial v^2} + (\sinh^2 u + \sin^2 v)\frac{\partial^2\phi}{\partial z^2}\right) \\ &= \frac{1}{\sinh^2 u + \sin^2 v}\left(\frac{\partial^2\phi}{\partial u^2} + \frac{\partial^2\phi}{\partial v^2}\right) + \frac{\partial^2\phi}{\partial z^2}, \end{aligned}$$

and so $\partial\phi/\partial t = c\Delta\phi$ can be written as

$$\frac{\partial\phi}{\partial t} = \frac{c}{\sinh^2 u + \sin^2 v}\left(\frac{\partial^2\phi}{\partial u^2} + \frac{\partial^2\phi}{\partial v^2}\right) + c\frac{\partial^2\phi}{\partial z^2}$$

in elliptic cylindrical coordinates.

29. If $U = U(\rho, \theta, \phi, t)$, then

$$\Delta U = \frac{\partial^2 U}{\partial \rho^2} + \frac{2}{\rho}\frac{\partial U}{\partial \rho} + \frac{1}{\rho^2 \sin^2\phi}\frac{\partial^2 U}{\partial \theta^2} + \frac{1}{\rho^2}\frac{\partial^2 U}{\partial \phi^2} + \frac{\cot\phi}{\rho^2}\frac{\partial U}{\partial \phi}.$$

(a) If U is independent of θ, then $\partial U/\partial\theta = 0$ and the equation $\partial U/\partial t = c\Delta U$ is of the form

$$\frac{\partial U}{\partial t} = c\left(\frac{\partial^2 U}{\partial \rho^2} + \frac{2}{\rho}\frac{\partial U}{\partial \rho} + \frac{1}{\rho^2}\frac{\partial^2 U}{\partial \phi^2} + \frac{\cot\phi}{\rho^2}\frac{\partial U}{\partial \phi}\right).$$

(b) In this case, $\partial U/\partial\theta = 0$ and $\partial U/\partial\phi = 0$, so

$$\frac{\partial U}{\partial t} = c\left(\frac{\partial^2 U}{\partial \rho^2} + \frac{2}{\rho}\frac{\partial U}{\partial \rho}\right).$$

(c) In this case, $\partial U/\partial t = 0$, $\partial U/\partial\theta = 0$ and $\partial U/\partial\phi = 0$, and the equation is of the form

$$\frac{\partial^2 U}{\partial \rho^2} + \frac{2}{\rho}\frac{\partial U}{\partial \rho} = 0.$$

(d) In this case, $\partial U/\partial t = 0$, and $\partial U/\partial\rho = 0$, so

$$\frac{1}{\rho^2 \sin^2\phi}\frac{\partial^2 U}{\partial \theta^2} + \frac{1}{\rho^2}\frac{\partial^2 U}{\partial \phi^2} + \frac{\cot\phi}{\rho^2}\frac{\partial U}{\partial \phi} = 0.$$

4.4 Differential Forms

1. The expression $f\beta$ is a 1-form, as a product of a scalar function and a 1-form. Its value is $f\beta = e^{x+y}(x^2+y^2)dz$.

3. The expression $\alpha \wedge \beta$ is the wedge product of two 1-forms. It is equal to the 2-form

$$\alpha\wedge\beta = (3xdx + yzdy)\wedge(x^2+y^2)dz = 3x(x^2+y^2)dxdz + yz(x^2+y^2)dydz.$$

5. Since β is a 1-form and μ is a 2-form their difference $\beta - \mu$ is not defined.

7. The form $\alpha\wedge\alpha$ is a 2-form, as a wedge product of 1-forms. Its value is

$$\begin{aligned}\alpha\wedge\alpha &= (3xdx + yzdy)\wedge(3xdx + yzdy)\\ &= 9x^2 dx\wedge dx + 3xyzdy\wedge dx + 3xyzdx\wedge dy + y^2z^2 dy\wedge dy = 0,\end{aligned}$$

since $dx\wedge dx = dy\wedge dy = 0$ and $dy\wedge dx = -dx\wedge dy$. (As a matter of fact, we could have avoided this computation: recall that the wedge product of any 1-form with itself is zero.)

9. The wedge product $\beta\wedge\gamma$ is a 3-form, and so the difference $\beta\wedge\gamma - \nu$ is defined. From

$$\begin{aligned}\beta\wedge\gamma &= (x^2+y^2)dz\wedge(2dxdy - x\cos y\, dydz)\\ &= 2(x^2+y^2)dzdxdy - x(x^2+y^2)\cos y\, dzdydz = 2(x^2+y^2)dxdydz\end{aligned}$$

it follows that

$$(\beta\wedge\gamma) - \nu = 2(x^2+y^2)dxdydz - \sin x\, dxdydz = (2x^2+2y^2-\sin x)dxdydz.$$

11. Since ν is a 3-form and α is a 1-form, their wedge product is a 4-form. In $\mathbb{R}^3$ there are no non-zero 4-forms, and so $\nu \wedge \alpha = 0$ (0 denotes a zero 4-form).

Alternatively, notice that each of the two terms in $\nu \wedge \alpha$ contains a wedge product of four basic 1-forms with a repeated factor (and so each term is zero).

13. In the computation we will immediately drop terms that involve a repeated basic 1-form, since all such products are zero.

(a) The left side is computed to be

$$\begin{aligned}\alpha \wedge (\beta \wedge \gamma) &= x^2 dx \wedge [(2dx - e^z dy) \wedge (dy + dz - zdx)]\\ &= x^2 dx \wedge (2dxdy + 2dxdz - e^z dydz + ze^z dydx)\\ &= x^2 dx \wedge ((2 - ze^z)dxdy + 2dxdz - e^z dydz)\\ &= -x^2 e^z dxdydz.\end{aligned}$$

Similarly,

$$\begin{aligned}(\alpha \wedge \beta) \wedge \gamma &= (x^2 dx \wedge (2dx - e^z dy)) \wedge (dy + dz - zdx)\\ &= -x^2 e^z dxdy \wedge (dy + dz - zdx) = -x^2 e^z dxdydz.\end{aligned}$$

(b) Since

$$\alpha \wedge \gamma = x^2 dx \wedge (dy + dz - zdx) = x^2 dxdy + x^2 dxdz$$

and

$$\gamma \wedge \alpha = (dy + dz - zdx) \wedge x^2 dx = x^2 dydx + x^2 dzdx = -x^2 dxdy - x^2 dxdz$$

it follows that $\alpha \wedge \gamma = -\gamma \wedge \alpha$.

15. The differential $d\alpha$ is the 1-form

$$\begin{aligned}d\alpha &= \frac{\partial}{\partial x}(e^{xyz})dx + \frac{\partial}{\partial y}(e^{xyz})dy + \frac{\partial}{\partial z}(e^{xyz})dz\\ &= yze^{xyz} dx + xze^{xyz} dy + xye^{xyz} dz = e^{xyz}(yzdx + xzdy + xydz).\end{aligned}$$

17. From $\alpha = x^2(dxdy + dydz)$ we get

$$\begin{aligned}d\alpha &= d(x^2) \wedge dxdy + d(x^2) \wedge dydz\\ &= 2xdx \wedge dxdy + 2xdx \wedge dydz = 2xdxdydz.\end{aligned}$$

19. By definition,

$$\begin{aligned}d\alpha &= d(\sinh x) \wedge dx + d(\sinh^2 x) \wedge dy - d(1) \wedge dz\\ &= \cosh x\, dx \wedge dx + 2 \sinh x \cosh x\, dx \wedge dy - 0 = 2 \sinh x \cosh x\, dxdy.\end{aligned}$$

21. By definition,

$$\begin{aligned}d\alpha &= d(z) \wedge dxdy + d(y) \wedge dydz + d(xy) \wedge dzdx\\ &= dz \wedge dxdy + dy \wedge dydz + (ydx + xdy) \wedge dzdx\end{aligned}$$

$$
\begin{aligned}
&= dxdydz + 0 + ydxdzdx + xdydzdx \\
&= dxdydz + xdxdydz = (1+x)dxdydz.
\end{aligned}
$$

23. By definition,

$$d(\arctan x) = \frac{\partial}{\partial x}(\arctan x)\,dx + \frac{\partial}{\partial y}(\arctan x)\,dy = \frac{1}{1+x^2}dx.$$

25. Instead of a straightforward computation, we use Theorem 4.6. By parts (a) and (c) of the theorem,

$$
\begin{aligned}
d\alpha &= d(dydz - dxdy) = d(dydz) - d(dxdy) \\
&= d(dy)\wedge dz - dy \wedge d(dz) - [d(dx)\wedge dy - dx \wedge d(dy)] = 0,
\end{aligned}
$$

since, by part (b) of the same theorem, the differential applied twice gives zero.

27. Since

$$
\begin{aligned}
d\alpha &= d(x^3dy - 2ydx + zdz) \\
&= 3x^2dx\wedge dy - 2dy\wedge dx + dz\wedge dz = (3x^2+2)dxdy,
\end{aligned}
$$

it follows that

$$d(d\alpha) = d((3x^2+2)dxdy) = d(3x^2+2)\wedge dxdy = 6xdx\wedge dxdy = 0.$$

29. By definition,

$$
\begin{aligned}
d(fg) &= d(xye^z(2x - y^2 - xz)) \\
&= [ye^z(2x-y^2-xz) + xye^z(2-z)]dx + [xe^z(2x-y^2-xz) + xye^z(-2y)]dy \\
&\quad + [xye^z(2x-y^2-xz) + xye^z(-x)]dz \\
&= e^z[(4xy - 2xyz - y^3)dx + (2x^2 - 3xy^2 - x^2z)dy + (x^2y - xy^3 - x^2yz)dz]
\end{aligned}
$$

The right side is computed to be

$$
\begin{aligned}
df\wedge g + f\wedge dg &= d(xye^z)\wedge(2x-y^2-xz) + (xye^z)\wedge d(2x-y^2-xz) \\
&= (ye^zdx + xe^zdy + xye^zdz)(2x-y^2-xz) + xye^z(2dx - 2ydy - zdx - xdz) \\
&= e^z[(y(2x-y^2-xz) + xy(2-z))dx + (x(2x-y^2-xz) + xy(-2y))dy \\
&\quad + (xy(2x-y^2-xz) + xy(-x))dz] \\
&= e^z[(4xy - 2xyz - y^3)dx + (2x^2 - 3xy^2 - x^2z)dy + (x^2y - xy^3 - x^2yz)dz].
\end{aligned}
$$

It follows that the corresponding coefficients of $d(fg)$ and $df\wedge g + f\wedge dg$ are equal.

31. By definition,

$$
\begin{aligned}
d\alpha &= d(f(x,y)dxdy + g(y,z)dydz + h(x,z)dzdx) \\
&= (f_xdx + f_ydy)\wedge dxdy + (g_ydy + g_zdz)\wedge dydz + (h_xdx + h_zdz)\wedge dzdx = 0,
\end{aligned}
$$

since each product involves a repeated basic 1-form.

33. By definition,

$$div\,\mathbf{F} = 2xyz - 2y + y = 2xyz - y.$$

On the other hand,

$$\begin{aligned} d\alpha &= d(yzdxdy - (x^2+y^2)dzdx + x^2yzdydz) \\ &= (zdy + ydz)\wedge dxdy - (2xdx + 2ydy)\wedge dzdx + (2xyzdx + x^2zdy + x^2ydz)\wedge dydz \\ &= ydzdxdy - 2ydydzdx + 2xyzdxdydz = (2xyz - y)dxdydz. \end{aligned}$$

35. By definition,

$$df = \frac{2x}{x^2+y^2+z^2+1}dx + \frac{2y}{x^2+y^2+z^2+1}dy + \frac{2z}{x^2+y^2+z^2+1}dz.$$

Since

$$\nabla f = \left(\frac{2x}{x^2+y^2+z^2+1}, \frac{2y}{x^2+y^2+z^2+1}, \frac{2z}{x^2+y^2+z^2+1}\right),$$

the corresponding components are equal. The identity $d(df) = 0$ represents $curl\,(\nabla f) = \mathbf{0}$.

4.5 Chapter Review

Solutions to Computer Projects

1. (a) The maximum rate of change of f at $(1,2)$ is given by $\|\nabla f(1,2)\|$ and the direction is that of the vector $\nabla f(1,2)$. From $f(x,y) = 4y^2e^{-x/2} - 3\pi\cos x$ it follows that

$$\begin{aligned} \nabla f(1,2) &= (-2y^2e^{-x/2} + 3\pi\sin x, 8ye^{-x/2})\Big|_{(1,2)} \\ &= (-8e^{-1/2} + 3\pi\sin 1, 16e^{-1/2}) \approx (3.07843, 9.70450) \end{aligned}$$

and

$$\|\nabla f(1,2)\| \approx 10.18105.$$

(b) The rate of change of f at $(1,2)$ in the direction of the unit vector $\mathbf{u}$ is $D_{\mathbf{u}}f(1,2) = \nabla f(1,2)\cdot\mathbf{u}$, and so

$$D_{\mathbf{u}}f(1,2) = \|\nabla f(1,2)\|\cos\alpha,$$

where α is the angle between $\nabla f(1,2)$ and $\mathbf{u}$.

From $D_{\mathbf{u}}f(1,2) < 3.26$ it follows that $\|\nabla f(1,2)\|\cos\alpha < 3.26$ and

$$\cos\alpha < \frac{3.26}{\|\nabla f(1,2)\|} \approx 0.32020.$$

If $\cos\alpha = 0.32020$, then $\alpha = \arccos 0.32020 \approx 1.24485$ rad. It follows that the rate of change of f at $(1,2)$ in a direction that makes an angle α, where $1.24485 < \alpha \le \pi$, is smaller than 3.26.

(c) Let $\mathbf{u} = (\cos\theta, \sin\theta)$. Then

$$g(\theta) = \nabla f(1,2) \cdot \mathbf{u} \approx (3.07843, 9.70450) \cdot (\cos\theta, \sin\theta) = 3.07843\cos\theta + 9.70450\sin\theta.$$

From $g'(\theta) = -3.07843\sin\theta + 9.70450\cos\theta = 0$ we get $\theta \approx 1.26362$ and $\theta \approx 1.26362 + \pi \approx 4.40521$. Since $g'(1.2) > 0$ and $g'(1.3) < 0$, it follows that $\theta = 1.26362$ gives maximum directional derivative (and that is the angle that $\nabla f(1,2)$ forms with respect to the positive x-axis). Since $g'(4.3) < 0$ and $g'(4.5) > 0$, it follows that $\theta = 4.40521$ gives minimum directional derivative of f at $(1,2)$.

(d) See Figure 4.5.1.

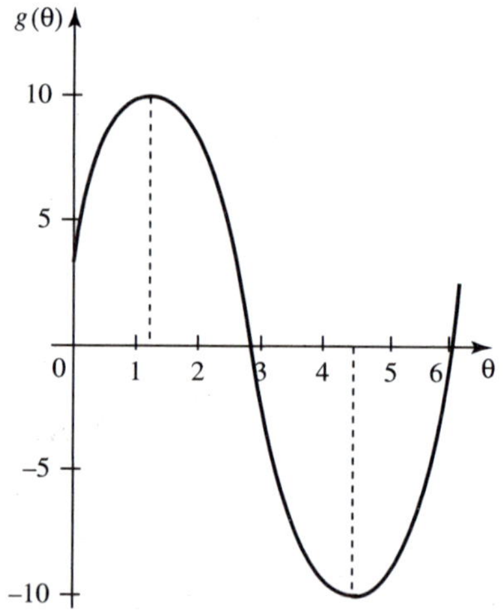

Figure 4.5.1

2. (a) By definition (or, using a computer),

$$\begin{aligned}
curl\,\mathbf{F} &= \begin{vmatrix} \mathbf{i} & \mathbf{j} & \mathbf{k} \\ \partial/\partial x & \partial/\partial y & \partial/\partial z \\ 3\sin(6x)\cos^2(4y)+1 & -4\sin^2(3x)\sin(8y)+1 & 0 \end{vmatrix} \\
&= \left(\frac{\partial}{\partial x}(-4\sin^2(3x)\sin(8y)+1) - \frac{\partial}{\partial y}(3\sin(6x)\cos^2(4y)+1)\right)\mathbf{k} \\
&= [-4(2)\sin(3x)\cos(3x)(3)\sin(8y) - 3\sin(6x)2\cos(4y)(-\sin(4y))4]\mathbf{k} \\
&= (-12\sin(6x)\sin(8y) + 12\sin(6x)\sin(8y))\mathbf{k} = \mathbf{0}.
\end{aligned}$$

(b) From $f_x = 3\sin(6x)\cos^2(4y) + 1$ it follows that

$$f(x,y) = -\tfrac{1}{2}\cos(6x)\cos^2(4y) + x + C(y).$$

The requirement $f_y = -4\sin^2(3x)\sin(8y) + 1$ combined with the formula for f that we recovered implies that

$$\cos(6x)\cos(4y)\sin(4y)\,4 + C'(y) = -4\sin^2(3x)\sin(8y) + 1.$$

Since $\cos(6x) = 1 - 2\sin^2(3x)$, it follows that

$$4(1 - 2\sin^2(3x))\cos(4y)\sin(4y) + C'(y) = -4\sin^2(3x)\sin(8y) + 1,$$

and

$$2\sin(8y) - 4\sin^2(3x)\sin(8y) + C'(y) = -4\sin^2(3x)\sin(8y) + 1;$$

hence $C'(y) = 1 - 2\sin(8y)$ and $C(y) = y + \cos(8y)/4 + C$. So

$$f(x, y) = -\tfrac{1}{2}\cos(6x)\cos^2(4y) + x + y + \tfrac{1}{4}\cos(8y) + C.$$

From $f(0, 0) = 2$ it follows that $C = 9/4$, and

$$f(x, y) = -\tfrac{1}{2}\cos(6x)\cos^2(4y) + x + y + \tfrac{1}{4}\cos(8y) + \tfrac{9}{4}.$$

(c) See Figure 4.5.2 (a).

(d) See Figure 4.5.2 (b).

(e) Since the gradient ∇f is perpendicular to the level curves of f, and $\mathbf{F} = \nabla f$, it follows that the flow lines of $\mathbf{F}$ are perpendicular to the level curves of f.

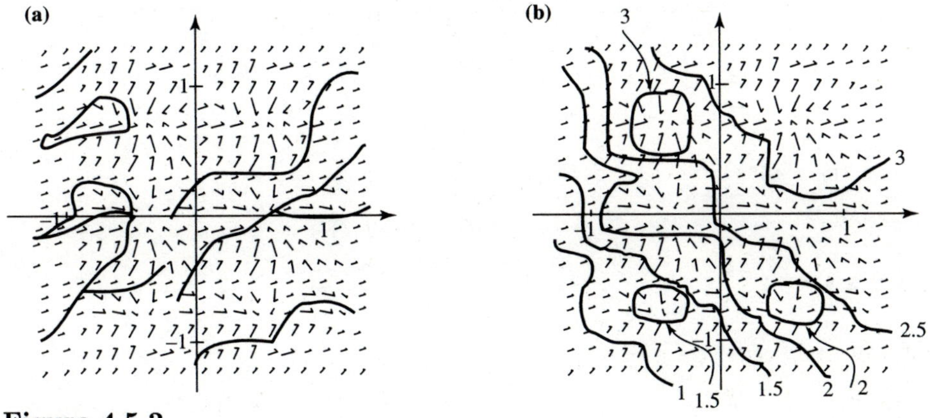

Figure 4.5.2

Solutions to Further Explorations

1. By a product rule for the divergence,

$$\mathit{div}\,(\|\mathbf{r}\|^{-2}\mathbf{r}) = \|\mathbf{r}\|^{-2}\mathit{div}\,\mathbf{r} + \nabla(\|\mathbf{r}\|^{-2})\cdot\mathbf{r}.$$

Since $\mathit{div}\,\mathbf{r} = \mathit{div}\,(x\mathbf{i} + y\mathbf{j} + z\mathbf{k}) = 3$ and

$$\begin{aligned}\nabla(\|\mathbf{r}\|^{-2}) &= \nabla(x^2 + y^2 + z^2)^{-1}\\ &= \frac{-2x}{(x^2+y^2+z^2)^2}\mathbf{i} + \frac{-2y}{(x^2+y^2+z^2)^2}\mathbf{j} + \frac{-2z}{(x^2+y^2+z^2)^2}\mathbf{k}\\ &= -2\|\mathbf{r}\|^{-4}\mathbf{r},\end{aligned}$$

it follows that

$$\mathit{div}\,(\|\mathbf{r}\|^{-2}\mathbf{r}) = 3\|\mathbf{r}\|^{-2} - 2\|\mathbf{r}\|^{-4}\mathbf{r}\cdot\mathbf{r} = 3\|\mathbf{r}\|^{-2} - 2\|\mathbf{r}\|^{-2} = \|\mathbf{r}\|^{-2},$$

since $\mathbf{r}\cdot\mathbf{r} = \|\mathbf{r}\|^2$. Using the expression for $\nabla(\|\mathbf{r}\|^{-2})$ computed above, we get

$$\begin{aligned}\Delta(div\,(\|\mathbf{r}\|^{-2}\mathbf{r})) &= \Delta\|\mathbf{r}\|^{-2} = div\,(\nabla\|\mathbf{r}\|^{-2})\\ &= div\,(-2\|\mathbf{r}\|^{-4}\mathbf{r}) = -2div\,(\|\mathbf{r}\|^{-4}\mathbf{r})\\ &= -2(\|\mathbf{r}\|^{-4}div\,\mathbf{r} + \nabla(\|\mathbf{r}\|^{-4})\cdot\mathbf{r}),\end{aligned}$$

again by the product rule for the divergence. From

$$\begin{aligned}\nabla(\|\mathbf{r}\|^{-4}) &= \nabla(x^2+y^2+z^2)^{-2}\\ &= \frac{-4x}{(x^2+y^2+z^2)^3}\mathbf{i} + \frac{-4y}{(x^2+y^2+z^2)^3}\mathbf{j} + \frac{-4z}{(x^2+y^2+z^2)^3}\mathbf{k}\\ &= -4\|\mathbf{r}\|^{-6}\mathbf{r},\end{aligned}$$

we finally get

$$\Delta\|\mathbf{r}\|^{-2} = -2(3\|\mathbf{r}\|^{-4} + -4\|\mathbf{r}\|^{-6}\mathbf{r}\cdot\mathbf{r}) = -2(-\|\mathbf{r}\|^{-4}) = 2\|\mathbf{r}\|^{-4}.$$

3. Let us compute the derivative of $f(t\mathbf{x}) = t^p f(\mathbf{x})$ with respect to t. The left side is (let $\mathbf{x} = (x,y,z)$) $f(t\mathbf{x}) = f(tx,ty,tz)$, and, by the chain rule,

$$\begin{aligned}\frac{\partial}{\partial t}f(tx,ty,tz) &= D_1f(tx,ty,tz)\,x + D_2f(tx,ty,tz)\,y + D_3f(tx,ty,tz)\,z\\ &= (D_1f(tx,ty,tz), D_2f(tx,ty,tz), D_3f(tx,ty,tz))\cdot(x,y,z)\\ &= \nabla f(tx,ty,tz)\cdot\mathbf{x}.\end{aligned}$$

The derivative on the right side is $pt^{p-1}f(x,y,z)$, and hence

$$\nabla f(tx,ty,tz)\cdot\mathbf{x} = pt^{p-1}f(x,y,z).$$

Now substitute $t=1$ to get

$$\nabla f(x,y,z)\cdot\mathbf{x} = pf(x,y,z);$$

i.e., $\mathbf{x}\cdot\nabla f(\mathbf{x}) = pf(\mathbf{x})$.

5. (a) The direction of the most rapid descent at $(10,10,390)$ is given by

$$-\nabla z(10,10) = 0.2x\mathbf{i} + 0.8y\mathbf{j}\Big|_{(10,10)} = 2\mathbf{i} + 8\mathbf{j}.$$

(b) The direction of the fastest ascent is $\nabla z(10,10) = -2\mathbf{i} - 8\mathbf{j}$. To answer the second part of the question, we have to find the angle between the xy-plane and the tangent line to the surface at the point $(10,10,390)$ that lies directly above the vector $-2\mathbf{i}-8\mathbf{j}$ (whose tail is at $(10,10)$). But that angle is equal to the angle between a normal to the surface and the z-axis. Rewrite the equation of the surface as $g(x,y,z) = 0.1x^2 + 0.4y^2 + z = 440$ to get a normal

$$\nabla g(10,10,390) = 0.2x\mathbf{i} + 0.8y\mathbf{j} + \mathbf{k}\Big|_{(10,10,390)} = 2\mathbf{i} + 8\mathbf{j} + \mathbf{k}.$$

The angle θ between the normal and the z-axis satisfies

$$\cos\theta = \frac{(2\mathbf{i}+8\mathbf{j}+\mathbf{k})\cdot\mathbf{k}}{\sqrt{69}} = \frac{1}{\sqrt{69}}.$$

Hence $\theta = \arccos(1/\sqrt{69}) \approx 1.4502$ rad.

(c) We have to find a unit vector $\mathbf{u} = u\mathbf{i} + v\mathbf{j}$ so that $D_{\mathbf{u}}z(10,10) = 0.1$. From

$$D_{\mathbf{u}}z(10,10) = \nabla z(10,10) \cdot \mathbf{u} = (-2\mathbf{i} - 8\mathbf{j}) \cdot (u\mathbf{i} + v\mathbf{j}) = 0.1$$

it follows that $2u + 8v = -0.1$. Since $\mathbf{u}$ is a unit vector, $u^2 + v^2 = 1$; it remains to solve the two equations for u and v. From $2u + 8v = -0.1$ we get $u = -0.05 - 4v$ and $(-0.05 - 4v)^2 + v^2 = 1$. It follows that $17v^2 + 0.4v - 0.9975 = 0$, and therefore $v_1 \approx 0.2308$ (hence $u_1 = -0.05 - 4v_1 \approx -0.9730$) and $v_2 \approx -0.2543$ (hence $u_2 = -0.05 - 4v_2 \approx -0.9671$). So there are two desired directions, (approximately equal to) $-0.9730\mathbf{i} + 0.2308\mathbf{j}$ and $-0.9671\mathbf{i} + 0.2543\mathbf{j}$.

7. We use the chain rule to compute the derivative of $f(\mathbf{c}(t))$:

$$\begin{aligned}\frac{d}{dt} f(\mathbf{c}(t)) &= Df(\mathbf{c}(t)) \cdot \mathbf{c}'(t) = grad\, f(\mathbf{c}(t)) \cdot \mathbf{c}'(t) \\ &= -\mathbf{F}(\mathbf{c}(t)) \cdot \mathbf{c}'(t) = -\mathbf{c}'(t) \cdot \mathbf{c}'(t) = -\|\mathbf{c}'(t)\|^2,\end{aligned}$$

since $\mathbf{F}(\mathbf{c}(t)) = \mathbf{c}'(t)$ by the definition of a flow line. It follows that the derivative of $f(\mathbf{c}(t))$ is always negative; i.e., f is a decreasing function.

9. (a) The direction of the largest increase in the voltage is given by $\nabla V(1,1,-3)$. Since $V = (1 + \sqrt{x^2 + y^2 + z^2})^{-1}$, it follows that

$$\begin{aligned}\frac{\partial V}{\partial x}(1,1,-3) &= -(1 + \sqrt{x^2 + y^2 + z^2})^{-2}\, \tfrac{1}{2} \frac{2x}{\sqrt{x^2 + y^2 + z^2}}\Bigg|_{(1,1,-3)} \\ &= -\frac{x}{(1 + \sqrt{x^2 + y^2 + z^2})^2 \sqrt{x^2 + y^2 + z^2}}\Bigg|_{(1,1,-3)} \\ &= -\frac{1}{(1 + \sqrt{11})^2 \sqrt{11}} \approx -0.01618.\end{aligned}$$

By interchanging x and y, we get $(\partial V/\partial y)(1,1,-3) \approx -0.01618$. Finally,

$$\begin{aligned}\frac{\partial V}{\partial z}(1,1,-3) &= -\frac{z}{(1 + \sqrt{x^2 + y^2 + z^2})^2 \sqrt{x^2 + y^2 + z^2}}\Bigg|_{(1,1,-3)} \\ &\approx 0.04854,\end{aligned}$$

and so $\nabla V(1,1,-3) \approx -0.01618\mathbf{i} - 0.01618\mathbf{j} + 0.04854\mathbf{k}$

(b) At $(1,1,-3)$, there is no change in V in directions perpendicular to $\nabla V(1,1,-3)$. All such vectors lie in the plane through $(1,1,-3)$ whose normal is given by $\nabla V(1,1,-3)$.

5. INTEGRATION ALONG PATHS

5.1 Paths and Parametrizations

1. $\phi(t)$ is bijective: its inverse $\phi^{-1}: [0, \ln 2] \to [0,1]$ is $\phi^{-1}(t) = e^t - 1$. The derivative $\phi'(t) = 1/(t+1)$ is defined and continuous on $[0,1]$, and therefore ϕ can be a reparametrization of a path.

3. Since $\phi'(t) = t^{-2/3}/3$, it follows that $\phi'(0)$ is not defined. Consequently, ϕ cannot be a reparametrization of a path.

5. $\phi(t) = e^t$ is bijective: its inverse $\phi^{-1}: [1, e] \to [0,1]$ is $\phi^{-1}(t) = \ln t$. The derivative $\phi'(t) = e^t$ is defined and continuous on $[0,1]$, and therefore ϕ can be a reparametrization of a path.

7. ϕ is not bijective, since, for example, $\phi(-1) = \phi(1) = 1$. It is not differentiable either, since $\phi'(0)$ does not exist. Either of these two facts proves that ϕ cannot be a reparametrization of a path.

9. $\mathbf{c}(t)$ is not one-to-one; for example, $\mathbf{c}(0) = (0,1,4\pi^2)$ and $\mathbf{c}(4\pi) = (0,1,4\pi^2)$. It follows that $\mathbf{c}(t)$ is not a simple curve. Since $\mathbf{c}(-2\pi) = (0,1,16\pi^2)$ and $\mathbf{c}(6\pi) = (0,1,16\pi^2)$, it follows that $\mathbf{c}(t)$ is closed. However, since $\mathbf{c}(0) = \mathbf{c}(4\pi)$, $\mathbf{c}$ is not one-to-one on $[-2\pi, 6\pi)$, and so it is not a simple closed curve.

11. From $\|\mathbf{c}(t)\| = \sqrt{t^2 \sin^2 t + t^2 \cos^2 t} = \sqrt{t^2} = |t| = t$ (since $t \ge 0$), it follows that the distance $\|\mathbf{c}(t) - \mathbf{0}\| = \|\mathbf{c}(t)\|$ between $\mathbf{c}(t)$ and the origin is t. Therefore, no two points $\mathbf{c}(t_1)$ and $\mathbf{c}(t_2)$, with $t_1 \neq t_2$, are at the same distance from the origin, and so $\mathbf{c}$ is simple (that is, it does not intersect itself). Since $\mathbf{c}(0) = (0,0)$ and $\mathbf{c}(2\pi) = (0, 2\pi)$, $\mathbf{c}$ is not closed. Consequently, $\mathbf{c}$ cannot be a simple closed curve.

13. Assume that $\mathbf{c}(t_1) = \mathbf{c}(t_2)$. Then

$$\left(t_1 - \frac{1}{t_1}, t_1 + \frac{1}{t_1}\right) = \left(t_2 - \frac{1}{t_2}, t_2 + \frac{1}{t_2}\right)$$

implies

$$t_1 - \frac{1}{t_1} = t_2 - \frac{1}{t_2} \qquad \text{and} \qquad t_1 + \frac{1}{t_1} = t_2 + \frac{1}{t_2}.$$

Adding the two equations we get $t_1 = t_2$. In other words, $\mathbf{c}(t)$ is one-to-one (whenever defined, and, in particular, on $[1,2]$) and the curve $\mathbf{c}$ is simple. Since $\mathbf{c}(1) = (0,2)$ and $\mathbf{c}(2) = (3/2, 5/2)$, $\mathbf{c}$ is not closed. In that case, it cannot be a simple closed curve.

15. Let $x = t$; then $\sqrt{t^2+1}$, and hence

$$\mathbf{c}(t) = (t, \sqrt{t^2+1}), \qquad t \in [-1,1]$$

is one possible parametrization of the given curve. It is continuous, and since $\mathbf{c}'(t) = (1, t/\sqrt{t^2+1})$, it is also differentiable and C^1.

17. (a) Since $\mathbf{c}_1'(t) = (1,0)$, it follows that $\mathbf{c}_1(t)$ is C^1, and hence also differentiable and continuous.

(b) The paths $\mathbf{c}_{21}(t) = (-t^2, 1)$, $t \in [-1,0]$, and $\mathbf{c}_{22}(t) = (t^2,1)$, $t \in [0,1]$, are continuous. Since $\mathbf{c}_{21}(0) = (0,1) = \mathbf{c}_{22}(0)$, it follows that $\mathbf{c}_2(t)$ is a continuous path. The paths $\mathbf{c}_{21}'(t) = (-2t, 0)$, $t \in [-1,0]$, and $\mathbf{c}_{22}'(t) = (2t,0)$, $t \in [0,1]$, are both continuous. Since $\mathbf{c}_{21}'(0) = (0,0) = \mathbf{c}_{22}'(0)$, it follows that $\mathbf{c}_2'(t)$ is a continuous path; therefore, $\mathbf{c}_2(t)$ is a C^1 path.

(c) $\mathbf{c}_3(t) = (t^{1/3}, 1)$ is continuous; however, it is not differentiable, since $\mathbf{c}_3'(t) = (t^{-2/3}/3, 0)$ is not defined at $t = 0$. For the same reason, $\mathbf{c}_3$ is not piecewise C^1 or C^1.

(d) Since $\mathbf{c}_4'(t) = (3t^2, 0)$ is continuous, it follows that $\mathbf{c}_4(t)$ is a C^1 path.

(e) $\mathbf{c}_5(t)$ traverses the given line segment as follows: it goes from $(-1,1)$ to $(0,1)$, then jumps to $(1,1)$ and goes back to $(0,1)$. So $\mathbf{c}_5$ is not continuous. Hovever, both $(t,1)$ and $(1-t,1)$ are differentiable and have continuous derivatives; so $\mathbf{c}_5$ is piecewise C^1.

19. Drop t for simplicity of notation. Let $\boldsymbol{\gamma} = \mathbf{c}(\phi) = (\phi^2, 2-\phi^2)$. We have to find ϕ so that $\|\boldsymbol{\gamma}'\| = C_1$ (C_1 is a constant). From $\boldsymbol{\gamma}' = (2\phi\phi', -2\phi\phi')$ it follows that

$$\|\boldsymbol{\gamma}'\| = \sqrt{4\phi^2(\phi')^2 + 4\phi^2(\phi')^2} = |\phi\phi'|\sqrt{8} = C_1;$$

i.e., $\phi\phi' = \pm C_1/\sqrt{8}$. Integrating both sides with respect to t, we get

$$\tfrac{1}{2}\phi^2 = \pm\frac{C_1 t}{\sqrt{8}} + D_1,$$

where D_1 is a constant. Therefore,

$$\phi = \pm\sqrt{\pm 2C_1 t/\sqrt{8} + 2D_1}$$

or (simplify the expressions involving constants; let $C = \pm 2C_1/\sqrt{8}$ and $D = 2D_1$) $\phi(t) = \pm\sqrt{Ct+D}$. It follows that the reparametrization of $\mathbf{c}$ by $\phi(t) = \pm\sqrt{Ct+D}$, $t \in [1,3]$, where C and D are chosen so that $\phi(t)$ is defined, will have constant speed.

21. Recall that the speed of the curve $\mathbf{c}(t) = (\cos(at), \sin(at))$, $a > 0$, is given by

$$\|\mathbf{c}'(t)\| = \sqrt{(-a\sin(at))^2 + (a\cos(at))^2} = a.$$

Therefore, to adjust the speed we have to choose the arguments of sin and cos correctly. Let $\mathbf{c}_1(t) = (\cos t, \sin t)$, $t \in [0, 2\pi]$; i.e., the speed of $\mathbf{c}_1$ is 1. Since $\mathbf{c}(0) = (1,0)$, and both components of $\mathbf{c}_1(t)$ are positive as t starts increasing from 0, it follows that $\mathbf{c}_1(t)$ is oriented counterclockwise. That answers the second part of the question.

To reverse the orientation, let

$$\mathbf{c}_2(t) = (\cos(-t), \sin(-t)) = (\cos t, -\sin t),$$

where $t \in [0, 2\pi]$. This time, $\mathbf{c}(0) = (1, 0)$, and, as t starts increasing from 0, the y-component is negative; so $\mathbf{c}_2$ is oriented clockwise. To get the correct speed, we use

$$\mathbf{c}_2(t) = (\cos(St), -\sin(St)), \qquad t \in [0, 2\pi/S].$$

This is the answer to the first part of the question.

5.2 Path Integrals of Real-Valued Functions

1. The values of f along $\mathbf{c}$ are

$$f(\mathbf{c}(t)) = 2(e^t + 1) - (e^t - 2) = e^t + 4.$$

From $\mathbf{c}'(t) = (e^t, e^t)$ it follows that $\|\mathbf{c}'(t)\| = \sqrt{e^{2t} + e^{2t}} = e^t\sqrt{2}$ and so

$$\begin{aligned}\int_{\mathbf{c}} (2x - y)ds &= \int_0^{\ln 2} (e^t + 4)e^t\sqrt{2}\,dt = \sqrt{2}\int_0^{\ln 2} (e^{2t} + 4e^t)\,dt \\ &= \sqrt{2}\left(\tfrac{1}{2}e^{2t} + 4e^t\right)\Big|_0^{\ln 2} = \frac{11\sqrt{2}}{2} \approx 7.7782.\end{aligned}$$

3. The values of f along $\mathbf{c}$ are $f(\mathbf{c}(t)) = (t^2 + t^2 + t^2)^{-1} = (3t^2)^{-1} = t^{-2}/3$. From $\mathbf{c}'(t) = (1, 1, 1)$ it follows that $\|\mathbf{c}'(t)\| = \sqrt{3}$. Let $\mathbf{c}_b(t) = (t, t, t)$, $1 \le t \le b$; then

$$\int_{\mathbf{c}_b} (x^2 + y^2 + z^2)^{-1} ds = \int_1^b \frac{t^{-2}}{3}\sqrt{3}\,dt = \frac{-\sqrt{3}}{3}\left(\frac{1}{t}\right)\Big|_1^b = \frac{\sqrt{3}}{3}\left(1 - \frac{1}{b}\right).$$

Therefore,

$$\int_{\mathbf{c}} (x^2 + y^2 + z^2)^{-1} ds = \lim_{b\to\infty} \int_{\mathbf{c}_b} (x^2 + y^2 + z^2)^{-1} ds = \frac{\sqrt{3}}{3}.$$

5. The values of f along $\mathbf{c}$ are $f(\mathbf{c}(t)) = f(t^2\mathbf{i} + \ln t\mathbf{j} + 2t\mathbf{k}) = \ln t - 4t^2$. From $\mathbf{c}'(t) = 2t\mathbf{i} + t^{-1}\mathbf{j} + 2\mathbf{k}$ we get

$$\|\mathbf{c}'(t)\| = \sqrt{4t^2 + 4 + t^{-2}} = \sqrt{(2t + t^{-1})^2} = 2t + t^{-1},$$

since $t \ge 0$. It follows that

$$\begin{aligned}\int_{\mathbf{c}} f\,ds &= \int_1^4 (\ln t - 4t^2)(2t + t^{-1})\,dt \\ &= 2\int_1^4 t\ln t\,dt - 8\int_1^4 t^3\,dt + \int_1^4 \frac{\ln t}{t}\,dt - 4\int_1^4 t\,dt.\end{aligned}$$

Using integration by parts with $u = \ln t$ and $dv = tdt$, we get ($du = dt/t$ and $v = t^2/2$)

$$\int t\ln t\,dt = \tfrac{1}{2}t^2\ln t - \int \tfrac{1}{2}t\,dt = \tfrac{1}{2}t^2\ln t - \tfrac{1}{4}t^2 + C,$$

and therefore

$$\int_1^4 t\ln t\,dt = \left(\tfrac{1}{2}t^2\ln t - \tfrac{1}{4}t^2\right)\Big|_1^4 = 8\ln 4 - \tfrac{15}{4}.$$

Using the substitution $u = \ln t$, $du = (1/t)dt$, we get

$$\int_1^4 \frac{\ln t}{t}\,dt = \int_0^{\ln 4} u\,du = \tfrac{1}{2}u^2\Big|_0^{\ln 4} = \tfrac{1}{2}(\ln 4)^2.$$

Hence

$$\begin{aligned}\int_{\mathbf{c}} f\,ds &= 2\left(8\ln 4 - \tfrac{15}{4}\right) - 2t^4\Big|_1^4 + \tfrac{1}{2}(\ln 4)^2 - 2t^2\Big|_1^4\\ &= 16\ln 4 + \tfrac{1}{2}(\ln 4)^2 - 547.5 \approx -524.3584.\end{aligned}$$

7. The values of f along $\mathbf{c}$ are $f(2t, 3t, 1-4t) = 6t^2 + 3t + 1 - 4t = 6t^2 - t + 1$. From $\mathbf{c}'(t) = (2, 3, -4)$ it follows that $\|\mathbf{c}'(t)\| = \sqrt{29}$ and so

$$\int_{\mathbf{c}} (xy + y + z)ds = \int_0^1 (6t^2 - t + 1)\sqrt{29}\,dt = \sqrt{29}\left(2t^3 - \tfrac{1}{2}t^2 + t\right)\Big|_0^1 = \tfrac{5}{2}\sqrt{29} \approx 13.4629.$$

9. The values of f along $\mathbf{c}$ are $f(2\sin t, 4t, 2\cos t) = (2\sin t)(4t)(2\cos t) = 16t\sin t\cos t = 8t\sin 2t$. From $\mathbf{c}'(t) = (2\cos t, 4, -2\sin t)$ it follows that $\|\mathbf{c}'(t)\| = \sqrt{4\cos^2 t + 16 + 4\sin^2 t} = \sqrt{20}$ and so

$$\int_{\mathbf{c}} xyz\,ds = \int_0^{6\pi} 8t\sin 2t\,\sqrt{20}\,dt = 8\sqrt{20}\int_0^{6\pi} t\sin 2t\,dt.$$

Using integration by parts with $u = t$ and $dv = \sin 2tdt$, we get $(du = dt$ and $v = -(\cos 2t)/2)$

$$\int t\sin 2t\,dt = -\tfrac{1}{2}t\cos 2t + \int \tfrac{1}{2}\cos 2t\,dt = -\tfrac{1}{2}t\cos 2t + \tfrac{1}{4}\sin 2t + C,$$

and therefore

$$\int_0^{6\pi} t\sin 2t\,dt = \left(-\tfrac{1}{2}t\cos 2t + \tfrac{1}{4}\sin 2t\right)\Big|_0^{6\pi} = -3\pi.$$

Finally,

$$\int_{\mathbf{c}} xyz\,ds = 8\sqrt{20}(-3\pi) = -48\pi\sqrt{5} \approx -337.1911$$

11. Parametrize $\mathbf{c}$ by

$$\mathbf{c}(t) = (0,0) + t(3,-4) = (3t, -4t), \qquad 0 \le t \le 1.$$

Then $\mathbf{c}'(t) = (3, -4)$, $\|\mathbf{c}'(t)\| = 5$ and

$$\int_{\mathbf{c}} e^{x+3y}\,ds = \int_0^1 5e^{-9t}\,dt = -\tfrac{5}{9}e^{-9t}\Big|_0^1 = -\tfrac{5}{9}(e^{-9} - 1) = \tfrac{5}{9}(1 - e^{-9}) \approx 0.5556$$

13. The curve $\mathbf{c}$ consists of two curves, parametrized by

$$\mathbf{c}_1(t) = (4,2,0) + t(-4,0,0) = (4-4t, 2, 0), \quad t \in [0,1],$$

and

$$\mathbf{c}_2(t) = (0, 2\cos t, 2\sin t), \quad t \in [0,\pi];$$

therefore,

$$\int_{\mathbf{c}} f\,ds = \int_{\mathbf{c}_1} f ds + \int_{\mathbf{c}_2} f ds.$$

From $f(\mathbf{c}_1(t)) = 4 - 4t - 8 = -4 - 4t$, $\mathbf{c}_1'(t) = (-4, 0, 0)$ and $\|\mathbf{c}_1'(t)\| = 4$ it follows that

$$\int_{\mathbf{c}_1} (x - 4y + z)\,ds = \int_0^1 (-4 - 4t)\,4\,dt = -16(t + \tfrac{1}{2}t^2)\Big|_0^1 = -24.$$

From $f(\mathbf{c}_2(t)) = -8\cos t + 2\sin t$, $\mathbf{c}_2'(t) = (0, -2\sin t, 2\cos t)$ and $\|\mathbf{c}_2'(t)\| = 2$ it follows that

$$\int_{\mathbf{c}_2} (x - 4y + z)\, ds = \int_0^\pi 2(-8\cos t + 2\sin t)\, dt = 2(-8\sin t - 2\cos t)\Big|_0^\pi = 2(2-(-2)) = 8.$$

It follows that $\int_{\mathbf{c}} f\, ds = -24 + 8 = -16$.

15. It is possible. The path integral of a real-valued function is independent of a parametrization of a curve; that is, it is independent of the way the curve is traversed. However, in general, it depends on the curve that is used to join the given points.

Consider the following intuitive example: let $A = (1,0)$ and $B = (2,0)$ and $f(x,y) = y^2$. The path integral along the straight line from A to B is zero, since the value of f at each point on that line is zero. On the other hand, the path integral along a curve in the first quadrant joining A and B is positive, since $f(x,y) = y^2 > 0$.

17. We are asked to find the area of the "fence" in the shape of the cylinder $x^2 + y^2 = 4$ whose height is determined by $z = y + 2$. Therefore, area $= \int_{\mathbf{c}} f\, ds$, where $f(x,y) = y + 2$ and $\mathbf{c}$ is the circle $x^2 + y^2 = 4$.

Parametrize $\mathbf{c}$ by $\mathbf{c}(t) = (2\cos t, 2\sin t)$, $t \in [0, 2\pi]$. Then $\mathbf{c}'(t) = (-2\sin t, 2\cos t)$, $\|\mathbf{c}'(t)\| = 2$ and

$$\text{area} = \int_{\mathbf{c}} (y+2)\, ds = \int_0^{2\pi} 2(2\sin t + 2)\, dt = 4(-\cos t + t)\Big|_0^{2\pi} = 8\pi.$$

19. We are asked to find the area of the "fence" in the shape of the graph of $y = \sin x$, whose height is determined by $z = \sin x \cos x$. Therefore, area $= \int_{\mathbf{c}} f\, ds$, where $f(x,y) = \sin x \cos x$ and $\mathbf{c}$ is the curve $y = \sin x$, $0 \le x \le \pi/2$.

Parametrize $\mathbf{c}$ by $\mathbf{c}(t) = (t, \sin t)$, $t \in [0, \pi/2]$. Then $\mathbf{c}'(t) = (1, \cos t)$, and its length is $\|\mathbf{c}'(t)\| = \sqrt{1 + \cos^2 t}$ and

$$\text{area} = \int_0^{\pi/2} \sin t \cos t \sqrt{1 + \cos^2 t}\, dt.$$

Using the substitution $u = 1 + \cos^2 t$, $du = -2\sin t \cos t dt$, we get

$$\text{area} = -\tfrac{1}{2} \int_2^1 u^{1/2}\, du = -\tfrac{1}{3} u^{3/2}\Big|_2^1 = \tfrac{1}{3}(2^{3/2} - 1) \approx 0.6095.$$

21. The function $f(x,y) = e^{x^2+y^2}$ is constant along the given circle; i.e., since $x^2 + y^2 = 16$, it follows that $f(x,y) = e^{16}$ for all points on that circle. The path integral $\int_{\mathbf{c}} e^{x^2+y^2}\, ds$ is the area of the "fence" defined by $x^2 + y^2 = 16$ of height e^{16}. In other words, it is the surface of the cylinder of radius 4 and height e^{16}. Its surface area is $2\pi(4)(e^{16}) = 8\pi e^{16}$. Therefore, $\int_{\mathbf{c}} e^{x^2+y^2}\, ds = 8\pi e^{16}$.

23. Since $f(x,y) \le 1$ for all x and y, it is not possible that the average value of f be any number greater than 1.

25. Since $\mathbf{c}$ is a circle of radius 3, its length is $\ell(\mathbf{c}) = 6\pi$. Hence

$$\overline{f} = \frac{1}{\ell(\mathbf{c})}\int_{\mathbf{c}} f\,ds = -\frac{1}{6\pi}\int_{\mathbf{c}} \sqrt{x^2+z^2}\,ds.$$

From $\mathbf{c}'(t) = -3\sin t\,\mathbf{j} + 3\cos t\,\mathbf{k}$ we get $\|\mathbf{c}'(t)\| = 3$ and

$$\overline{f} = -\frac{1}{6\pi}\int_0^{2\pi} 3\sqrt{9\cos^2 t}\,dt = -\frac{3}{2\pi}\int_0^{2\pi} |\cos t|\,dt.$$

Since

$$\begin{aligned}\int_0^{2\pi} |\cos t|\,dt &= \int_0^{\pi/2} |\cos t|\,dt + \int_{\pi/2}^{3\pi/2} |\cos t|\,dt + \int_{3\pi/2}^{2\pi} |\cos t|\,dt \\ &= \int_0^{\pi/2} \cos t\,dt - \int_{\pi/2}^{3\pi/2} \cos t\,dt + \int_{3\pi/2}^{2\pi} \cos t\,dt \\ &= 1+2+1 = 4,\end{aligned}$$

it follows that $\overline{f} = -6/\pi$.

5.3 Path Integrals of Vector Fields

1. Parametrize $\mathbf{c}$ by $\mathbf{c}(t) = t\mathbf{i} + t^2\mathbf{j}$, $t \in [-1,1]$. Then $\mathbf{c}'(t) = \mathbf{i} + 2t\mathbf{j}$, $\mathbf{F}(\mathbf{c}(t)) = \mathbf{F}(t\mathbf{i}+t^2\mathbf{j}) = t^4\mathbf{i} - t^2\mathbf{j}$, and

$$\int_{\mathbf{c}} (y^2\mathbf{i} - x^2\mathbf{j})\cdot d\mathbf{s} = \int_{-1}^{1} (t^4\mathbf{i} - t^2\mathbf{j})\cdot(\mathbf{i}+2t\mathbf{j})\,dt = \int_{-1}^{1} (t^4 - 2t^3)\,dt = \left(\tfrac{1}{5}t^5 - \tfrac{1}{2}t^4\right)\Big|_{-1}^{1} = \tfrac{2}{5}.$$

3. Parametrize the segment from $(0,0)$ to $(0,1)$ by $\mathbf{c}_1(t) = (0,t)$, $0 \le t \le 1$. Then $\mathbf{c}_1'(t) = (0,1)$ and

$$\int_{\mathbf{c}_1} (e^{x+y}\mathbf{i} - \mathbf{j})\cdot d\mathbf{s} = \int_0^1 (e^t, -1)\cdot(0,1)\,dt = \int_0^1 (-1)\,dt = -1.$$

Parametrize the segment from $(0,1)$ to $(1,1)$ by $\mathbf{c}_2(t) = (t,1)$, $0 \le t \le 1$. Then $\mathbf{c}_2'(t) = (1,0)$ and

$$\int_{\mathbf{c}_2} (e^{x+y}\mathbf{i} - \mathbf{j})\cdot d\mathbf{s} = \int_0^1 (e^{t+1}, -1)\cdot(1,0)\,dt = \int_0^1 e^{t+1}\,dt = e^{t+1}\Big|_0^1 = e^2 - e.$$

Parametrize the segment from $(1,1)$ to $(1,0)$ by $\mathbf{c}_3(t) = (1, 1-t)$, $0 \le t \le 1$. Then $\mathbf{c}_3'(t) = (0,-1)$ and

$$\int_{\mathbf{c}_3} (e^{x+y}\mathbf{i} - \mathbf{j})\cdot d\mathbf{s} = \int_0^1 (e^{2-t}, -1)\cdot(0,-1)\,dt = \int_0^1 dt = 1.$$

Finally, parametrize the segment from $(1,0)$ to $(0,0)$ by $\mathbf{c}_4(t) = (1-t, 0)$, $0 \le t \le 1$. Then $\mathbf{c}_4'(t) = (-1,0)$ and

$$\int_{\mathbf{c}_4} (e^{x+y}\mathbf{i} - \mathbf{j})\cdot d\mathbf{s} = \int_0^1 (e^{1-t}, -1)\cdot(-1,0)\,dt = \int_0^1 (-e^{1-t})\,dt = e^{1-t}\Big|_0^1 = 1 - e.$$

It follows that

$$\int_{\mathbf{c}}(e^{x+y}\mathbf{i}-\mathbf{j})\cdot d\mathbf{s} = \int_{\mathbf{c}_1}(e^{x+y}\mathbf{i}-\mathbf{j})\cdot d\mathbf{s} + \int_{\mathbf{c}_2}(e^{x+y}\mathbf{i}-\mathbf{j})\cdot d\mathbf{s} + \int_{\mathbf{c}_3}(e^{x+y}\mathbf{i}-\mathbf{j})\cdot d\mathbf{s}$$
$$+\int_{\mathbf{c}_4}(e^{x+y}\mathbf{i}-\mathbf{j})\cdot d\mathbf{s}$$
$$= -1+e^2-e+1+1-e = e^2-2e+1.$$

5. The value of $\mathbf{F}$ along $\mathbf{c}$ is $\mathbf{F}(\mathbf{c}(t)) = \mathbf{F}(\sin t, \cos t, 2t) = (\cos t, -\sin t, 1)$. Since $\mathbf{c}'(t) = (\cos t, -\sin t, 2)$, it follows that

$$\int_{\mathbf{c}}(y,-x,1)\cdot d\mathbf{s} = \int_0^{\pi}(\cos t, -\sin t, 1)\cdot(\cos t, -\sin t, 2)\,dt = \int_0^{\pi} 3\,dt = 3\pi.$$

7. The value of $\mathbf{F}$ along $\mathbf{c}$ is $\mathbf{F}(\mathbf{c}(t)) = \mathbf{F}(\sin t, \cos t, t^2) = (\sin^2 t, \sin t\cos t, 2t^4)$. Since $\mathbf{c}'(t) = (\cos t, -\sin t, 2t)$, it follows that

$$\int_{\mathbf{c}}(x^2, xy, 2z^2)\cdot d\mathbf{s} = \int_0^{\pi/2}(\sin^2 t, \sin t\cos t, 2t^4)\cdot(\cos t, -\sin t, 2t)\,dt$$
$$= \int_0^{\pi/2} 4t^5\,dt = \tfrac{2}{3}t^6\Big|_0^{\pi/2} = \tfrac{2}{3}\left(\tfrac{\pi}{2}\right)^6 \approx 10.0145.$$

9. The value of $\mathbf{F}$ along $\mathbf{c}$ is $\mathbf{F}(\mathbf{c}(t)) = \mathbf{F}(4t^3\mathbf{i}+t^2\mathbf{j}) = 8t^5\mathbf{i}+e^{t^2}\mathbf{j}$. Since $\mathbf{c}'(t) = 12t^2\mathbf{i}+2t\mathbf{j}$, we get

$$\int_{\mathbf{c}}(2xy\mathbf{i}+e^y\mathbf{j})\cdot d\mathbf{s} = \int_0^1(8t^5\mathbf{i}+e^{t^2}\mathbf{j})\cdot(12t^2\mathbf{i}+2t\mathbf{j})\,dt$$
$$= \int_0^1\left(96t^7+2te^{t^2}\right)dt = \left(12t^8+e^{t^2}\right)\Big|_0^1 = 12+e-1 = 11+e$$

(we used the substitution $u=t^2$, $du = 2tdt$, to evaluate $\int 2te^{t^2}dt = e^{t^2}$).

11. The work is given by $W = \int_{\mathbf{c}}\mathbf{F}\cdot d\mathbf{s}$, where $\mathbf{F}(x,y) = x^3\mathbf{i}+(x+y)\mathbf{j}$ and $\mathbf{c}(t) = \sin t\,\mathbf{i}+t^2\mathbf{j}$, $t\in[0,\pi/2]$. Since $\mathbf{F}(\mathbf{c}(t)) = \sin^3 t\,\mathbf{i}+(\sin t+t^2)\mathbf{j}$ and $\mathbf{c}'(t) = \cos t\,\mathbf{i}+2t\mathbf{j}$, it follows that

$$W = \int_0^{\pi/2}(\sin^3 t\,\mathbf{i}+(\sin t+t^2)\mathbf{j})\cdot(\cos t\,\mathbf{i}+2t\mathbf{j})\,dt$$
$$= \int_0^{\pi/2}(\sin^3 t\cos t+2t\sin t+2t^3)\,dt.$$

Using the substitution $u=\sin t$, $du = \cos t dt$, we get

$$\int_0^{\pi/2}\sin^3 t\cos t\,dt = \int_0^1 u^3\,du = \tfrac{1}{4}.$$

Using integration by parts with $u=t$ and $dv = \sin t dt$, we get ($du = dt$, $v = -\cos t$)

$$\int_0^{\pi/2} t\sin t\,dt = -t\cos t\Big|_0^{\pi/2} + \int_0^{\pi/2}\cos t\,dt = \int_0^{\pi/2}\cos t\,dt = \sin t\Big|_0^{\pi/2} = 1.$$

Finally,

$$\int_0^{\pi/2} 2t^3\,dt = \tfrac{1}{2}t^4\Big|_0^{\pi/2} = \tfrac{1}{32}\pi^4,$$

and so

$$W = \tfrac{1}{4} + 2 + \tfrac{1}{32}\pi^4 = \tfrac{9}{4} + \tfrac{1}{32}\pi^4 \approx 5.2940.$$

13. There is no need to parametrize a circle or to compute a path integral. Just notice that the direction of **F** is radial (i.e., along lines through the origin), and therefore orthogonal to any circle centered at the origin. Consequently, the work must be zero.

15. Use W_a, W_b, etc. to denote the work along the curves $\mathbf{c}_a$, $\mathbf{c}_b$, etc. in parts (a), (b), etc.

(a) The work is

$$W_a = \int_{\mathbf{c}_{a1}} \mathbf{F} \cdot d\mathbf{s} + \int_{\mathbf{c}_{a2}} \mathbf{F} \cdot d\mathbf{s},$$

where $\mathbf{c}_{a1}(t) = (t, 0)$, $t \in [0,1]$ and $\mathbf{c}_{a2}(t) = (1, t)$, $t \in [0,1]$. Since $\mathbf{c}'_{a1}(t) = (1,0)$ and $\mathbf{c}'_{a2}(t) = (0,1)$, it follows that

$$W_a = \int_{\mathbf{c}_{a1}} \mathbf{F} \cdot d\mathbf{s} + \int_{\mathbf{c}_{a2}} \mathbf{F} \cdot d\mathbf{s} = \int_0^1 (0,0) \cdot (1,0)\, dt + \int_0^1 (t,0) \cdot (0,1)\, dt = 0 + 0 = 0.$$

So the work along $\mathbf{c}_a$ is zero.

(b) Parametrize the given path by $\mathbf{c}_b(t) = (t, t^2)$, $t \in [0,1]$. Then $\mathbf{c}'_b(t) = (1, 2t)$, $\mathbf{F}(\mathbf{c}_b(t)) = (t^2, 0)$ and

$$W_b = \int_0^1 (t^2, 0) \cdot (1, 2t)\, dt = \int_0^1 t^2\, dt = \tfrac{1}{3}.$$

(c) Use a parametrization $\mathbf{c}_c(t) = (t, t^4)$, $t \in [0,1]$. Then $\mathbf{c}'_c(t) = (1, 4t^3)$, $\mathbf{F}(\mathbf{c}_c(t)) = (t^4, 0)$ and

$$W_c = \int_{\mathbf{c}_c} \mathbf{F} \cdot d\mathbf{s} = \int_0^1 (t^4, 0) \cdot (1, 4t^3)\, dt = \int_0^1 t^4\, dt = \tfrac{1}{5}.$$

(d) Parametrize the given path by $\mathbf{c}_d(t) = (t, t)$, $t \in [0,1]$. Then $\mathbf{c}'_d(t) = (1,1)$, $\mathbf{F}(\mathbf{c}_d(t)) = (t, 0)$ and

$$W_d = \int_{\mathbf{c}_d} \mathbf{F} \cdot d\mathbf{s} = \int_0^1 (t, 0) \cdot (1,1)\, dt = \int_0^1 t\, dt = \tfrac{1}{2}.$$

(e) Use $\mathbf{c}_e(t) = (t, \sin(\pi t/2))$, $t \in [0,1]$. Then $\mathbf{c}'_e(t) = (1, (\pi/2)\cos(\pi t/2))$, $\mathbf{F}(\mathbf{c}_e(t)) = (\sin(\pi t/2), 0)$ and

$$\begin{aligned} W_e &= \int_{\mathbf{c}_e} \mathbf{F} \cdot d\mathbf{s} = \int_0^1 (\sin(\pi t/2), 0) \cdot (1, (\pi/2)\cos(\pi t/2))\, dt \\ &= \int_0^1 \sin(\pi t/2)\, dt = -\tfrac{2}{\pi}\cos(\pi t/2)\Big|_0^1 = \tfrac{2}{\pi} \approx 0.6366. \end{aligned}$$

(f) The work is

$$W_f = \int_{\mathbf{c}_{f1}} \mathbf{F} \cdot d\mathbf{s} + \int_{\mathbf{c}_{f2}} \mathbf{F} \cdot d\mathbf{s},$$

where $\mathbf{c}_{f1}(t) = (0, t)$, $t \in [0,1]$ and $\mathbf{c}_{f2}(t) = (t, 1)$, $t \in [0,1]$. Then $\mathbf{c}'_{f1}(t) = (0,1)$, $\mathbf{c}'_{f2}(t) = (1,0)$, and

$$W_f = \int_{\mathbf{c}_{f1}} \mathbf{F} \cdot d\mathbf{s} + \int_{\mathbf{c}_{f2}} \mathbf{F} \cdot d\mathbf{s} = \int_0^1 (t,0) \cdot (0,1)\, dt + \int_0^1 (1,0) \cdot (1,0)\, dt = 0 + 1 = 1.$$

See Figure 5.3.15; the vector field $\mathbf{F}$ becomes stronger as the y-coordinate increases. The curves $\mathbf{c}_f$, $\mathbf{c}_e$ and $\mathbf{c}_d$ are "aligned better" than $\mathbf{c}_a$, $\mathbf{c}_b$ and $\mathbf{c}_c$ with $\mathbf{F}$ in regions where $\mathbf{F}$ is stronger. That is why in these cases the corresponding work (i.e., W_f, W_e and W_d) is larger. Using similar argument we can compare any pair of paths.

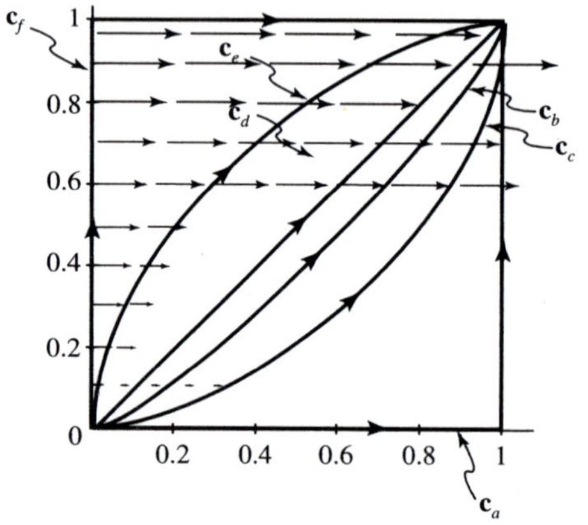

Figure 5.3.15

17. By definition (from $\mathbf{c}(t) = (t, 2t^3)$ we get $x = t$ and $y = 2t^3$),

$$\int_{\mathbf{c}} xydy = \int_0^1 \left(xy\frac{dy}{dt}\right)\, dt = \int_0^1 t \cdot 2t^3 \cdot 6t^2\, dt = 12\int_0^1 t^6\, dt = \tfrac{12}{7}.$$

19. From $\mathbf{c}(t) = (\sqrt{t}, t^2, t^3)$ it follows that $x = \sqrt{t}$, $y = t^2$ and $z = t^3$, and therefore

$$\begin{aligned}\int_{\mathbf{c}} (ydy - xyzdz) &= \int_0^1 \left(y\frac{dy}{dt} - xyz\frac{dz}{dt}\right)\, dt \\ &= \int_0^1 (2t^3 - 3t^{15/2})\, dt = \left(\tfrac{1}{2}t^4 - \tfrac{6}{17}t^{17/2}\right)\Big|_0^1 = \tfrac{1}{2} - \tfrac{6}{17} = \tfrac{5}{34}.\end{aligned}$$

21. Parametrize $\mathbf{c}$ by $\mathbf{c}(t) = (\sin t, \cos t)$, $t \in [0, 2\pi]$ ($\mathbf{c}$ starts and ends at $(0, 1)$ and is oriented clockwise). Then

$$\begin{aligned}\int_{\mathbf{c}} \alpha &= \int_0^{2\pi} \left(\frac{1}{1+x^2}\frac{dx}{dt} + \frac{1}{1+y^2}\frac{dy}{dt}\right)\, dt \\ &= \int_0^{2\pi} \left(\frac{1}{1+\sin^2 t}\cos t + \frac{1}{1+\cos^2 t}(-\sin t)\right)\, dt \\ &= \arctan(\sin t) + \arctan(\cos t)\Big|_0^{2\pi} = \left(0 + \tfrac{1}{4}\pi\right) - \left(0 + \tfrac{1}{4}\pi\right) = 0.\end{aligned}$$

Both integrals are solved by substitution. For example, using $u = \sin t$, $du = \cos t dt$, we get

$$\int \frac{1}{1+\sin^2 t}\cos t\, dt = \int \frac{1}{1+u^2}\, du = \arctan u = \arctan(\sin t).$$

The other integral is computed similarly, using the substitution $u = \cos t$.

23. Parametrize $\mathbf{c}$ by $\mathbf{c}(t) = (2\cos t, 2\sin t)$, $t \in [0, 2\pi]$. Then $x^2 + y^2 = 4$ and

$$\int_{\mathbf{c}} \frac{ydx + xdy}{x^2 + y^2} = \tfrac{1}{4}\int_0^{2\pi} (2\sin t(-2\sin t) + 2\cos t(2\cos t))\, dt$$
$$= \int_0^{2\pi} (\cos^2 t - \sin^2 t)\, dt = \int_0^{2\pi} \cos 2t\, dt = \tfrac{1}{2}\sin 2t\Big|_0^{2\pi} = 0.$$

25. (a) Since $x = t$ and $y = t$, it follows that

$$\int_{\mathbf{c}_1} (xydx + 2ydy) = \int_0^1 \left(xy\frac{dx}{dt} + 2y\frac{dy}{dt}\right) dt$$
$$= \int_0^1 (t^2 + 2t)\, dt = \left(\tfrac{1}{3}t^3 + t^2\right)\Big|_0^1 = \tfrac{4}{3}.$$

(b) In this case, $x = \sin t$ and $y = \sin t$, and

$$\int_{\mathbf{c}_2} (xydx + 2ydy) = \int_0^{\pi/2} \left(xy\frac{dx}{dt} + 2y\frac{dy}{dt}\right) dt$$
$$= \int_0^{\pi/2} (\sin^2 t\cos t + 2\sin t\cos t)\, dt$$
$$= \left(\tfrac{1}{3}\sin^3 t + \sin^2 t\right)\Big|_0^{\pi/2} = \tfrac{1}{3} + 1 = \tfrac{4}{3}.$$

(The substitution $u = \sin t$ is used to solve both integrals.)

(c) From $x = \cos t$ and $y = \cos t$ it follows that

$$\int_{\mathbf{c}_3} (xydx + 2ydy) = \int_0^{\pi/2} \left(xy\frac{dx}{dt} + 2y\frac{dy}{dt}\right) dt$$
$$= \int_0^{\pi/2} (\cos^2 t(-\sin t) + 2\cos t(-\sin t))\, dt$$
$$= \left(\tfrac{1}{3}\cos^3 t + \cos^2 t\right)\Big|_0^{\pi/2} = -1 - \tfrac{1}{3} = -\tfrac{4}{3}.$$

(The substitution $u = \cos t$ is used to solve both integrals.)

As a path integral of a vector field, a path integral of a 1-form depends only on the orientation of the parametrization that is used. The paths $\mathbf{c}_1$ and $\mathbf{c}_2$ are oriented from $(0,0)$ to $(1,1)$, whereas $\mathbf{c}_3$ is oriented from $(1,1)$ to $(0,0)$. This is the reason why the results in (a) and (b) are the same, and the result in (c) is the negative of the results in (a) and (b).

27. Let $\mathbf{c}\colon [a,b] \to \mathbb{R}^2$ (or $\mathbb{R}^3$) Assume, for simplicity, that $\mathbf{c}$ consists of two C^1 paths (the proof in the general case of n C^1 paths is analogous); so assume that the restriction $\mathbf{c}_1 = \mathbf{c}|^{[a,d]}$ of $\mathbf{c}$ to the interval $[a,d]$ and the restriction $\mathbf{c}_2 = \mathbf{c}|^{[d,b]}$ of $\mathbf{c}$ to the interval $[d,b]$ are C^1 paths. It follows that

$$\int_{\mathbf{c}} \mathbf{F}\cdot d\mathbf{s} = \int_{\mathbf{c}_1} \mathbf{F}\cdot d\mathbf{s} + \int_{\mathbf{c}_2} \mathbf{F}\cdot d\mathbf{s}.$$

Let $\boldsymbol{\gamma}(t) = \mathbf{c}(\phi(t))$ be a reparametrization of $\mathbf{c}(t)$, and assume that it preserves its orientation. By Theorem 5.3,

$$\int_{\mathbf{c}_1} \mathbf{F} \cdot d\mathbf{s} = \int_{\boldsymbol{\gamma}_1} \mathbf{F} \cdot d\mathbf{s} \quad \text{and} \quad \int_{\mathbf{c}_2} \mathbf{F} \cdot d\mathbf{s} = \int_{\boldsymbol{\gamma}_2} \mathbf{F} \cdot d\mathbf{s},$$

where $\boldsymbol{\gamma}_1 = \mathbf{c}_1(\phi(t))$ and $\boldsymbol{\gamma}_2 = \mathbf{c}_2(\phi(t))$. Therefore,

$$\int_{\mathbf{c}} \mathbf{F} \cdot d\mathbf{s} = \int_{\mathbf{c}_1} \mathbf{F} \cdot d\mathbf{s} + \int_{\mathbf{c}_2} \mathbf{F} \cdot d\mathbf{s} = \int_{\boldsymbol{\gamma}_1} \mathbf{F} \cdot d\mathbf{s} + \int_{\boldsymbol{\gamma}_2} \mathbf{F} \cdot d\mathbf{s} = \int_{\boldsymbol{\gamma}} \mathbf{F} \cdot d\mathbf{s}.$$

The proof in the orientation-reversing case is analogous.

29. Parametrize the circle by $\mathbf{c}(t) = r\cos t\,\mathbf{i} + r\sin t\,\mathbf{j}$, $t \in [0, 2\pi]$. Then $\mathbf{c}'(t) = -r\sin t\,\mathbf{i} + r\cos t\,\mathbf{j}$ and the circulation of $\mathbf{F}$ around $\mathbf{c}$ is

$$\begin{aligned}\int_{\mathbf{c}} \mathbf{F} \cdot d\mathbf{s} &= \int_0^{2\pi} \mathbf{F}(\mathbf{c}(t)) \cdot \mathbf{c}'(t)\, dt \\ &= \int_0^{2\pi} (r\cos t\,\mathbf{i} + r\sin t\,\mathbf{j}) \cdot (-r\sin t\,\mathbf{i} + r\cos t\,\mathbf{j})\, dt = \int_0^{2\pi} 0\, dt = 0.\end{aligned}$$

Next, we need the outward unit normal vector $\mathbf{n}$. The dot product of the vector $\mathbf{N} = \cos t\,\mathbf{i} + \sin t\,\mathbf{j}$ with $\mathbf{c}'(t)$ is zero, hence $\mathbf{N}$ is normal to $\mathbf{c}$. Since $\mathbf{N}$ is a unit vector, it follows that $\mathbf{n} = \pm(\cos t\,\mathbf{i} + \sin t\,\mathbf{j})$. In the first quadrant, $\mathbf{n}$ has positive $\mathbf{i}$ and $\mathbf{j}$ components, and so $\mathbf{n} = \cos t\,\mathbf{i} + \sin t\,\mathbf{j}$ (check that this choice works for all quadrants; for example, in the third quadrant both components of $\mathbf{n}$ are negative — and the outward normal also has negative components). The flux of $\mathbf{F}$ across $\mathbf{c}$ is

$$\int_{\mathbf{c}} \mathbf{F} \cdot \mathbf{n}\, ds = \int_0^{2\pi} (r\cos t\,\mathbf{i} + r\sin t\,\mathbf{j}) \cdot (\cos t\,\mathbf{i} + \sin t\,\mathbf{j})\, dt = \int_0^{2\pi} r\, dt = 2\pi r.$$

31. Parametrize the semi-circular part of $\mathbf{c}$ by $\mathbf{c}_1(t) = r\cos t\,\mathbf{i} + r\sin t\,\mathbf{j}$, $t \in [0, \pi]$, and the straight line part by $\mathbf{c}_2(t) = t\mathbf{i}$, $t \in [-r, r]$. Then $\mathbf{c}_1'(t) = -r\sin t\,\mathbf{i} + r\cos t\,\mathbf{j}$ and $\mathbf{c}_2'(t) = \mathbf{i}$; so

$$\int_{\mathbf{c}_1} \mathbf{F} \cdot d\mathbf{s} = \int_{-r}^{r} (r\cos t\,\mathbf{i} + r\sin t\,\mathbf{j}) \cdot (-r\sin t\,\mathbf{i} + r\cos t\,\mathbf{j})\, dt = \int_{-r}^{r} 0\, dt = 0$$

and

$$\int_{\mathbf{c}_2} \mathbf{F} \cdot d\mathbf{s} = \int_{-r}^{r} t\mathbf{i} \cdot \mathbf{i}\, dt = \int_{-r}^{r} t\, dt = 0.$$

It follows that the circulation of $\mathbf{F}$ along $\mathbf{c}$ is

$$\int_{\mathbf{c}} \mathbf{F} \cdot d\mathbf{s} = \int_{\mathbf{c}_1} \mathbf{F} \cdot d\mathbf{s} + \int_{\mathbf{c}_2} \mathbf{F} \cdot d\mathbf{s} = 0.$$

The outward unit normal to $\mathbf{c}_1$ is $\mathbf{n}_1 = \cos t\,\mathbf{i} + \sin t\,\mathbf{j}$ (see the solution of Exercise 29). Therefore,

$$\int_{\mathbf{c}_1} \mathbf{F} \cdot \mathbf{n}_1\, ds = \int_0^{\pi} (r\cos t\,\mathbf{i} + r\sin t\,\mathbf{j}) \cdot (\cos t\,\mathbf{i} + \sin t\,\mathbf{j})\, dt = \int_0^{\pi} r\, dt = \pi r.$$

The outward unit normal to $\mathbf{c}_2$ is $\mathbf{n}_2 = -\mathbf{j}$. It follows that

$$\int_{\mathbf{c}_2} \mathbf{F} \cdot \mathbf{n}_2\, ds = \int_0^{\pi} t\mathbf{i} \cdot (-\mathbf{j})\, dt = \int_0^{\pi} 0\, dt = 0.$$

Consequently, the flux of $\mathbf{F}$ across $\mathbf{c}$ is

$$\int_{\mathbf{c}} \mathbf{F} \cdot \mathbf{n}\, ds = \int_{\mathbf{c}_1} \mathbf{F} \cdot \mathbf{n}_1\, ds + \int_{\mathbf{c}_2} \mathbf{F} \cdot \mathbf{n}_2\, ds = \pi r.$$

5.4 Path Integrals Independent of Path

1. Let $U = \mathbb{R}^2 - \{(x,y) \mid x^2 + y^2 = 1\}$. U is not connected, since any continuous curve joining a point inside to a point outside the circle $x^2 + y^2 = 1$ would have to cross the circle, and so would not lie entirely in U. Since U is not connected, it cannot be simply-connected. It is not star-shaped either; if it were star-shaped, it would have been connected.

3. Let U denote the given set and let $\mathbf{c}$ be the circle $x^2 + y^2 = 1$ in the plane $z = 1$; then $U = \mathbb{R}^3 - \mathbf{c}$. The set U is connected: any two points in U can be joined by a continuous curve that is completely contained in U (compare with Exercise 1: this time, we have a lot of "space" to avoid the points on the circle $\mathbf{c}$). U is not simply-connected; see Figure 5.4.3. The curve $\mathbf{c}_1$ that goes around $\mathbf{c}$ cannot be deformed (without breaking) to a point. U is not star-shaped either: let A be any point and let ℓ be the line through A and any point B on $\mathbf{c}$; see Figure 5.4.3. The line segment $\overline{AP}$, where P is a point on the half-line of ℓ starting at B that does not contain A is not completely contained in U, since B does not belong to U.

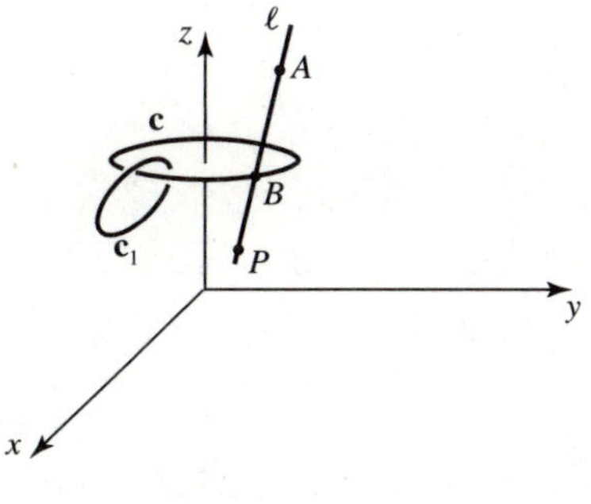

Figure 5.4.3

5. Let U be the given set. U is not connected: any continuous curve connecting a point inside the sphere to a point outside of it would have to cross the sphere, and would not be completely contained in U. Since U is not connected, it cannot be simply-connected or star-shaped either.

Notice that this exercise is a higher-dimensional analogue of Exercise 1.

7. Let $U = \mathbb{R}^3 - \mathbf{c}$. The set U is connected: there is a continuous curve joining any two points in U. U is also simply-connected: any simple closed curve can be continuously deformed to a point; see Figure 5.4.7. U is not star-shaped: take any point A and let ℓ be the line through

A and through any point B on $\mathbf{c}$. A point P on the half-line of ℓ starting at B that does not contain A cannot be joined to A by a straight line that is entirely in U; see Figure 5.4.7.

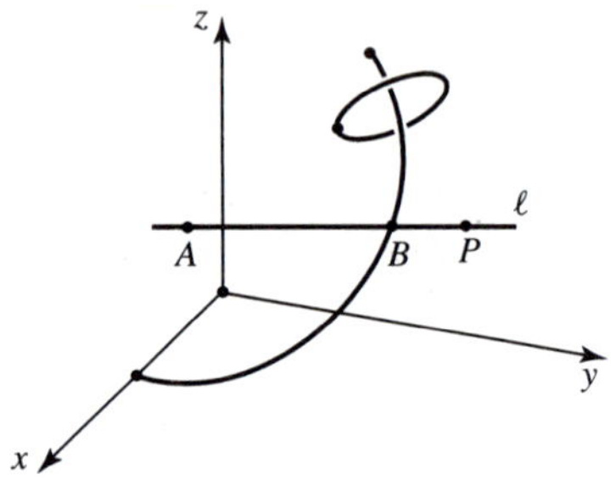

Figure 5.4.7

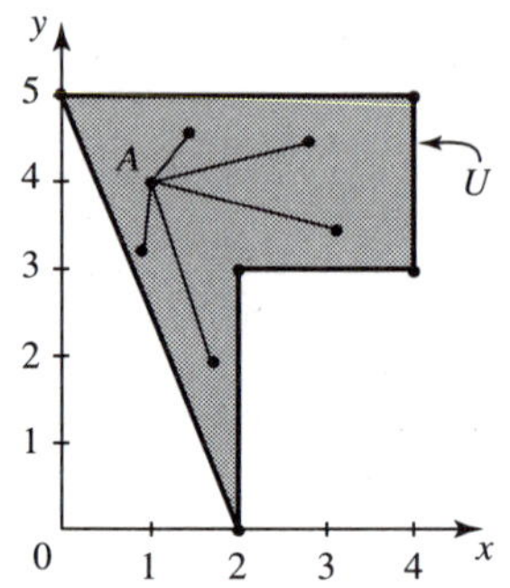

Figure 5.4.9

9. Let U be the given region, and let $A = (1, 4)$. Any point P that belongs to U can be joined with A by a straight line segment that is in U. So U is star-shaped; see Figure 5.4.9. Since U is star-shaped, it is also connected. Since any simple closed curve can be deformed to a point (there are no "holes" in U), it follows that U is simply-connected.

11. The domain of $\mathbf{F}$ is $\mathbb{R}^2$, which is an open and simply-connected set. Since

$$curl\,\mathbf{F} = \begin{vmatrix} \mathbf{i} & \mathbf{j} & \mathbf{k} \\ \partial/\partial x & \partial/\partial y & \partial/\partial z \\ y^2 \cos x & 2y \sin x & 0 \end{vmatrix} = (2y\cos x - 2y\cos x)\mathbf{k} = \mathbf{0},$$

it follows from Theorem 5.4 that $\mathbf{F}$ is a gradient vector field. Consequently, any path starting at $(1, 1)$ and ending at $(1, 3)$ can be used in the evaluation of the path integral of $\mathbf{F}$. Let us take the straight line segment and parametrize it as $\mathbf{c}(t) = \mathbf{i} + t\mathbf{j}$, $t \in [1, 3]$. Then $\mathbf{c}'(t) = \mathbf{j}$ and

$$\int_{\mathbf{c}} \mathbf{F} \cdot d\mathbf{s} = \int_1^3 (t^2 \cos 1\,\mathbf{i} + 2t \sin 1\,\mathbf{j}) \cdot \mathbf{j}\, dt = 2 \sin 1 \int_1^3 t\, dt = 8 \sin 1 \approx 6.7318.$$

Here is an alternative: first, we find a function f such that $\nabla f = \mathbf{F}$. From $y^2 \cos x\mathbf{i} + 2y \sin x\,\mathbf{j} = f_x\mathbf{i} + f_y\mathbf{j}$ it follows that $f_x = y^2 \cos x$ and $f_y = 2y \sin x$. Integrate the first equation with respect to x to obtain $f = y^2 \sin x + C(y)$. Now substitute this into the expression for f_y, thus getting $2y \sin x + C'(y) = 2y \sin x$; i.e., $C'(y) = 0$ and $C(y) = C$ (where C is a real number). Hence $f = y^2 \sin x + C$. Now use the Fundamental Theorem 5.5:

$$\int_{\mathbf{c}} \mathbf{F} \cdot d\mathbf{s} = \int_{\mathbf{c}} \nabla f \cdot d\mathbf{s} = f(1, 3) - f(1, 1) = 9 \sin 1 - \sin 1 = 8 \sin 1.$$

13. By definition,

$$curl\,\mathbf{F} = \begin{vmatrix} \mathbf{i} & \mathbf{j} & \mathbf{k} \\ \partial/\partial x & \partial/\partial y & \partial/\partial z \\ -y/(x^2+y^2) & x/(x^2+y^2) & 1 \end{vmatrix} = \left(\frac{x^2 + y^2 - 2x^2}{(x^2+y^2)^2} + \frac{x^2 + y^2 - 2y^2}{(x^2+y^2)^2} \right) \mathbf{k} = \mathbf{0}.$$

From $\mathbf{c}_1(t) = (\cos t, \sin t, 0)$ it follows that $\mathbf{c}_1'(t) = (-\sin t, \cos t, 0)$ and

$$\int_{\mathbf{c}_1} \mathbf{F} \cdot d\mathbf{s} = \int_0^{\pi} (-\sin t, \cos t, 1) \cdot (-\sin t, \cos t, 0)\, dt = \int_0^{\pi} 1\, dt = \pi.$$

From $\mathbf{c}_2(t) = (\cos t, -\sin t, 0)$, we get $\mathbf{c}_2'(t) = (-\sin t, -\cos t, 0)$ and

$$\int_{\mathbf{c}_2} \mathbf{F} \cdot d\mathbf{s} = \int_0^{\pi} (\sin t, \cos t, 0) \cdot (-\sin t, -\cos t, 0)\, dt = \int_0^{\pi} (-1)\, dt = -\pi.$$

Obviously, the path integral of $\mathbf{F}$ depends on the path, so $\mathbf{F}$ cannot be conservative. The domain of $\mathbf{F}$ is the set $\mathbb{R}^3 - \{(0,0,0)\}$, which is not simply-connected; consequently, Theorem 5.8 does not apply and there is no contradiction.

15. Since

$$\mathit{curl}\,\mathbf{F} = \begin{vmatrix} \mathbf{i} & \mathbf{j} & \mathbf{k} \\ \partial/\partial x & \partial/\partial y & \partial/\partial z \\ 4x^2 - 4y^2 + x & 7xy + \ln y & 0 \end{vmatrix} = (7y + 8y)\mathbf{k} \neq \mathbf{0},$$

it follows that $\mathbf{F}$ is not a gradient vector field.

17. The *curl* of $\mathbf{F}$ is computed to be

$$\mathit{curl}\,\mathbf{F} = \begin{vmatrix} \mathbf{i} & \mathbf{j} & \mathbf{k} \\ \partial/\partial x & \partial/\partial y & \partial/\partial z \\ 2x\ln y & 2y + x^2/y & 0 \end{vmatrix} = \left(\frac{2x}{y} - \frac{2x}{y}\right)\mathbf{k} = \mathbf{0}.$$

The domain of $\mathbf{F}$ is the set $\{(x,y) \mid y > 0\}$; i.e., it is the upper half-plane (without the x-axis). It is simply-connected, and consequently, $\mathbf{F}$ is a gradient vector field.

From $\mathbf{F} = \nabla f$ it follows that

$$2x \ln y\, \mathbf{i} + 2y + \frac{x^2}{y}\mathbf{j} = f_x\mathbf{i} + f_y\mathbf{j};$$

i.e., $f_x = 2x\ln y$ and $f_y = 2y + x^2/y$. Integrating the equation for f_x with respect to x we get $f = x^2 \ln y + C(y)$; the equation for f_y then implies that $x^2/y + C'(y) = 2y + x^2/y$, so that $C'(y) = 2y$ and $C(y) = y^2 + C$. So all functions of the form

$$f(x,y) = x^2 \ln y + y^2 + C,$$

where C is a constant, satisfy $\mathbf{F} = \nabla f$.

19. The *curl* of $\mathbf{F}$ is computed to be

$$\begin{aligned} \mathit{curl}\,\mathbf{F} &= \begin{vmatrix} \mathbf{i} & \mathbf{j} & \mathbf{k} \\ \partial/\partial x & \partial/\partial y & \partial/\partial z \\ y\cos(xy) & x\cos(xy) - z\sin y & \cos y \end{vmatrix} \\ &= (-\sin y + \sin y)\mathbf{i} + (\cos(xy) - xy\sin(xy) - \cos(xy) + xy\sin(xy))\mathbf{k} = \mathbf{0}. \end{aligned}$$

The domain of $\mathbf{F}$ is $\mathbb{R}^3$, which is a simply-connected set. Hence, $\mathbf{F}$ is a gradient vector field.

From $\mathbf{F} = \nabla f$ it follows that

$$y\cos(xy)\mathbf{i} + (x\cos(xy) - z\sin y)\mathbf{j} + \cos y\mathbf{k} = f_x\mathbf{i} + f_y\mathbf{j} + f_z\mathbf{k};$$

i.e., $f_x = y\cos(xy)$, $f_y = x\cos(xy) - z\sin y$ and $f_z = \cos y$. Integrating the equation for f_x with respect to x we get $f = \sin(xy) + C(y,z)$; the equation for f_y implies that

$$x\cos(xy) + \frac{\partial C(y,z)}{\partial y} = x\cos(xy) - z\sin y,$$

so that $\partial C(y,z)/\partial y = -z\sin y$; it follows that $C(y,z) = z\cos y + C(z)$. Hence

$$f = \sin(xy) + z\cos y + C(z).$$

Finally, substituting this formula for f into the expression for f_z, we get $\cos y + C'(z) = \cos y$; i.e., $C'(z) = 0$ and $C(z) = C$. So all functions of the form

$$f(x,y,z) = \sin(xy) + z\cos y + C,$$

where C is a constant, satisfy $\mathbf{F} = \nabla f$.

21. The given integral can be interpreted as $\int_{\mathbf{c}} \mathbf{F}\cdot d\mathbf{s}$, where $\mathbf{F} = (\cos x\tan z, 1, \sin x\sec^2 z)$, and $\mathbf{c}$ is a curve joining $(0,0,0)$ and $(\pi, \pi/2, \pi/3)$. The curl of $\mathbf{F}$ is computed to be

$$\mathit{curl}\,\mathbf{F} = \begin{vmatrix} \mathbf{i} & \mathbf{j} & \mathbf{k} \\ \partial/\partial x & \partial/\partial y & \partial/\partial z \\ \cos x\tan z & 1 & \sin x\sec^2 z \end{vmatrix} = -(\cos x\sec^2 z - \cos x\sec^2 z)\mathbf{j} = \mathbf{0}.$$

Take the domain of $\mathbf{F}$ to be the set $U = \{(x,y,z) \mid -\pi/2 < z < \pi/2\}$, since both endpoints $(0,0,0)$ and $(\pi,\pi/2,\pi/3)$ belong to it. U is a simply-connected set and so $\mathbf{F}$ is a gradient vector field on U.

From $\mathbf{F} = \nabla f$ it follows that

$$(\cos x\tan z, 1, \sin x\sec^2 z) = (f_x, f_y, f_z);$$

i.e., $f_x = \cos x\tan z$, $f_y = 1$ and $f_z = \sin x\sec^2 z$. Integrating the equation for f_x with respect to x we get $f = \sin x\tan z + C(y,z)$; substituting f into the equation for f_y we get $\partial C(y,z)/\partial y = 1$, so that $C(y,z) = y + C(z)$. Hence

$$f = \sin x\tan z + y + C(z).$$

Finally, substituting this formula for f into the expression for f_z, we get $\sin x\sec^2 z + C'(z) = \sin x\sec^2 z$; i.e., $C'(z) = 0$ and $C(z) = C$. So $\mathbf{F} = \nabla f$, where $f(x,y,z) = \sin x\tan z + y + C$, and C is a constant. By the Fundamental Theorem 5.5,

$$\int_{\mathbf{c}} \mathbf{F}\cdot d\mathbf{s} = \int_{\mathbf{c}} \nabla f\cdot d\mathbf{s} = f(\pi,\pi/2,\pi/3) - f(0,0,0) = \sin\pi\tan(\pi/3) + \frac{\pi}{2} - 0 = \frac{\pi}{2}.$$

23. (a) The *curl* of $\mathbf{F}$ is

$$\mathit{curl}\,\mathbf{F} = \begin{vmatrix} \mathbf{i} & \mathbf{j} & \mathbf{k} \\ \partial/\partial x & \partial/\partial y & \partial/\partial z \\ -(1+x)ye^x & -xe^x & 0 \end{vmatrix} = (-e^x - xe^x + (1+x)e^x)\mathbf{k} = \mathbf{0}.$$

Since the domain of $\mathbf{F}$ is $\mathbb{R}^3$ (which is simply-connected), it follows that $\mathbf{F}$ is conservative.

(b) By the Fundamental Theorem (see Theorem 5.5),

$$\int_{\mathbf{c}} \mathbf{F} \cdot d\mathbf{s} = -\int_{\mathbf{c}} \nabla V \cdot d\mathbf{s} = -V(x,y)\Big|_{(0,0)}^{(x,0)} = -V(x,0) + V(0,0) = -V(x,0),$$

since $V(0,0) = 0$ by assumption. Now let $\mathbf{c}$ be the straight line segment from $(0,0)$ to $(x,0)$; parametrize it by $\mathbf{c}(t) = t\mathbf{i}$, $t \in [0,x]$. Then $\mathbf{F}(\mathbf{c}(t)) = -te^t\mathbf{j}$, $\mathbf{c}'(t) = \mathbf{i}$ and hence

$$\int_{\mathbf{c}} \mathbf{F} \cdot d\mathbf{s} = \int_0^x \mathbf{F}(\mathbf{c}(t)) \cdot \mathbf{c}'(t)\, dt = \int_0^x (-te^t\mathbf{j}) \cdot \mathbf{i}\, dt = \int_0^x 0\, dt = 0.$$

Comparing the two expressions for $\int_{\mathbf{c}} \mathbf{F} \cdot d\mathbf{s}$, we get that $V(x,0) = 0$ for every x.

(c) The computation is analogous to (b). By the Fundamental Theorem 5.5,

$$\int_{\mathbf{c}} \mathbf{F} \cdot d\mathbf{s} = -\int_{\mathbf{c}} \nabla V \cdot d\mathbf{s} = -V(x,y)\Big|_{(0,0)}^{(0,y)} = -V(0,y) + V(0,0) = -V(0,y),$$

since $V(0,0) = 0$. Take $\mathbf{c}$ to be the straight line segment from $(0,0)$ to $(0,y)$; parametrize it by $\mathbf{c}(t) = t\mathbf{j}$, $t \in [0,y]$. Then $\mathbf{F}(\mathbf{c}(t)) = -t\mathbf{i}$, $\mathbf{c}'(t) = \mathbf{j}$ and

$$\int_{\mathbf{c}} \mathbf{F} \cdot d\mathbf{s} = \int_0^x \mathbf{F}(\mathbf{c}(t)) \cdot \mathbf{c}'(t)\, dt = \int_0^y (-t\mathbf{i}) \cdot \mathbf{j}\, dt = \int_0^y 0\, dt = 0.$$

Comparing the two expressions for $\int_{\mathbf{c}} \mathbf{F} \cdot d\mathbf{s}$, we get that $V(0,y) = 0$ for every y.

The reader is invited to check that $V(x,y) = xye^x$.

25. Parametrize $\mathbf{c}$ by $\mathbf{c}(t) = 3t\mathbf{i} - 2t\mathbf{j}$, $t \in [0,1]$. Then

$$\mathbf{F}(\mathbf{c}(t)) = \mathbf{F}(3t\mathbf{i} - 2t\mathbf{j}) = -12t^2e^{-2t}\mathbf{i} + (9t^2 - 18t^3)e^{-2t}\mathbf{j},$$

$\mathbf{c}'(t) = 3\mathbf{i} - 2\mathbf{j}$ and

$$\int_{\mathbf{c}} \mathbf{F} \cdot d\mathbf{s} = \int_0^1 \mathbf{F}(\mathbf{c}(t)) \cdot \mathbf{c}'(t)\, dt = \int_0^1 (-12t^2e^{-2t}\mathbf{i} + (9t^2 - 18t^3)e^{-2t}\mathbf{j}) \cdot (3\mathbf{i} - 2\mathbf{j})\, dt$$
$$= \int_0^1 (-54t^2e^{-2t} + 36t^3e^{-2t})\, dt.$$

From integration tables (or, using integration by parts) we get

$$\int_0^1 t^2e^{-2t}\, dt = -\tfrac{1}{4}e^{-2t}(2t^2 + 2t + 1)\Big|_0^1 = \tfrac{1}{4}(1 - 5e^{-2}).$$

Similarly,

$$\int_0^1 t^3e^{-2t}\, dt = -\tfrac{1}{8}e^{-2t}(4t^3 + 6t^2 + 6t + 3)\Big|_0^1 = \tfrac{1}{8}(3 - 19e^{-2}).$$

Finally,

$$\int_{\mathbf{c}} \mathbf{F} \cdot d\mathbf{s} = -54\left(\tfrac{1}{4}(1 - 5e^{-2})\right) + 36\left(\tfrac{1}{8}(3 - 19e^{-2})\right) = -18e^{-2} \approx -2.4360.$$

(b) Since $curl\,\mathbf{F} = \mathbf{0}$ and the domain of $\mathbf{F}$ is $\mathbb{R}^2$, which is a simply-connected set, it follows that $\mathbf{F}$ is a gradient vector field. From $\mathbf{F} = \nabla f$ it follows that

$$2xye^y\mathbf{i} + x^2e^y(1+y)\mathbf{j} = f_x\mathbf{i} + f_y\mathbf{j};$$

i.e., $f_x = 2xye^y$ and $f_y = x^2e^y(1+y)$. Integrating the equation for f_x with respect to x we get $f = x^2ye^y + C(y)$; the equation for f_y implies that $x^2(e^y + ye^y) + C'(y) = x^2e^y(1+y)$, so that $C'(y) = 0$ and $C(y) = C$. So all functions of the form $f(x,y) = x^2ye^y + C$, where C is a constant, satisfy $\mathbf{F} = \nabla f$. Hence, by the Fundamental Theorem,

$$\int_{\mathbf{c}} \mathbf{F}\cdot d\mathbf{s} = \int_{\mathbf{c}} \nabla f \cdot d\mathbf{s} = f(3,-2) - f(0,0) = -18e^{-2}.$$

27. By a product rule for *curl*,

$$curl\,\mathbf{F}(\mathbf{r}) = curl\,(\|\mathbf{r}\|^2\mathbf{r}) = \|\mathbf{r}\|^2 curl\,\mathbf{r} + grad\,\|\mathbf{r}\|^2 \times \mathbf{r}.$$

Now

$$curl\,\mathbf{r} = \begin{vmatrix} \mathbf{i} & \mathbf{j} & \mathbf{k} \\ \partial/\partial x & \partial/\partial y & \partial/\partial z \\ x & y & z \end{vmatrix} = \mathbf{0}$$

and

$$\nabla\|\mathbf{r}\|^2 = \nabla(x^2+y^2+z^2) = 2x\mathbf{i} + 2y\mathbf{j} + 2z\mathbf{k} = 2\mathbf{r},$$

and therefore

$$curl\,\mathbf{F}(\mathbf{r}) = \|\mathbf{r}\|^2\,\mathbf{0} + 2\mathbf{r}\times\mathbf{r} = \mathbf{0}.$$

From $\mathbf{F} = \nabla f$ it follows that

$$((x^2+y^2+z^2)x, (x^2+y^2+z^2)y, (x^2+y^2+z^2)z) = (f_x, f_y, f_z);$$

i.e., $f_x = (x^2+y^2+z^2)x$, $f_y = (x^2+y^2+z^2)y$ and $f_z = (x^2+y^2+z^2)z$. Integrating the equation for f_x with respect to x we get

$$f = \tfrac{1}{4}x^4 + \tfrac{1}{2}x^2(y^2+z^2) + C(y,z);$$

substituting f into the equation for f_y we get $yx^2 + \partial C(y,z)/\partial y = yx^2 + y^3 + z^2y$, so that $\partial C(y,z)/\partial y = y^3 + z^2y$, and $C(y,z) = y^4/4 + y^2z^2/2 + C(z)$. Hence

$$f = \tfrac{1}{4}x^4 + \tfrac{1}{4}y^4 + \tfrac{1}{2}x^2y^2 + \tfrac{1}{2}x^2z^2 + \tfrac{1}{2}y^2z^2 + C(z).$$

Finally, substituting this formula for f into the expression for f_z, we get $C'(z) = z^3$; i.e., $C(z) = z^4/4 + C$. So $\mathbf{F} = \nabla f$, where

$$\begin{aligned} f(x,y,z) &= \tfrac{1}{4}(x^4+y^4+z^4) + \tfrac{1}{2}(x^2y^2 + x^2z^2 + y^2z^2) + C \\ &= \tfrac{1}{4}(x^2+y^2+z^2)^2 + C = \tfrac{1}{4}\|\mathbf{r}\|^4 + C. \end{aligned}$$

29. The *curl* of $\mathbf{F}$ is

$$\begin{aligned} curl\,\mathbf{F} &= \begin{vmatrix} \mathbf{i} & \mathbf{j} & \mathbf{k} \\ \partial/\partial x & \partial/\partial y & \partial/\partial z \\ \ln(x+y^2) + x/(x+y^2) & 2xy/(x+y^2) & 0 \end{vmatrix} \\ &= \left(\frac{2y(x+y^2) - 2xy}{(x+y^2)^2} - \frac{2y}{(x+y^2)} + \frac{2xy}{(x+y^2)^2}\right)\mathbf{k} = \mathbf{0}. \end{aligned}$$

The domain of $\mathbf{F}$ is the set $U = \{(x,y) \mid x + y^2 > 0\}$. We do not have to identify it precisely, but only note that it contains the first quadrant. That part is simply-connected, and contains the curve; it follows that $\mathbf{F}$ is a gradient vector field on that set.

From $\mathbf{F} = \nabla f$ it follows that

$$(\ln(x+y^2) + x/(x+y^2))\mathbf{i} + (2xy/(x+y^2))\mathbf{j} = f_x\mathbf{i} + f_y\mathbf{j};$$

i.e., $f_x = \ln(x+y^2) + x/(x+y^2)$ and $f_y = 2xy/(x+y^2)$. Integrating the equation for f_y with respect to y we get

$$f = x\ln(x+y^2) + C(x);$$

the equation for f_x implies that

$$\ln(x+y^2) + \frac{x}{x+y^2} + C'(x) = \ln(x+y^2) + \frac{x}{x+y^2},$$

so that $C'(x) = 0$ and $C(x) = C$. Hence $f(x,y) = x\ln(x+y^2) + C$, and, by the Fundamental Theorem,

$$\int_{\mathbf{c}} \mathbf{F}\cdot d\mathbf{s} = \int_{\mathbf{c}} \nabla f \cdot d\mathbf{s} = x\ln(x+y^2)\Big|_{(1,1)}^{(2,8)} = 2\ln 66 - \ln 2 = \ln 2178 \approx 7.6862.$$

31. Recall that the path integral $\int_{\mathbf{c}} \alpha$ of a 1-form α is equal to the path integral of the vector field $\mathbf{F}$ whose components are the same as the components of α. Since $\alpha = df$ it follows that $\mathbf{F} = \nabla f$ (the differential corresponds to the gradient). Therefore, by the Fundamental Theorem,

$$\int_{\mathbf{c}} \alpha = \int_{\mathbf{c}} df = \int_{\mathbf{c}} \nabla f = f(\mathbf{c}(b)) - f(\mathbf{c}(a)).$$

Here is an alternative proof: since α is exact, there exists a differentiable function f such that $\alpha = df$. Therefore

$$\int_{\mathbf{c}} \alpha = \int_{\mathbf{c}} df = \int_{\mathbf{c}} (f_x dx + f_y dy + f_z dz) = \int_a^b \left(f_x \tfrac{dx}{dt} + f_y \tfrac{dy}{dt} + f_z \tfrac{dz}{dt}\right) dt$$

$$= \int_a^b (f_x, f_y, f_z)\cdot\left(\tfrac{dx}{dt}, \tfrac{dy}{dt}, \tfrac{dz}{dt}\right) dt = \int_{\mathbf{c}} \nabla f \cdot d\mathbf{s} = f(\mathbf{c}(b)) - f(\mathbf{c}(a)),$$

by the Fundamental Theorem of Calculus.

5.5 Chapter Review

Solutions to Computer Projects

1. (a) The direction of the line is given by the vector $(\cos\theta, \sin\theta)$; so it can be parametrized as (fix θ — we use t as the parameter) $\mathbf{c}_\theta(t) = (t\cos\theta, t\sin\theta)$, $t \in [0,1]$ (since $\mathbf{c}_\theta(0) = (0,0)$ and $\mathbf{c}_\theta(1) = (\cos\theta, \sin\theta)$, the distance between $\mathbf{c}_\theta(0)$ and $\mathbf{c}_\theta(1)$ is 1, as needed). From

$$\mathbf{F}(\mathbf{c}_\theta(t)) = \mathbf{F}(t\cos\theta, t\sin\theta) = t\sin\theta e^{t^2\sin\theta\cos\theta}\mathbf{i} + t^2\sin^2\theta\mathbf{j}$$

and $\mathbf{c}'_\theta(t) = (\cos\theta, \sin\theta)$ it follows that (the work depends on θ, so call it $W(\theta)$)

$$\begin{aligned} W(\theta) &= \int_0^1 \mathbf{F}(\mathbf{c}_\theta) \cdot \mathbf{c}'_\theta(t)\, dt \\ &= \int_0^1 t\sin\theta\cos\theta e^{t^2\sin\theta\cos\theta} + t^2\sin^3\theta\, dt \\ &= \left(\tfrac{1}{2}e^{t^2\sin\theta\cos\theta} + \tfrac{1}{3}t^3\sin^3\theta\right)\Big|_0^1 \\ &= \tfrac{1}{2}(e^{\sin\theta\cos\theta} - 1) + \tfrac{1}{3}\sin^3\theta. \end{aligned}$$

(b) Figure 5.5.1 (a) shows the plot of $W(\theta)$ on $[0, 2\pi]$. Using a computer, we will now locate the critical points that give maximum and minimum work. Solving $W'(\theta) = 0$, we get $\theta \approx 1.03951$ rad (global maximum on $[0, 2\pi]$), $\theta \approx 2.66062$ rad (local (but not global) minimum), $\theta \approx 3.75451$ rad (local (but not global) maximum) and $\theta \approx 5.04291$ rad (global minimum on $[0, 2\pi]$). So $\theta \approx 1.03951$ rad is the direction along which the work of $\mathbf{F}$ is the largest; $\theta \approx 5.04291$ rad is the direction of the smallest work.

Clearly, $W(0) = W(2\pi) = 0$. Looking at the plot in Figure 5.5.1 (a), we see that there are two more directions along which the work is zero: $\theta \approx 2.17310$ rad and $\theta \approx 4.25790$ rad.

(c) The vector field $\mathbf{F}$ does maximum work along a path that is "best aligned" with it. In our case (we considered only straight lines, and not all possible paths) such a path is a line that makes an angle of approximately 1.03951 rad with respect to the positive x-axis. $\mathbf{F}$ does minimum work along a path that is "worst aligned" with its direction (i.e., that runs in the opposite direction). In our case, that direction is given by $\theta \approx 5.04291$ rad (again, measured with respect to the positive x-axis). $\mathbf{F}$ does no work along the x-axis (since $\mathbf{F} = \mathbf{0}$ for points on the x-axis) and along two directions, $\theta \approx 2.17310$ rad and $\theta \approx 4.25790$ rad; these are the directions along which $\mathbf{F}$ is the "most perpendicular."

(d) See Figure 5.5.1 (b) (drawn is a unit vector field, so that the picture looks better).

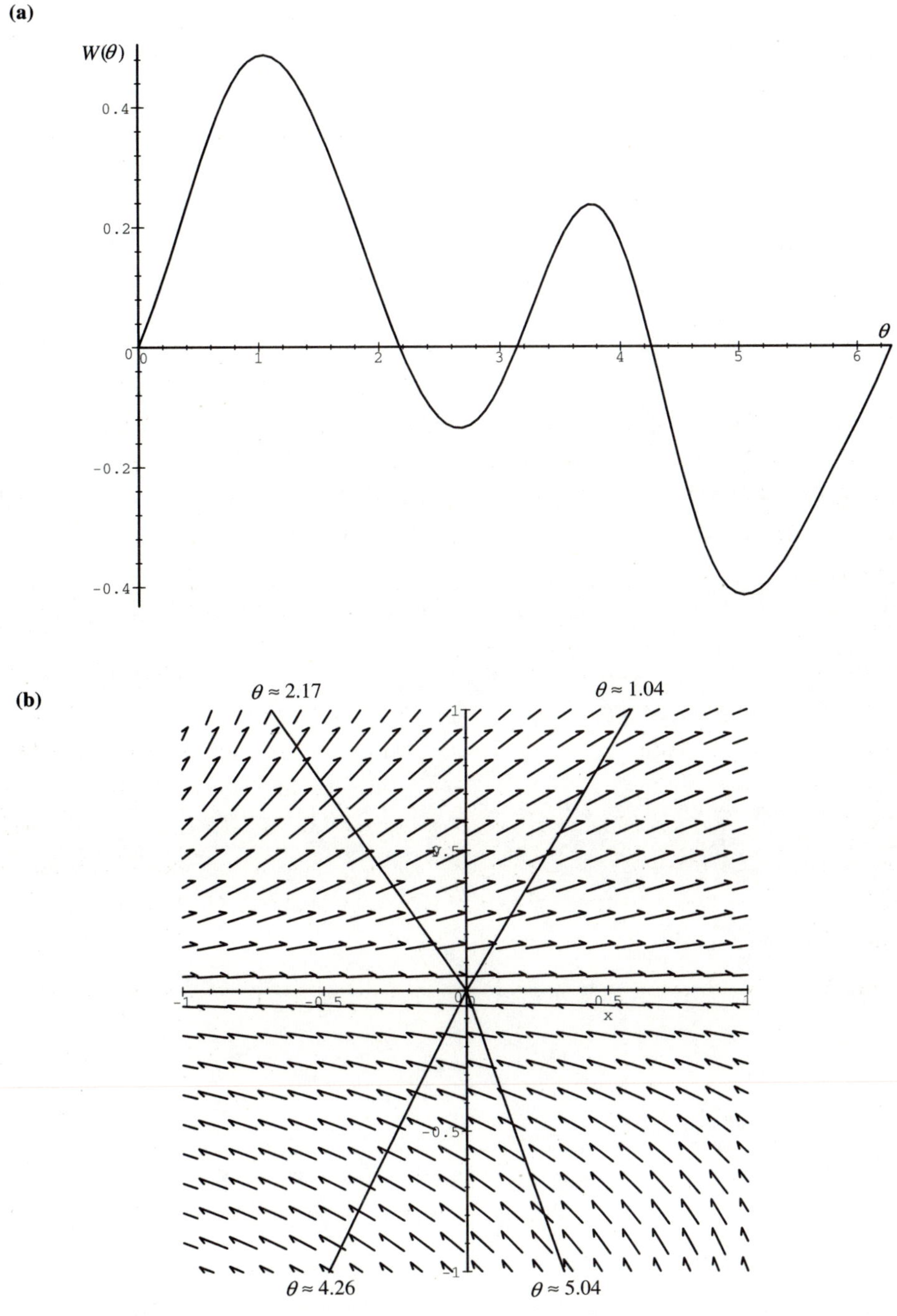

Figure 5.5.1

2. (a) Using the angle θ with respect to the positive x-axis, we can use the parametrization $\mathbf{c}_\theta = (t\cos\theta, t\sin\theta)$, $t \in [0,1]$. This line segment starts at the origin and ends at the point

$(\cos\theta, \sin\theta)$ on the unit circle.

(b) Since

$$f(\mathbf{c}_\theta(t)) = t^2\cos^2\theta + 3t\sin\theta e^{-2t^2\cos^2\theta - 2t^2\sin^2\theta} = t^2\cos^2\theta + 3t\sin\theta e^{-2t^2}$$

and $\|\mathbf{c}_\theta(t)\| = \|(\cos\theta, \sin\theta)\| = 1$, it follows that

$$\int_{\mathbf{c}_\theta} f ds = \int_0^1 f(\mathbf{c}_\theta(t))\|\mathbf{c}_\theta(t)\|\, dt = \int_0^1 (t^2\cos^2\theta + 3t\sin\theta e^{-2t^2})\, dt$$
$$= \tfrac{1}{3}\cos^2\theta \left(t^3\Big|_0^1\right) - \tfrac{3}{4}\sin\theta\left(e^{-2t^2}\Big|_0^1\right) = \tfrac{1}{3}\cos^2\theta - \tfrac{3}{4}(e^{-2}-1)\sin\theta.$$

(c) Let

$$I(\theta) = \tfrac{1}{3}\cos^2\theta - \tfrac{3}{4}(e^{-2}-1)\sin\theta$$

be the value of the path integral of f along the line segment $\mathbf{c}_\theta$. Figure 5.5.2 (a). shows the plot of $I(\theta)$ on $[0, 2\pi]$. It is not clear where the maximum is: so zoom in on the "interesting" part: for example, select the interval $[1.2, 2]$; see Figure 5.5.2 (b). The situation is much clearer now. Solving $I'(\theta) = 0$ using a computer, we get $\theta \approx 1.33690$ rad and $I(\theta) \approx 0.64875$ (a candidate for the maximum); $\theta \approx 1.57080$ rad (local minimum — we do not need it); $\theta \approx 1.80479$ rad and $I(\theta) \approx 0.64875$ (a candidate for the maximum); $\theta \approx 4.71239$ rad (global minimum — we do not need it). So the path integral is largest along the lines defined by $\theta \approx 1.33690$ rad and $\theta \approx 1.80479$ rad.

(d) Using a computer, we find that $I(\theta) = 0.1$ when $\theta \approx 3.45711$ rad and $\theta \approx 5.96767$ rad. From Figure 5.5.2 (a) it follows that if $3.45711 < \theta < 5.96767$, then the path integral $I(\theta)$ is less than 0.1.

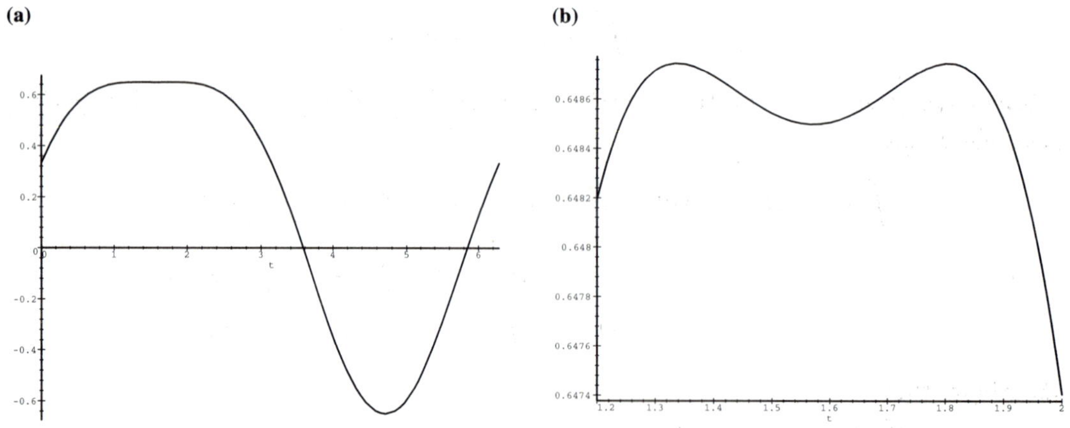

Figure 5.5.2

Solutions to Further Explorations

1. Define $\boldsymbol{\gamma}(t) = \mathbf{c}(\phi(t)) = (\phi(t)^{3/2}, \phi(t))$. Drop t for simplicity of notation. Then $\boldsymbol{\gamma}' = (3\phi^{1/2}\phi'/2, \phi')$, and from $\|\boldsymbol{\gamma}'\| = C$ ($C > 0$ is a constant) it follows that

$$\sqrt{\left(\tfrac{3}{2}\phi^{1/2}\phi'\right)^2 + (\phi')^2} = C;$$

i.e.,

$$|\phi'|\sqrt{\tfrac{9}{4}\phi + 1} = C.$$

Since we are looking for an orientation-preserving parametrization, $\phi' > 0$ and hence $|\phi'| = \phi'$; so we have to solve the differential equation

$$\phi'\sqrt{\tfrac{9}{4}\phi + 1} = C,$$

where $C > 0$. Integrating both sides, we get

$$\tfrac{2}{3}\left(\tfrac{9}{4}\phi + 1\right)^{3/2}\tfrac{4}{9} = Ct + D$$

and

$$\left(\tfrac{9}{4}\phi + 1\right)^{3/2} = \tfrac{27}{8}(Ct + D).$$

Solving for ϕ, we get $\phi(t) = (Ct + D)^{2/3} - 4/9$.

3. Parametrize $\mathbf{c}_\epsilon$ by

$$\mathbf{c}_\epsilon(t) = \mathbf{i} + \mathbf{j} + t((\epsilon - 1)\mathbf{i} + (\epsilon - 1)\mathbf{j}) = (1 + t\epsilon - t)\mathbf{i} + (1 + t\epsilon - t)\mathbf{j}, \qquad t \in [0, 1].$$

Then $\mathbf{F}(\mathbf{c}_\epsilon(t)) = \ln(1 + t(\epsilon - 1))\mathbf{i} + \mathbf{j}$, $\mathbf{c}'_\epsilon(t) = (\epsilon - 1)\mathbf{i} + (\epsilon - 1)\mathbf{j}$ and

$$\begin{aligned}\int_{\mathbf{c}_\epsilon} \mathbf{F} \cdot d\mathbf{s} &= \int_0^1 [\ln(1 + t(\epsilon - 1))\mathbf{i} + \mathbf{j}] \cdot [(\epsilon - 1)\mathbf{i} + (\epsilon - 1)\mathbf{j}]\, dt \\ &= (\epsilon - 1)\int_0^1 [\ln(1 + t(\epsilon - 1)) + 1]\, dt.\end{aligned}$$

Using integration by parts with $u = \ln(1+t(\epsilon-1))$ and $dv = dt$ (hence $du = (\epsilon-1)/(1+t(\epsilon-1))$ and $v = t$) we get

$$\begin{aligned}\int \ln(1 + t(\epsilon - 1))\, dt &= t\ln(1 + t(\epsilon - 1)) - \int \frac{t(\epsilon - 1)}{1 + t(\epsilon - 1)}\, dt \\ &= t\ln(1 + t(\epsilon - 1)) - \int 1 - \frac{1}{1 + t(\epsilon - 1)}\, dt \\ &= t\ln(1 + t(\epsilon - 1)) - t - \tfrac{1}{\epsilon - 1}\ln(1 + t(\epsilon - 1)).\end{aligned}$$

It follows that

$$\begin{aligned}\int_{\mathbf{c}_\epsilon} \mathbf{F} \cdot d\mathbf{s} &= (\epsilon - 1)\left[t\ln(1 + t(\epsilon - 1)) - t - \tfrac{1}{\epsilon - 1}\ln(1 + t(\epsilon - 1)) + t\right]\Big|_0^1 \\ &= (\epsilon - 1)\left(\ln\epsilon + \frac{1}{\epsilon - 1}\ln\epsilon\right) = \epsilon\ln\epsilon.\end{aligned}$$

By L'Hôpital's rule,

$$\lim_{\epsilon\to 0}\epsilon\ln\epsilon = \lim_{\epsilon\to 0}\frac{\ln\epsilon}{\epsilon^{-1}} = \lim_{\epsilon\to 0}\frac{1/\epsilon}{-\epsilon^{-2}} = \lim_{\epsilon\to 0}(-\epsilon^2) = 0$$

and so as $\epsilon \to 0$, $\int_{\mathbf{c}_\epsilon} \mathbf{F} \cdot d\mathbf{s} \to 0$.

5. We will compute the work of $\mathbf{F}$ in two different ways. Since $\mathbf{F}$ is conservative, we can choose any path joining $(0,0,0)$ and (x,y,z). Take the line $\mathbf{c}(t) = (tx, ty, tz)$, $t \in [0,1]$. Then

$$\mathbf{F}(\mathbf{c}(t)) = a\|\mathbf{c}(t)\|\mathbf{c}(t) = a\sqrt{t^2(x^2+y^2+z^2)}\ (tx,ty,tz) = at^2\sqrt{x^2+y^2+z^2}\ (x,y,z)$$

and $\mathbf{c}'(t) = (x,y,z)$, and therefore

$$\begin{aligned} W = \int_{\mathbf{c}} \mathbf{F} \cdot d\mathbf{s} &= \int_0^1 at^2\sqrt{x^2+y^2+z^2}\ (x,y,z)\cdot(x,y,z)\,dt \\ &= a(x^2+y^2+z^2)^{3/2}\ \tfrac{1}{3}t^3\Big|_0^1 = \tfrac{1}{3}a(x^2+y^2+z^2)^{3/2}. \end{aligned}$$

Since $\mathbf{F}$ is conservative, there is a potential function V such that $\mathbf{F} = -\nabla V$. Hence

$$W = \int_{\mathbf{c}} \mathbf{F} \cdot d\mathbf{s} = -\int_{\mathbf{c}} \nabla V \cdot d\mathbf{s} = -V\Big|_{(0,0,0)}^{(x,y,z)} = -V(x,y,z) + V(0,0,0) = -V(x,y,z),$$

since, by assumption, $V(0,0,0) = 0$. Comparing the two expressions for the work, we get

$$V(x,y,z) = -\tfrac{1}{3}a(x^2+y^2+z^2)^{3/2};$$

i.e., $V(\mathbf{r}) = -a\|\mathbf{r}\|^3/3$.

7. Let $E(t) = f(\mathbf{c}(t)) + m\|\mathbf{c}'(t)\|^2/2$. We will show that $E'(t) = 0$ (and consequently, that $E(t)$ is a constant). By the chain rule,

$$\frac{d}{dt}(f(\mathbf{c}(t))) = Df(\mathbf{c}(t)) \cdot D\mathbf{c}(t) = \nabla f(\mathbf{c}(t)) \cdot \mathbf{c}'(t).$$

Since $\|\mathbf{c}'(t)\|^2 = \mathbf{c}'(t)\cdot\mathbf{c}'(t)$, it follows that $(d/dt)\|\mathbf{c}'(t)\|^2 = 2\mathbf{c}'(t)\cdot\mathbf{c}''(t)$. Therefore,

$$\frac{d}{dt}E(t) = \nabla f(\mathbf{c}(t)) \cdot \mathbf{c}'(t) + m\mathbf{c}'(t)\cdot\mathbf{c}''(t).$$

Since $\mathbf{F} = -\nabla f$ and $\mathbf{F}(t) = m\mathbf{c}''(t)$ by Newton's Second Law, it follows that

$$\frac{d}{dt}E(t) = -\mathbf{F}(\mathbf{c}(t))\cdot\mathbf{c}'(t) + \mathbf{c}'(t)\cdot\mathbf{F}(\mathbf{c}(t)) = 0.$$

$E(t)$ is the sum of the kinetic energy and the potential energy. The fact that $E'(t) = 0$ is the Law of Conservation of Energy (i.e., total energy remains constant throughout the motion).

6. DOUBLE AND TRIPLE INTEGRALS

6.1 Double Integrals: Definition and Properties

1. Form the partition of R into 9 rectangles $R_{11}, R_{12}, \ldots, R_{33}$ using vertical lines $x = 0$, $x = 1$, $x = 2$ and $x = 3$ and horizontal lines $y = 0$, $y = 2/3$, $y = 4/3$ and $y = 2$. Taking (x_i^*, y_j^*) to be the upper-left corner of a subrectangle R_{ij}, $i, j = 1, 2, 3$, we get

$$\begin{aligned}\mathcal{R}_3 &= f(0,2/3)\Delta A_{11} + f(0,4/3)\Delta A_{12} + f(0,2)\Delta A_{13} + f(1,2/3)\Delta A_{21} + f(1,4/3)\Delta A_{22} \\ &\quad + f(1,2)\Delta A_{23} + f(2,2/3)\Delta A_{31} + f(2,4/3)\Delta A_{32} + f(2,2)\Delta A_{33},\end{aligned}$$

where $\Delta A_{ij} = 2/3$ (for all i and j) is the area of R_{ij}. It follows that (factor out $\Delta A_{ij} = 2/3$)

$$\begin{aligned}\mathcal{R}_3 &= \tfrac{2}{3}\left(\tfrac{7}{9} + \tfrac{55}{9} + 15 + \tfrac{16}{9} + \tfrac{64}{9} + 16 + \tfrac{25}{9} + \tfrac{73}{9} + 17\right) \\ &= \tfrac{2}{3}\left(\tfrac{240}{9} + 48\right) = \tfrac{448}{9} \approx 49.7778.\end{aligned}$$

3. Form the partition of R into 4 rectangles $R_{11}, R_{12}, R_{21}, R_{22}$, using vertical lines $x = 0$, $x = 5/2$ and $x = 5$ and horizontal lines $y = 0$, $y = 3/2$ and $y = 3$. Taking (x_i^*, y_j^*) to be the upper left corner of each subrectangle R_{ij}, $i, j = 1, 2$, we get

$$\mathcal{R}_2 = f(0,3/2)\Delta A_{11} + f(0,3)\Delta A_{12} + f(5/2,3/2)\Delta A_{21} + f(5/2,3)\Delta A_{22},$$

where $\Delta A_{ij} = (3/2)(5/2) = 15/4$ (for all i, j) is the area of R_{ij}. It follows that (factor out $\Delta A_{ij} = 15/4$)

$$\mathcal{R}_2 = \tfrac{15}{4}\left(-\tfrac{9}{4} - 9 + 4 - \tfrac{11}{4}\right) = \tfrac{15}{4}(-10) = -75/2.$$

5. Form the partition of R into 4 rectangles $R_{11}, R_{12}, R_{21}, R_{22}$, using vertical lines $x = -1$, $x = 0$ and $x = 1$ and horizontal lines $y = 0$, $y = 1/2$ and $y = 1$. Taking (x_i^*, y_j^*) to be the lower right corner of each subrectangle R_{ij}, $i, j = 1, 2$, we get

$$\mathcal{R}_2 = f(0,0)\Delta A_{11} + f(0,1/2)\Delta A_{12} + f(1,0)\Delta A_{21} + f(1,1/2)\Delta A_{22},$$

where $\Delta A_{ij} = 1/2$ (for all i, j) is the area of R_{ij}. It follows that

$$\mathcal{R}_2 = \tfrac{1}{2}\left(0 - \tfrac{1}{2} + 1 + \tfrac{1}{2}e^{1/2}\right) = \tfrac{1}{4}(1 + e^{1/2}) \approx 0.6622.$$

7. Let $f(x, y) = 1$. Then

$$\iint_R 1\,dA = \lim_{n\to\infty} \sum_{i=1}^{n}\sum_{j=1}^{n} f(x_i^*, y_j^*)\Delta A_{ij} = \lim_{n\to\infty} \sum_{i=1}^{n}\sum_{j=1}^{n} \Delta A_{ij},$$

where ΔA_{ij} is the area of the subrectangle R_{ij}, $i, j = 1, \ldots, n$. So, for every n,

$$\sum_{i=1}^{n}\sum_{j=1}^{n} \Delta A_{ij} = \text{area}(R)$$

and hence

$$\iint_R 1\,dA = \lim_{n\to\infty} \text{area}(R) = \text{area}(R).$$

9. Since $z = 3 - x \geq 0$ on $R = [0,1] \times [0,2]$, it follows that the double integral $\iint_R (3-x)\,dA$ is the volume of the solid V under $z = 3 - x$ and over $[0,1] \times [0,2]$; see Figure 6.1.9. The solid V consists of the rectangular box over $[0,1] \times [0,2]$ of height 2 with a triangular prism on top (the triangle is the right triangle with sides 1) of length 2. Therefore, the volume of V is $4 + 1 = 5$, and so $\iint_R (3-x)\,dA = 5$.

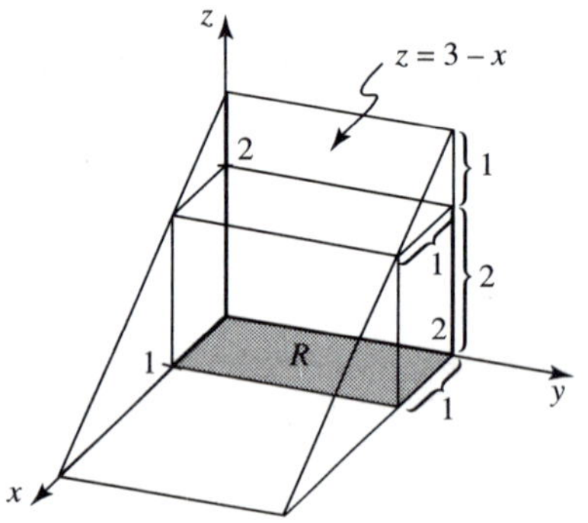

Figure 6.1.9

11. Since $-1 \leq x, y \leq 1$, it follows that $0 \leq x^2+y^2 \leq 2$, and so $1 \leq e^{x^2+y^2} \leq e^2$. Consequently,

$$\iint_R 1\,dA \leq \iint_R e^{x^2+y^2}\,dA \leq \iint_R e^2\,dA.$$

Since $\iint_R 1\,dA = \text{area}(R) = 4$ and $\iint_R e^2\,dA = e^2 \iint_R 1\,dA = 4e^2$, we get the desired inequality.

13. Since $\sin a \leq 1$ and $\cos a \leq 1$ for all $a \in R$, it follows that $\sin(2x)\cos(5y) \leq 1$ and therefore

$$\iint_R \sin(2x)\cos(5y)\,dA \leq \iint_R 1\,dA = \text{area}(R) = 2\pi^2,$$

since the dimensions of R are 2π and π.

15. The function f is continuous at all points in $R = [-1,1] \times [0,2]$ (notice that $\lim_{x\to 0} f(x,y) = 3$ for all y), and is hence integrable on R.

Since $f(x,y) > 0$ on R, we can interpret $\iint_R f\,dA$ as the volume of the solid V below $f(x,y)$ and above R. The part of V over $[0,1] \times [0,2]$ is a rectangular box of height 3; its volume is 6. The part of V over $[-1,0] \times [0,2]$ consists of a rectangular box of height 2 with a triangular prism on top (the triangle is the right triangle with sides 1), of height (or length) 2. The volume of the box and the prism is $4 + 1 = 5$. It follows that $\iint_R f\,dA = 11$.

17. The inequality $x^2 + y^2 \leq x^3 + y^3$ does not hold on $[0,1] \times [0,1]$; take, for example, $x = y = 1/2$: then $x^2 + y^2 = 1/2$, but $x^3 + y^3 = 1/4$. As a matter of fact, if $0 \leq x, y \leq 1$, then

$x^2 + y^2 \geq x^3 + y^3$, and $\iint_R (x^2 + y^2)\, dA > \iint_R (x^3 + y^3)\, dA$, if $R = [0,1] \times [0,1]$.

However, if $x, y \geq 1$, then $x^2 \leq x^3$ and $y^2 \leq y^3$, and so $x^2 + y^2 \leq x^3 + y^3$, and the inequality $\iint_R (x^2 + y^2)\, dA \leq \iint_R (x^3 + y^3)\, dA$ holds for $R = [1,2] \times [2,3]$.

19. The function $f(x,y) = e^{1/(x^2+y^2)}$ is not bounded on $[-1,1] \times [-1,1]$; its value approaches ∞ as $(x,y) \to (0,0)$. Consequently, f is not integrable on $[-1,1] \times [-1,1]$.

6.2 Double Integrals over General Regions

1. Let $g(x,y) = -f(x,y)$. Then $g(x,y) \geq 0$ and the solid region under g and above D is the mirror image (i.e., symmetric with respect to the xy-plane) of the solid region below D and above $f(x,y)$. Since

$$\iint_D g\, dA = -\iint_D f\, dA,$$

it follows that $\iint_D f\, dA$ is the negative of the volume of the solid region below D and above f.

If $f(x,y) \geq g(x,y)$ and $f(x,y) \geq 0$, $g(x,y) \geq 0$, then

$$\iint_D (f - g)\, dA = \iint_D f\, dA - \iint_D g\, dA$$

is the difference of the volume of the solid under f and above D and the volume of the solid under g and above D. In other words, $\iint_D (f - g)\, dA$ is the volume of the solid region between f and g and inside the vertical "cylinder" whose cross-sections all look like $D \subseteq \mathbb{R}^2$; see Figure 6.2.1.

Notice that this interpretation holds no matter whether f and g are positive or not (but as long as $f(x,y) \geq g(x,y)$). Here is a reason: by definition, (R_{ij}, $i, j = 1, \ldots, n$, is a rectangle forming a subdivision and ΔA_{ij} is its area)

$$\iint_D (f - g)\, dA = \lim_{n \to \infty} \sum_{i=1}^{n} \sum_{j=1}^{n} (f(x_i^*, y_j^*) - g(x_i^*, y_j^*)) \Delta A_{ij}.$$

Since $f(x,y) \geq g(x,y)$ (no additional assumptions!), $f(x_i^*, y_j^*) - g(x_i^*, y_j^*) \geq 0$, and

$$(f(x_i^*, y_j^*) - g(x_i^*, y_j^*)) \Delta A_{ij}$$

represents the volume of the rectangular box that approximates the volume of the solid between f and g and above R_{ij}. Adding up all such rectanglular boxes and taking the limit as R_{ij}'s shrink to a point, we get the volume of the solid between f and g defined over $D \subseteq \mathbb{R}^2$.

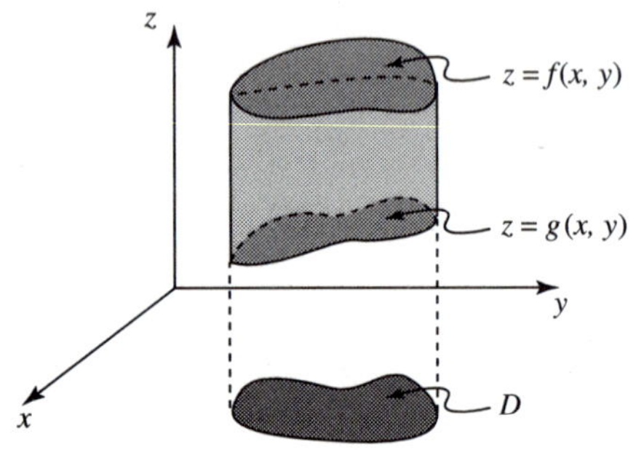

Figure 6.2.1

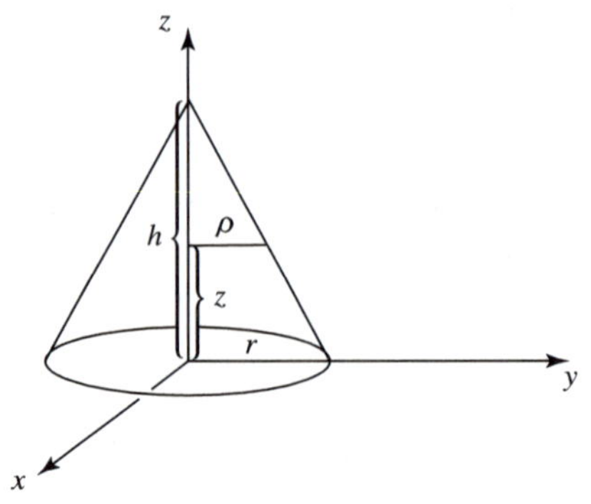

Figure 6.2.3

3. Place the cone so that its base lies in the xy-plane and its axis of rotation coincides with the z-axis. At a height z, $0 \le z \le h$, the cross-section is a circle of radius $\rho = r(h-z)/h$; see Figure 6.2.3. The radius was determined from the similar triangles ratio $h/r = (h-z)/\rho$. It follows that the cross-sectional area at height z is $\pi\rho^2 = \pi r^2(h-z)^2/h^2$ and the volume of the cone is equal to the "sum" of all such cross-sections from $z=0$ (where the base of the cone is) to $z=h$ (where its vertex is); i.e.,

$$\begin{aligned} v(V) &= \int_0^h \pi \frac{r^2(h-z)^2}{h^2}\, dz = \frac{\pi r^2}{h^2}\int_0^h (h-z)^2\, dz \\ &= \frac{\pi r^2}{h^2}\left(-\tfrac{1}{3}\right)(h-z)^3\Big|_0^h = \frac{\pi r^2}{h^2}\frac{h^3}{3} = \tfrac{1}{3}\pi r^2 h. \end{aligned}$$

5. Consider cross-sections parallel to the yz-plane. At x, $0 \le x \le 2$, the cross-section is a rectangle of length 1 and height $e^{-x+2}-1$. The volume of V is the "sum" of all such rectangles as x takes on values between 0 and 2; so

$$v(V) = \int_0^2 (e^{-x+2}-1)\, dx = \left(-e^{-x+2}-x\right)\Big|_0^2 = e^2 - 3.$$

7. Integrating with respect to y first, we get

$$\begin{aligned} \int_0^2 \left(\int_{2-x}^{x+1} (xe^y - 2y - 1)dy\right) dx &= \int_0^2 \left(xe^y - y^2 - y\right)\Big|_{2-x}^{x+1} dx \\ &= \int_0^2 \left(xe^{x+1} - (x+1)^2 - (x+1)\right) - \left(xe^{2-x} - (2-x)^2 - (2-x)\right) dx \\ &= \int_0^2 (xe^{x+1} - xe^{2-x} - 8x + 4)\, dx. \end{aligned}$$

Using integration by parts with $u = x$ and $dv = e^x dx$, we get

$$\int_0^2 xe^{x+1}\, dx = e\int_0^2 xe^x\, dx = e(x-1)e^x\Big|_0^2 = e^3 + e.$$

Similarly (by integration by parts with $u = x$ and $dv = e^{-x}dx$),

$$\int_0^2 xe^{2-x}\, dx = e^2\int_0^2 xe^{-x}\, dx = e^2(-x-1)e^{-x}\Big|_0^2 = e^2 - 3.$$

Since

$$\int_0^2 (-8x+4)\,dx = (-4x^2+4x)\Big|_0^2 = -8,$$

it follows that the value of the integral is $e^3 + e - (e^2 - 3) - 8 = e^3 - e^2 + e - 5 \approx 10.4148$.

9. Integrating with respect to ρ first, we get

$$\begin{aligned}\int_0^\pi \left(\int_0^{\cos\theta} \rho^2 \sin\theta d\rho\right) d\theta &= \int_0^\pi \left(\tfrac{1}{3}\rho^3 \sin\theta\right)\Big|_0^{\cos\theta} d\theta \\ &= \tfrac{1}{3}\int_0^\pi \cos^3\theta \,\sin\theta\, d\theta \\ &= \tfrac{1}{3}\left(-\tfrac{1}{4}\cos^4\theta\right)\Big|_0^\pi = -\tfrac{1}{12}(1-1) = 0.\end{aligned}$$

(To solve $\int \cos^3\theta \sin\theta d\theta$, use the substitution $u = \cos\theta$, $du = -\sin\theta d\theta$.)

11. Integrating with respect to x first, we get

$$\begin{aligned}\int_0^{\pi/2} \left(\int_0^{\sin y} x\cos y dx\right) dy &= \int_0^{\pi/2} \left(\tfrac{1}{2}x^2 \cos y\right)\Big|_0^{\sin y} dy \\ &= \tfrac{1}{2}\int_0^{\pi/2} \sin^2 y\, \cos y\, dy = \tfrac{1}{6}\sin^3 y\Big|_0^{\pi/2} = \tfrac{1}{6}.\end{aligned}$$

(To solve $\int \sin^2 y \cos y dy$, use the substitution $u = \sin y$, $du = \cos y dy$.)

13. The given integral can be expressed as the iterated integral

$$\begin{aligned}\iint_D f\,dA &= \int_3^4 \left(\int_0^2 (xy^{-1} - x^2y^2)dx\right) dy \\ &= \int_3^4 \left(\tfrac{1}{2}x^2y^{-1} - \tfrac{1}{3}x^3y^2\right)\Big|_0^2 dy = \int_3^4 \left(2y^{-1} - \tfrac{8}{3}y^2\right) dy \\ &= \left(2\ln y - \tfrac{8}{9}y^3\right)\Big|_3^4 = \left(2\ln 4 - \tfrac{512}{9}\right) - (2\ln 3 - 24) \\ &= 2\ln(4/3) - \tfrac{296}{9} \approx -32.3135.\end{aligned}$$

15. The given integral can be evalauted as the iterated integral

$$\begin{aligned}\iint_D f\,dA &= \int_0^1 \left(\int_{-y}^{1+y} (2xy - y)dx\right) dy \\ &= \int_0^1 \left(x^2y - xy\right)\Big|_{-y}^{1+y} dy \\ &= \int_0^1 \left[\,((1+y)^2y - (1+y)y) - (y^3 - (-y)y)\,\right] dy \\ &= \int_0^1 0\,dy = 0.\end{aligned}$$

17. Combining $y = x^2$ and $y = 4 - x^2$ we get $x^2 = 4 - x^2$ or $x^2 = 2$ and $x = \pm\sqrt{2}$. Therefore, the given integral can be evalauted as the iterated integral

$$\begin{aligned}\iint_D f\,dA &= \int_{-\sqrt{2}}^{\sqrt{2}} \left(\int_{x^2}^{4-x^2} x^{-2/3} dy \right) dx \\ &= \int_{-\sqrt{2}}^{\sqrt{2}} x^{-2/3} y \Big|_{x^2}^{4-x^2} dx = \int_{-\sqrt{2}}^{\sqrt{2}} \left(4x^{-2/3} - 2x^{4/3}\right) dx \\ &= \left(12x^{1/3} - \tfrac{6}{7}x^{7/3}\right) \Big|_{-\sqrt{2}}^{\sqrt{2}} = 24\,(2)^{1/6} - \tfrac{12}{7}\,(2)^{7/6} \approx 23.0906.\end{aligned}$$

19. Let $f(x, y) = e^{-x-y}$. Since $-1 \le x \le 1$ and $0 \le y \le 2$, it follows that $-1 \le -x \le 1$ and $-2 \le -y \le 0$ and so $-3 \le -x - y \le 1$. The fact that e^a is an increasing function of a implies that $e^{-3} \le e^{-x-y} \le e^1$. Consequently,

$$\iint_D e^{-3}\,dA \le \iint_D e^{-x-y}\,dA \le \iint_D e\,dA.$$

The area of D is 4, and so $\iint_D e^{-3}\,dA = 4e^{-3}$ and $\iint_D e\,dA = 4e$ and

$$4e^{-3} \le \iint_D e^{-x-y}\,dA \le 4e.$$

21. The plane can be viewed as the graph of the function $z = 6 - x - y/2$. For points (x, y) in the rectangle $R = [-1, 1] \times [0, 2]$, z is positive, and so the volume is equal to the double integral $\iint_R (6 - x - y/2)\,dA$. Using Fubini's Theorem,

$$\begin{aligned}\iint_R \left(6 - x - \tfrac{1}{2}y\right) dA &= \int_0^2 \left(\int_{-1}^1 \left(6 - x - \tfrac{1}{2}y\right) dx \right) dy \\ &= \int_0^2 \left(6x - \tfrac{1}{2}x^2 - \tfrac{1}{2}xy\right) \Big|_{-1}^1 dy = \int_0^2 (12 - y)\, dy \\ &= \left(12y - \tfrac{1}{2}y^2\right) \Big|_0^2 = 24 - 2 = 22.\end{aligned}$$

23. Write the equations of the given planes as $z = 1 - x - y$ and $z = (1 - x - y)/2$. The two planes intersect when $1 - x - y = (1 - x - y)/2$; i.e., when $x + y = 1$. It follows that the solid in question is bounded from above by $z = 1 - x - y$, from below by $z = (1 - x - y)/2$, and is located over the triangular region D in the xy-plane bounded by the coordinate axes and the line $x + y = 1$. Its volume is

$$\begin{aligned}v(V) &= \iint_D \left(1 - x - y - \tfrac{1}{2}(1 - x - y)\right) dA = \tfrac{1}{2} \int_0^1 \left(\int_0^{1-y} (1 - x - y) dx \right) dy \\ &= \tfrac{1}{2} \int_0^1 \left(x - \tfrac{1}{2}x^2 - xy\right) \Big|_0^{1-y} dy = \tfrac{1}{2} \int_0^1 \left(\tfrac{1}{2} - y + \tfrac{1}{2}y^2\right) dy \\ &= \tfrac{1}{2} \left(\tfrac{1}{2}y - \tfrac{1}{2}y^2 + \tfrac{1}{6}y^3\right) \Big|_0^1 = \tfrac{1}{12}.\end{aligned}$$

25. Let $h = f - g$; the assumption can be written as $\iint_D h\,dA = 0$, and so the average value of h over D is

$$\overline{h} = \frac{1}{\text{area}(D)} \iint_D h\,dA = 0.$$

By Theorem 6.5 and the remark following (6.2), it follows that there must be a point (x_0, y_0) in D where $h(x_0, y_0) = 0$. Since $h = f - g$, it follows that $f(x_0, y_0) = g(x_0, y_0)$.

27. The equation of the ellipse with semi-axes a and b is $x^2/a^2 + y^2/b^2 = 1$, or $y^2 = b^2(1 - x^2/a^2)$. We will compute the area in the first quadrant and then multiply by 4. Let D be the region in the first quadrant bounded by $y = b\sqrt{1 - x^2/a^2}$. Its area $A(D)$ can be computed as the double integral

$$A(D) = \iint_D 1\, dA = \int_0^1 \left(\int_0^{b\sqrt{1-x^2/a^2}} 1 dy \right) dx$$
$$= b \int_0^a \sqrt{1 - x^2/a^2}\, dx = \frac{b}{a} \int_0^a \sqrt{a^2 - x^2}\, dx.$$

We proceed using the formula for $\int \sqrt{a^2 - x^2} dx$ from integration tables (one could use the trigonometric substitution $x = a \sin t$ to compute this integral):

$$A(D) = \frac{b}{a} \left(\frac{x}{2} \sqrt{a^2 - x^2} + \frac{a^2}{2} \arcsin(x/a) \right) \Big|_0^a$$
$$= \frac{b}{a} \frac{a^2}{2} (\arcsin 1 - \arcsin 0) = \frac{ab}{2} \frac{\pi}{2} = \frac{ab\pi}{4}.$$

It follows that the area of the given ellipse in $ab\pi$.

29. Combining $y = x^2$ and $y = 4 - x^2$ we get $x^2 = 2$ and $x = \pm\sqrt{2}$. So D is the region in the xy-plane bounded by the y-axis on the left, by $y = 4 - x^2$ from above and by $y = x^2$ from below. Its area is

$$A(D) = \iint_D 1\, dA = \int_0^{\sqrt{2}} \left(\int_{x^2}^{4-x^2} 1 dy \right) dx$$
$$= \int_0^{\sqrt{2}} (4 - 2x^2)\, dx$$
$$= \left(4x - \tfrac{2}{3}x^3\right) \Big|_0^{\sqrt{2}} = 4\sqrt{2} - \tfrac{2}{3}(\sqrt{2})^3 = \tfrac{8}{3}\sqrt{2}.$$

31. The average value of f on R is

$$\overline{f} = \frac{1}{A(R)} \iint_R k(x^2 + y^2)\, dA$$
$$= \tfrac{1}{2}k \int_0^2 \left(\int_0^1 (x^2 + y^2) dx \right) dy = \tfrac{1}{2}k \int_0^2 \left(\tfrac{1}{3}x^3 + xy^2\right) \Big|_0^1 dy$$
$$= \tfrac{1}{2}k \int_0^2 \left(\tfrac{1}{3} + y^2\right) dy = \tfrac{1}{2}k \left(\tfrac{1}{3}y + \tfrac{1}{3}y^3\right) \Big|_0^2 = \tfrac{1}{2}k\tfrac{10}{3} = \tfrac{5}{3}k.$$

All points where $f(x, y) = 5k/3$ must satisfy $k(x^2 + y^2) = 5k/3$, or $x^2 + y^2 = 5/3$. Therefore, all points in R that also lie on the circle centered at the origin of radius $\sqrt{5/3}$ satisfy the conclusion of the Mean Value Theorem.

6.3 Examples and Techniques of Evaluation of Double Integrals

1. The given integral can be evaluated as the iterated integral

$$\begin{aligned}\iint_D f\,dA &= \int_0^1 \left(\int_0^{y/2} ye^x\,dx\right) dy \\ &= \int_0^1 ye^x\Big|_0^{y/2} dy = \int_0^1 \left(ye^{y/2} - y\right) dy \\ &= \left((2y-4)e^{y/2} - \tfrac{1}{2}y^2\right)\Big|_0^1 = \left(-2e^{1/2} - \tfrac{1}{2}\right) - (-4) = \tfrac{7}{2} - 2e^{1/2}.\end{aligned}$$

We used the formula $\int ye^{y/2}dy = (2y-4)e^{y/2}$, that can be found in tables of integrals; alternatively, this integral can be computed using integration by parts with $u = y$ and $dv = e^{y/2}dy$.

3. Since $y = 2x$ and $y = 5x$ intersect at the origin, D is the region in the first quadrant: it is bounded from above by the line $y = 5x$, from below by the line $y = 2x$, and is located over the interval $[0, 2]$. Hence

$$\begin{aligned}\iint_D f\,dA &= \int_0^2 \left(\int_{2x}^{5x} y^2 dy\right) dx \\ &= \int_0^1 \tfrac{1}{3}y^3\Big|_{2x}^{5x} dx = \tfrac{1}{3}\int_0^2 117x^3\,dx = \tfrac{39}{4}x^4\Big|_0^2 = 156.\end{aligned}$$

5. The solid in question is the solid below $z = \sqrt{4-y^2}$ and above the triangle D in the xy-plane that is bounded by the lines $x = 0$, $y = 0$ and $x + y = 2$. Its volume is

$$\begin{aligned}v(V) &= \iint_D \sqrt{4-y^2}\,dA = \int_0^2 \left(\int_0^{2-y} \sqrt{4-y^2}dx\right) dy \\ &= \int_0^2 x\sqrt{4-y^2}\Big|_0^{2-y} dy = \int_0^2 (2-y)\sqrt{4-y^2}\,dy \\ &= \int_0^2 2\sqrt{4-y^2}\,dy - \int_0^2 y\sqrt{4-y^2}\,dy.\end{aligned}$$

From a table of integrals (or using the trigonometric substitution $y = 2\sin\theta$), we get

$$\int_0^2 \sqrt{4-y^2}\,dy = \left(\tfrac{y}{2}\sqrt{4-y^2} + 2\arcsin(y/2)\right)\Big|_0^2 = 2\arcsin 1 = \pi.$$

Using the substitution $u = 4 - y^2$, we get

$$\int_0^2 y\sqrt{4-y^2}\,dy = -\tfrac{1}{3}(4-y^2)^{3/2}\Big|_0^2 = \tfrac{1}{3}4^{3/2} = \tfrac{8}{3}.$$

Therefore, the volume of V is $2\pi - 8/3 \approx 3.6165$.

7. The plane $x + y + z = 4$ intersects the xy-plane (i.e., $z = 0$) along the line $x + y = 4$. Combining $x + y = 4$ and $y = 3x$ we get $x = 1$ and $y = 3$. It follows that the solid in question (call it V) is below the plane $z = 4 - x - y$, and above the triangular region D in the xy-plane defined by $x = 0$, $y = 3x$ and $x + y = 4$; see Figure 6.3.7. The volume of V is

$$\begin{aligned} v(V) &= \iint_D (4 - x - y)\, dA = \int_0^1 \left(\int_{3x}^{4-x} (4 - x - y) dy \right) dx \\ &= \int_0^1 \left(4y - xy - \tfrac{1}{2}y^2\right) \Big|_{3x}^{4-x} dx = \int_0^1 (8x^2 - 16x + 8)\, dx = 8 \int_0^1 (x - 1)^2\, dx \\ &= \tfrac{8}{3}(x - 1)^3 \Big|_0^1 = \tfrac{8}{3}. \end{aligned}$$

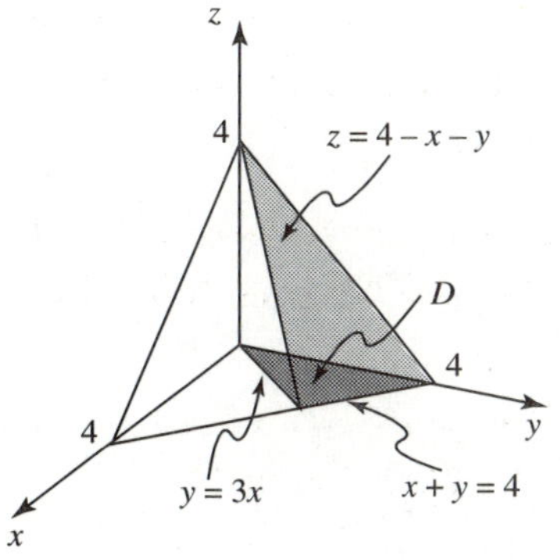

Figure 6.3.7

9. The given triangle (call it D) is bounded by $y = 1$, $y = x + 1$ and $x = 1$. The volume of the solid is

$$\begin{aligned} v &= \iint_D xy\, dA = \int_0^1 \left(\int_1^{x+1} xy dy \right) dx \\ &= \tfrac{1}{2} \int_0^1 xy^2 \Big|_1^{x+1} dx = \tfrac{1}{2} \int_0^1 (x^3 + 2x^2)\, dx \\ &= \tfrac{1}{2} \left(\tfrac{1}{4}x^4 + \tfrac{2}{3}x^3 \right) \Big|_0^1 = \tfrac{11}{24}. \end{aligned}$$

11. By separation of variables,

$$\begin{aligned} \int_0^1 \left(\int_0^2 (1 - x - y + xy) dx \right) dy &= \int_0^1 \left(\int_0^2 (1 - x)(1 - y) dx \right) dy \\ &= \left(\int_0^1 (1 - y)\, dy \right) \left(\int_0^2 (1 - x)\, dx \right) \\ &= \left(\left(y - \tfrac{1}{2}y^2\right) \Big|_0^1 \right) \left(\left(x - \tfrac{1}{2}x^2\right) \Big|_0^2 \right) = \tfrac{1}{2}(0) = 0. \end{aligned}$$

A direct computation gives

$$\begin{aligned} \int_0^1 \left(\int_0^2 (1 - x - y + xy) dx \right) dy &= \int_0^1 \left(x - \tfrac{1}{2}x^2 - xy + \tfrac{1}{2}x^2 y\right) \Big|_0^2 dy \\ &= \int_0^1 (-2y + 2y)\, dy = 0. \end{aligned}$$

13. The given integral is equal to $\iint_D (y^2 - x)\,dA$, where the region of integration D is given by $0 \le x \le 1$ and $\arctan x \le y \le \pi/4$; see Figure 6.3.13 (recall that $\arctan(1) = \pi/4$). Reversing the order of integration, we get (from $y = \arctan x$ it follows that $x = \tan y$)

$$\iint_D (y^2 - x)\,dA = \int_0^{\pi/4} \left(\int_0^{\tan y} (y^2 - x)dx \right) dy.$$

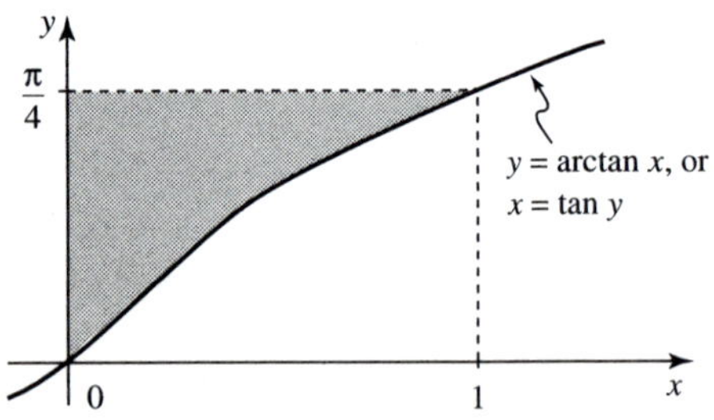

Figure 6.3.13

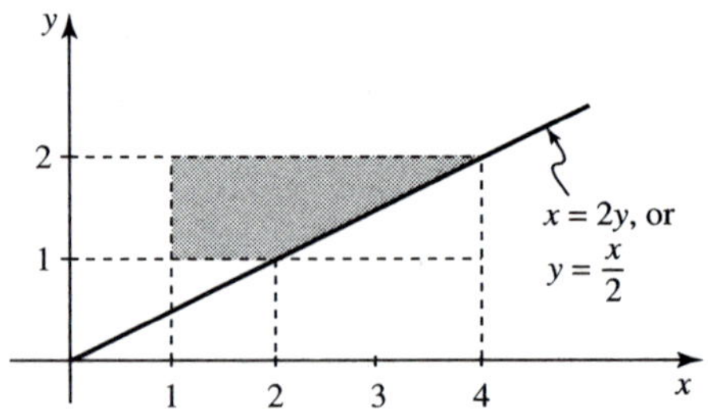

Figure 6.3.15

15. The given integral is equal to $\iint_D (\ln x/x)\,dA$, where the region of integration D is given by $1 \le y \le 2$ and $1 \le x \le 2y$; see Figure 6.3.15. The integral in the reversed order of integration has to be broken up into two integrals:

$$\iint_D \frac{\ln x}{x}\,dA = \int_1^2 \left(\int_1^2 \frac{\ln x}{x} dy \right) dx + \int_2^4 \left(\int_{x/2}^2 \frac{\ln x}{x} dy \right) dx.$$

17. The integral in question is equal to the double integral $\iint_D x\,dA$, where the D is the region in the xy-plane defined by $0 \le y \le 1$ and $0 \le x \le \arccos y$; in other words (recall that $x = \arccos y$ is the same curve as $y = \cos x$), D is the region under $y = \cos x$ on $[0, \pi/2]$. Reversing the order of integration, we get

$$\iint_D x\,dA = \int_0^{\pi/2} \left(\int_0^{\cos x} x dy \right) dx = \int_0^{\pi/2} \left(xy \Big|_0^{\cos x} \right) dx$$

$$= \int_0^{\pi/2} x \cos x\,dx = (x \sin x + \cos x) \Big|_0^{\pi/2} = \frac{\pi}{2} - 1.$$

The integral $\int x \cos x dx$ is solved using integration by parts, with $u = x$ and $dv = \cos x dx$.

19. The integral in question is equal to the double integral $\iint_D x \cos(2y^2)\,dA$, where D is the region in the xy-plane defined by $0 \le x \le 3$ and $x^2 \le y \le 9$; in other words, it is the region below the horizontal line $y = 9$ and above the parabola $y = x^2$, over the interval $[0, 3]$ on the x-axis. Reversing the order of integration, we get

$$\iint_D x \cos(2y^2)\,dA = \int_0^9 \left(\int_0^{\sqrt{y}} x \cos(2y^2) dx \right) dy$$

$$= \int_0^9 \tfrac{1}{2} x^2 \cos(2y^2) \Big|_0^{\sqrt{y}} dy = \tfrac{1}{2} \int_0^9 y \cos(2y^2)\,dy$$

$$= \frac{1}{2}\frac{1}{4}\sin(2y^2)\Big|_0^9 = \frac{1}{8}\sin 162 \approx -0.1223.$$

The integral $\int y\cos(2y^2)dy$ is solved using the substitution $u = 2y^2$.

21. The given integral is equal to the double integral $\iint_D 5\,dA$, where D consists of regions $1 \le y \le 2$ and $1 \le x \le \sqrt{y}$ and $2 \le y \le 4$ and $y/2 \le x \le \sqrt{y}$; see Figure 6.3.21. In other words, D is the region over the interval $[1,2]$ on the x-axis bounded from above by the line $y = 2x$ and from below by the parabola $y = x^2$. Hence

$$\iint_D 5\,dA = \int_1^2 \left(\int_{x^2}^{2x} 5dy\right) dx$$

$$= 5\int_1^2 \left(y\Big|_{x^2}^{2x}\right) dx = 5\int_1^2 (2x - x^2)\,dx$$

$$= 5\left(x^2 - \tfrac{1}{3}x^3\right)\Big|_1^2 = 5\left(\tfrac{4}{3} - \tfrac{2}{3}\right) = \tfrac{10}{3}.$$

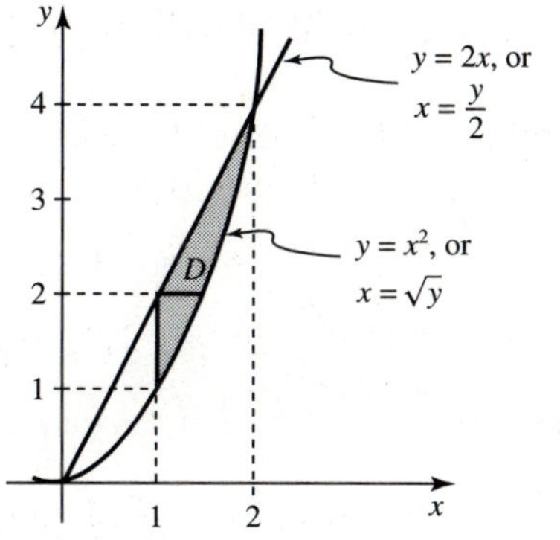

Figure 6.3.21

23. The region of integration D is the region below the parabola $y = x^2$ and over the interval $[0,1]$ on the x-axis. Hence

$$\iint_D e^{x^3}\,dA = \int_0^1 \left(\int_0^{x^2} e^{x^3}\,dy\right) dx$$

$$= \int_0^1 ye^{x^3}\Big|_0^{x^2} dx = \int_0^1 x^2 e^{x^3}\,dx = \tfrac{1}{3}e^{x^3}\Big|_0^1 = \tfrac{1}{3}(e-1).$$

The integral $\int x^2 e^{x^3} dx$ is solved by the substitution $u = x^3$, $du = 3x^2 dx$.

Note that in the reversed order of integration

$$\iint_D e^{x^3}\,dA = \int_0^1 \left(\int_{\sqrt{y}}^1 e^{x^3}\,dx\right) dy$$

we obtain the integral $\int e^{x^3}\,dx$, that cannot be solved as a compact formula (i.e., a numerical method or a power series method is needed).

6.4 Change of Variables in a Double Integral

1. The given integral is equal to the double integral $\iint_D e^{x^2+y^2}\,dA$, where D is the region defined by $0 \le x \le 2$ and $0 \le y \le \sqrt{4-x^2}$; it is the quarter-disk of radius 2 in the first quadrant. Consequently,

$$\iint_D e^{x^2+y^2}\,dA = \iint_{D^*} e^{r^2} r\,dA^*,$$

where D^* is defined by $0 \le r \le 2$ and $0 \le \theta \le \pi/2$. Therefore,

$$\begin{aligned}\iint_D e^{x^2+y^2}\,dA &= \int_0^{\pi/2}\left(\int_0^2 re^{r^2}\,dr\right)d\theta \\ &= \int_0^{\pi/2}\left(\tfrac{1}{2}e^{r^2}\right)\Big|_0^2\,d\theta = \tfrac{1}{2}(e^4-1)\int_0^{\pi/2}\,d\theta = \frac{\pi(e^4-1)}{4}.\end{aligned}$$

3. The given integral is equal to the double integral $\iint_D \arctan(y/x)\,dA$, where D is the region defined by $-1 \le x \le 1$ and $0 \le y \le \sqrt{1-x^2}$; it is the upper half-disk of radius 1. Since $\arctan(y/x) = \theta$, it follows that

$$\iint_D \arctan(y/x)\,dA = \iint_{D^*} \theta r\,dA^*,$$

where D^* is defined by $0 \le r \le 1$ and $0 \le \theta \le \pi$. So

$$\begin{aligned}\iint_D \arctan(y/x)\,dA &= \int_0^{\pi}\left(\int_0^1 \theta r dr\right)d\theta \\ &= \int_0^{\pi}\left(\tfrac{1}{2}\theta r^2\right)\Big|_0^1\,d\theta = \tfrac{1}{2}\int_0^{\pi}\theta\,d\theta = \frac{\pi^2}{4}.\end{aligned}$$

5. The disk $x^2+y^2 \le 1$ is represented in polar coordinates as the region D^* given by $0 \le r \le 1$ and $0 \le \theta \le 2\pi$. It follows that

$$\begin{aligned}\iint_D \sqrt{2x^2+2y^2+3}\,dA &= \iint_{D^*} r\sqrt{2r^2+3}\,dA^* \\ &= \int_0^{2\pi}\left(\int_0^1 r\sqrt{2r^2+3}\;dr\right)d\theta \\ &= \int_0^{2\pi}\left(\tfrac{1}{6}(2r^2+3)^{3/2}\right)\Big|_0^1\,d\theta \\ &= \tfrac{1}{6}(5^{3/2}-3^{3/2})\int_0^{2\pi}\,d\theta = \frac{\pi}{3}(5^{3/2}-3^{3/2}).\end{aligned}$$

The integral $\int r\sqrt{2r^2+3}dr$ is solved by the change of variables $u = 2r^2+3$.

7. Since the solid is inside the cylinder $x^2+y^2=5$, the region of integration $D \subseteq \mathbb{R}^2$ is the disk $x^2+y^2 \le 5$; so the volume of the solid in question is $v = \iint_D (x^2+y^2)\,dA$. Passing to polar coordinates we get

$$v = \iint_D (x^2+y^2)\,dA = \iint_{D^*} r^2\,r\,dA^*,$$

where D^* is defined by $0 \le r \le \sqrt{5}$ and $0 \le \theta \le 2\pi$. It follows that

$$v = \int_0^{2\pi} \left(\int_0^{\sqrt{5}} r^3 dr \right) d\theta = \left(\int_0^{2\pi} d\theta \right) \left(\int_0^{\sqrt{5}} r^3\, dr \right) = 2\pi \left(\tfrac{1}{4} r^4 \right) \Big|_0^{\sqrt{5}} = \tfrac{25}{2}\pi.$$

9. The solid in question is bounded from above by the ellipsoid $z = \sqrt{64 - 4x^2 - 4y^2}$, on the sides by the cylinder $x^2 + y^2 = 4$ (hence the region D of integration is the disk $x^2 + y^2 \le 4$ in the xy-plane), and from below by the xy-plane. Its volume is equal to $v = \iint_D \sqrt{64 - 4x^2 - 4y^2}\, dA$. Passing to polar coordinates, we get

$$v = \iint_D \sqrt{64 - 4x^2 - 4y^2}\, dA = \iint_{D^*} \sqrt{64 - 4r^2}\; r\, dA^*,$$

where D^* is defined by $0 \le r \le 2$ and $0 \le \theta \le 2\pi$. It follows that

$$\begin{aligned} v &= \int_0^{2\pi} \left(\int_0^2 2r\sqrt{16 - r^2}\, dr \right) d\theta \\ &= 2 \left(\int_0^{2\pi} d\theta \right) \left(\int_0^2 r\sqrt{16 - r^2}\, dr \right) \\ &= 4\pi \left(-\tfrac{1}{3}(16 - r^2)^{3/2} \right) \Big|_0^2 = -\frac{4\pi}{3}(12^{3/2} - 16^{3/2}) \approx 93.9578. \end{aligned}$$

11. Assume that T is a function of two variables (the proof in the case of n variables is analogous). Let

$$A = \begin{bmatrix} a_{11} & a_{12} \\ a_{21} & a_{22} \end{bmatrix} \quad \text{and} \quad \mathbf{b} = \begin{bmatrix} b_1 \\ b_2 \end{bmatrix}.$$

Then

$$T(x, y) = \begin{bmatrix} a_{11} & a_{12} \\ a_{21} & a_{22} \end{bmatrix} \begin{bmatrix} x \\ y \end{bmatrix} + \begin{bmatrix} b_1 \\ b_2 \end{bmatrix};$$

i.e., $T(x, y) = (a_{11}x + a_{12}y + b_1, a_{21}x + a_{22}y + b_2)$. The derivative of T is

$$DT = \begin{bmatrix} a_{11} & a_{12} \\ a_{21} & a_{22} \end{bmatrix} = A,$$

and therefore the Jacobian of T (which is equal to $\det(DT)$), is equal to $\det(A)$.

13. The area of the given rectangle (call it R^*) and the area of its image R are related by

$$\begin{aligned} A(R) &\approx \left| \det(DT)_{at\ (2,4)} \right| A(R^*) \\ &= \left| \det \begin{bmatrix} u(u^2 + v^2)^{-1/2} & v(u^2 + v^2)^{-1/2} \\ v & u \end{bmatrix}_{at\ (2,4)} \right| \cdot 0.05 \\ &= \left| u^2(u^2 + v^2)^{-1/2} - v^2(u^2 + v^2)^{-1/2} \right|_{at\ (2,4)} \cdot 0.05 \\ &= \left| 20^{-1/2}(-12) \right| \cdot 0.05 \approx 0.1342. \end{aligned}$$

15. The map T is an expansion by the factor a in the u-direction and a translation by b in the v-direction. Since translation does not change areas, it follows that the area of the image of D is equal to $a \cdot \text{area}(D)$.

17. Passing to polar coordinates, we get

$$\iint_D (5x + y^2 + x^2)\, dA = \iint_{D^*} (5r\cos\theta + r^2) r\, dA^*,$$

where D^* is the region in the $r\theta$-plane defined by $1 \le r \le 2$ and $0 \le \theta \le \pi$. So

$$\begin{aligned}\iint_D (5x + y^2 + x^2)\, dA &= \int_0^{\pi} \left(\int_1^2 (5r^2\cos\theta + r^3) dr \right) d\theta \\ &= \int_0^{\pi} \left(\tfrac{5}{3}\cos\theta r^3 + \tfrac{1}{4} r^4\right)\Big|_1^2 d\theta = \int_0^{\pi} \left(\tfrac{35}{3}\cos\theta + \tfrac{15}{4}\right) d\theta \\ &= \left(\tfrac{35}{3}\sin\theta + \tfrac{15}{4}\theta\right)\Big|_0^{\pi} = \frac{15\pi}{4}.\end{aligned}$$

19. Let D denote the region inside $r = 1 - \sin\theta$. Its area is given by

$$A(D) = \iint_D dA = \iint_{D^*} r\, dA^*,$$

where D^* is the region in the $r\theta$-plane defined by $0 \le \theta \le 2\pi$ and $0 \le r \le 1 - \sin\theta$. So

$$\begin{aligned}A(D) &= \int_0^{2\pi} \left(\int_0^{1-\sin\theta} r dr \right) d\theta \\ &= \int_0^{2\pi} \left(\tfrac{1}{2} r^2\right)\Big|_0^{1-\sin\theta} d\theta = \tfrac{1}{2}\int_0^{2\pi} (1 - \sin\theta)^2\, d\theta \\ &= \tfrac{1}{2}\int_0^{2\pi} (1 - 2\sin\theta + \sin^2\theta)\, d\theta\end{aligned}$$

(using $\sin^2\theta = (1 - \cos 2\theta)/2$)

$$\begin{aligned}&= \tfrac{1}{2}\int_0^{2\pi} \left(\tfrac{3}{2} - 2\sin\theta - \tfrac{1}{2}\cos 2\theta\right) d\theta \\ &= \tfrac{1}{2}\left(\tfrac{3}{2}\theta + 2\cos\theta - \tfrac{1}{4}\sin 2\theta\right)\Big|_0^{2\pi} = \frac{3\pi}{2}.\end{aligned}$$

21. It is a good idea to sketch the region. Let us consider $r = \theta$ first. When $\theta = 0$, $r = 0$. As θ increases, the distance r from a point on the curve to the origin increases at the same rate as θ; plot a few points, for example $(\pi/6, \pi/6)$, $(\pi/4, \pi/4)$, $(\pi/3, \pi/3)$, $(\pi/2, \pi/2)$, etc. Such a curve is called a spiral. The curve $r = 2\theta$ is another spiral: this time, the distance r increases twice as fast as the angle, and the spiral "opens up" faster; see Figure 6.4.21. The area of the region D in the first quadrant bounded by $r = \theta$ and $r = 2\theta$ is

$$A(D) = \iint_D dA = \iint_{D^*} r\, dA^*,$$

where D^* is the region in the $r\theta$-plane defined by $0 \le \theta \le \pi/2$ and $\theta \le r \le 2\theta$. So

$$A(D) = \int_0^{\pi/2} \left(\int_{\theta}^{2\theta} r dr \right) d\theta$$

$$= \int_0^{\pi/2} \left(\tfrac{1}{2}r^2\right)\Big|_\theta^{2\theta} d\theta = \tfrac{3}{2}\int_0^{\pi/2} \theta^2\, d\theta = \tfrac{3}{2}\tfrac{1}{3}\theta^3\Big|_0^{\pi/2} = \tfrac{1}{2}\left(\tfrac{\pi}{2}\right)^3.$$

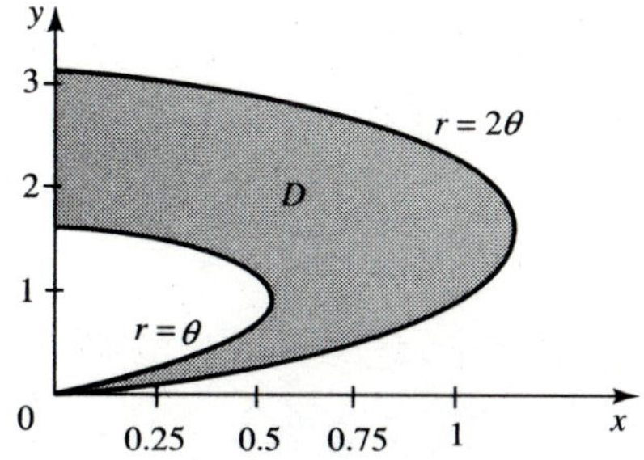

Figure 6.4.21

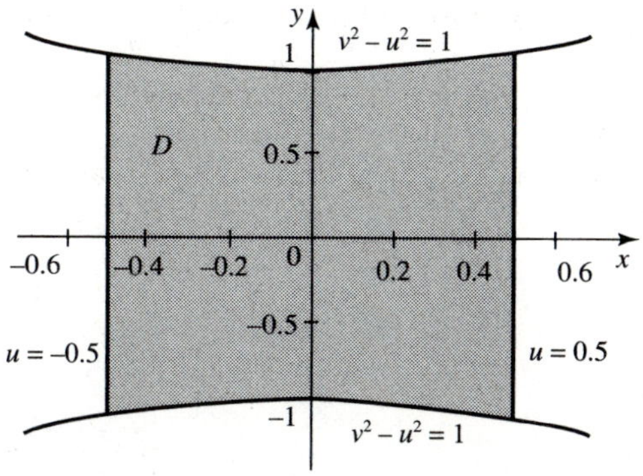

Figure 6.4.23

23. The change of variables is given by $T(u,v) = (u+v, -u+v)$. T is a linear map, and therefore the corresponding matrix $A = \begin{bmatrix} 1 & 1 \\ -1 & 1 \end{bmatrix}$ is equal to DT. In any case, the Jacobian of the change of variables T is

$$\frac{\partial(x,y)}{\partial(u,v)} = \det(A) = 2$$

and so (since $x^2 - y^2 = (u+v)^2 - (-u+v)^2 = 4uv$)

$$\iint_D (x^2 - y^2)\, dA = \iint_{D^*} 4uv\,|2|\, dA^*,$$

where D^* is the region that maps to D under T. From $y = x-1$ it follows that $-u+v = u+v-1$; i.e., $2u = 1$ or $u = 1/2$. Similarly, from $y = x+1$ it follows that $-u+v = u+v+1$; i.e., $2u = -1$ or $u = -1/2$. Finally, $xy = 1$ implies $(-u+v)(u+v) = 1$, i.e., $v^2 - u^2 = 1$; see Figure 6.4.23. It follows that (from $v^2 - u^2 = 1$ we get $v = \pm\sqrt{1+u^2}$)

$$\iint_D (x^2 - y^2)\, dA = 8\iint_{D^*} uv\, dA^* = 8\int_{-1/2}^{1/2} \left(\int_{-\sqrt{1+u^2}}^{\sqrt{1+u^2}} uv dv \right) du$$

$$= 8\int_{-1/2}^{1/2} \tfrac{1}{2}uv^2 \Big|_{-\sqrt{1+u^2}}^{\sqrt{1+u^2}} du$$

$$= 4\int_{-1/2}^{1/2} \left(u(1+u^2) - u(1+u^2)\right)\, du = 0.$$

25. See Figure 6.4.25. The volume of the wedge is $v = \iint_D (y+3)\, dA$, where D is the disk $x^2 + y^2 \le 9$. Changing to polar coordinates, we get

$$v = \iint_D (y+3)\, dA = \iint_{D^*} (r\sin\theta + 3) r\, dA^*,$$

where D^* is the region in the $r\theta$-plane defined by $0 \le \theta \le 2\pi$ and $0 \le r \le 3$. So

$$v = \int_0^{2\pi} \left(\int_0^3 (r^2 \sin\theta + 3r) dr \right) d\theta = \int_0^{2\pi} \left(\tfrac{1}{3} r^3 \sin\theta + \tfrac{3}{2} r^2 \right) \Big|_0^3 d\theta$$

$$= \int_0^{2\pi} \left(9 \sin\theta + \tfrac{27}{2}\right) d\theta = \left(-9\cos\theta + \tfrac{27}{2}\theta\right) \Big|_0^{2\pi} = 27\pi.$$

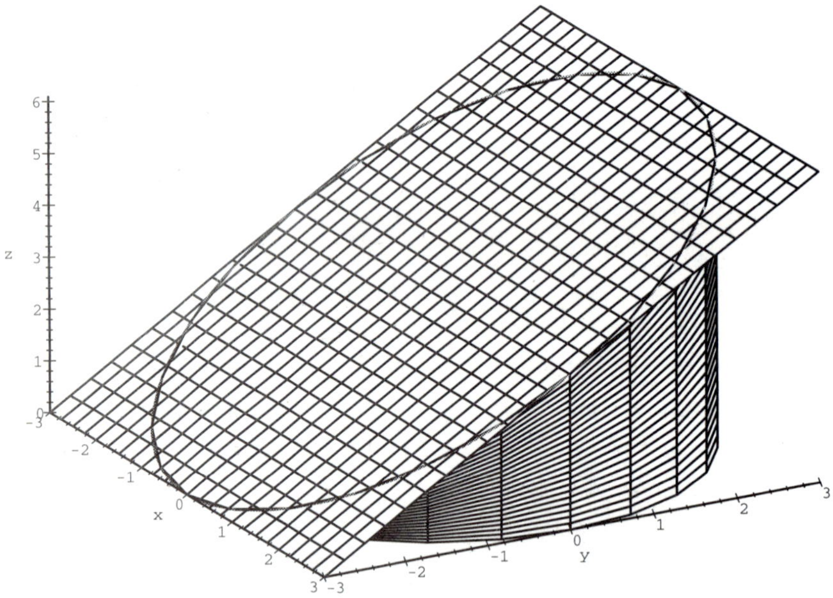

Figure 6.4.25

27. Represent the change of variables as the map $T(u, v) = (v, u/v)$. Then

$$DT(u, v) = \begin{bmatrix} 0 & 1 \\ 1/v & -u/v^2 \end{bmatrix},$$

and the Jacobian is

$$\frac{\partial(x, y)}{\partial(u, v)} = \det(DT) = -\frac{1}{v}.$$

From $xy^3 = u^3/v^2$ it follows that

$$\iint_D xy^3 \, dA = \iint_{D^*} \frac{u^3}{v^2} \left| -\frac{1}{v} \right| dA^*,$$

where D^* is the region that maps to D under T. From $x = 1$ it follows that $v = 1$; from $x = 2$ it follows that $v = 2$. Replacing x and y in $xy = 1$ by u and v, we get $v(u/v) = 1$; i.e., $u = 1$. Similarly, $xy = 3$ implies that $u = 3$. In words, D^* is the rectangle $[1, 3] \times [1, 2]$ in the uv-plane. Therefore (since $v > 0$, replace $|v|$ by v)

$$\iint_D xy^3 \, dA = \int_1^2 \left(\int_1^3 \frac{u^3}{v^3} du \right) dv = \tfrac{1}{4} \int_1^2 \frac{1}{v^3} u^4 \Big|_1^3 dv$$

$$= 20 \int_1^2 \frac{1}{v^3}\, dv = 20\left(-\tfrac{1}{2}v^{-2}\right)\Big|_1^2 = \tfrac{15}{2}.$$

29. Let $u = 3x - 2y$ and $v = x + y$. Solving for x and y we get

$$x = \tfrac{1}{5}u + \tfrac{2}{5}v \qquad \text{and} \qquad y = -\tfrac{1}{5}u + \tfrac{3}{5}v.$$

So the change of variables can be represented as a linear map

$$T(u,v) = \left(\tfrac{1}{5}u + \tfrac{2}{5}v, -\tfrac{1}{5}u + \tfrac{3}{5}v\right).$$

It follows that

$$DT = \begin{bmatrix} 1/5 & 2/5 \\ -1/5 & 3/5 \end{bmatrix}$$

and so

$$\frac{\partial(x,y)}{\partial(u,v)} = \det(DT) = \tfrac{3}{25} + \tfrac{2}{25} = \tfrac{1}{5}.$$

By the Change of Variables Theorem,

$$\iint_D 5(x+y)\, dA = \iint_{D^*} 5v\tfrac{1}{5}\, dA^*,$$

where D^* is the region that maps to D under T. The new variables u and v were chosen so that D^* is simple. Indeed, $3x - 2y = 5$ implies that $u = 5$ and $3x - 2y = -2$ gives $u = -2$. The remaining two equations imply $v = -2$ and $v = 1$. In other words, D^* is the rectangle $[-2,5] \times [-2,1]$ in the uv-plane. Consequently,

$$\iint_D 5(x+y)\, dA = \int_{-2}^5 \left(\int_{-2}^1 v dv\right) du = \int_{-2}^5 \tfrac{1}{2}v^2\Big|_{-2}^1 du$$

$$= -\tfrac{3}{2}\int_{-2}^5 du = -\tfrac{3}{2}(7) = -\tfrac{21}{2}.$$

31. Let $u = x + y$ and $v = y/x$. The corresponding region D^* in the uv-plane is given by $u = 0$, $u = 2$, $v = 1$ and $v = 2$; i.e., it is the rectangle $[0,2] \times [1,2]$. Solving $u = x + y$ and $v = y/x$ for x and y, we get (for example, substitute $y = u - x$ into $v = y/x$ and solve for x)

$$x = \frac{u}{1+v} \qquad \text{and} \qquad y = \frac{uv}{1+v}.$$

Therefore, the change of variables is

$$T(u,v) = \left(\frac{u}{1+v}, \frac{uv}{1+v}\right);$$

its derivative is the matrix

$$DT = \begin{bmatrix} 1/(1+v) & -u/(1+v)^2 \\ v/(1+v) & u/(1+v)^2 \end{bmatrix}$$

and so

$$\frac{\partial(x,y)}{\partial(u,v)} = \det(DT) = \frac{u + uv}{(1+v)^3} = \frac{u}{(1+v)^2}.$$

By the Change of Variables Theorem,

$$\iint_D e^x\, dA = \iint_{D^*} e^{u/(1+v)} \left|\frac{u}{(1+v)^2}\right| dA^*$$

(since $0 \le u \le 2$, $|u| = u$)

$$= \int_0^2 \left(\int_1^2 e^{u/(1+v)} \frac{u}{(1+v)^2} dv \right) du.$$

Use the substitution $t = u/(1+v)$, $dt = -udv/(1+v)^2$ for the inner integration:

$$\int_1^2 e^{u/(1+v)} \frac{u}{(1+v)^2}\, dv = -\int_{u/2}^{u/3} e^t\, dt = -e^t \Big|_{u/2}^{u/3} = e^{u/2} - e^{u/3}.$$

Therefore,

$$\iint_D e^x\, dA = \int_0^2 (e^{u/2} - e^{u/3})\, du = \left(2e^{u/2} - 3e^{u/3}\right)\Big|_0^2$$
$$= (2e - 3e^{2/3}) - (2 - 3) = 1 + 2e - 3e^{2/3}.$$

33. Let $u = x + y$ and $v = x - y$. Solving for x and y we get $x = (u+v)/2$ and $y = (u-v)/2$. So this change of variables is represented by the linear map

$$T(u,v) = \left(\tfrac{1}{2}(u+v), \tfrac{1}{2}(u-v)\right).$$

Since

$$DT = \begin{bmatrix} 1/2 & 1/2 \\ 1/2 & -1/2 \end{bmatrix},$$

it follows that the Jacobian is

$$\frac{\partial(x,y)}{\partial(u,v)} = \det(DT) = -\tfrac{1}{2}.$$

By the Change of Variables Theorem,

$$\iint_D \sin\left(\tfrac{x+y}{x-y}\right)\, dA = \iint_{D^*} \sin(u/v)\tfrac{1}{2}\, dA^*,$$

where D^* is the region that maps to D under T. From $x - y = 1$ and $x - y = 5$ it follows that $v = 1$ and $v = 5$. The equation $x = 0$ implies $(u+v)/2 = 0$; i.e., $u + v = 0$. Similarly, $y = 0$ implies $(u-v)/2 = 0$; i.e., $u - v = 0$. It follows that D^* is the part of the horizontal strip in the uv-plane defined by $v = 1$ and $v = 5$, bounded by the lines $v = u$ and $v = -u$. So

$$\iint_D \sin\left(\tfrac{x+y}{x-y}\right)\, dA = \tfrac{1}{2}\iint_{D^*} \sin(u/v)\, dA^*$$
$$= \tfrac{1}{2}\int_1^5 \left(\int_{-v}^{v} \sin(u/v) du \right) dv = \tfrac{1}{2}\int_1^5 (-v\cos(u/v))\Big|_{-v}^{v} dv$$
$$= -\tfrac{1}{2}\int_1^5 ((v\cos 1) - (v\cos(-1))\, dv = 0,$$

since $\cos 1 = \cos(-1)$.

6.5 Triple Integrals

1. We evaluate the given triple integral as the following iterated integral:

$$\begin{aligned}\iiint_V (2x-y-z)\,dV &= \int_0^3\left(\int_0^2\left(\int_0^2 (2x-y-z)dz\right)dy\right)dx\\ &= \int_0^3\left(\int_0^2 \left(2xz - zy - \tfrac{1}{2}z^2\right)\Big|_0^2 dy\right)dx\\ &= \int_0^3\left(\int_0^2 (4x-2y-2)dy\right)dx\\ &= \int_0^3 \left(4xy - y^2 - 2y\right)\Big|_0^2 dx\\ &= \int_0^3 (8x-8)\,dx = (4x^2-8x)\Big|_0^3 = 12.\end{aligned}$$

3. V can be viewed as a type-1(3D) region: the corresponding elementary (2D) region D is the triangle bounded by $x=0$, $y=0$ and $x+y=1$ (the last equation is obtained by substituting $z=0$ into $x+y+z=1$). The bottom and the top surfaces are $z=0$ and $z=1-x-y$. It follows that

$$\begin{aligned}\iiint_V y^2\,dV &= \iint_D\left(\int_0^{1-x-y} y^2dz\right)dA\\ &= \int_0^1\left(\int_0^{1-y}\left(\int_0^{1-x-y} y^2dz\right)dx\right)dy\\ &= \int_0^1\left(\int_0^{1-y} (y^2z)\Big|_0^{1-x-y} dx\right)dy\\ &= -\int_0^1\left(\int_0^{1-y} y^2(x+y-1)dx\right)dy\\ &= -\int_0^1 y^2\left(\tfrac{1}{2}x^2 + xy - x\right)\Big|_0^{1-y} dy\\ &= \int_0^1 \tfrac{1}{2}y^2(1-2y+y^2)\,dy = \left(\tfrac{1}{10}y^5 - \tfrac{1}{4}y^4 + \tfrac{1}{6}y^3\right)\Big|_0^1 = \tfrac{1}{60}.\end{aligned}$$

5. Combining $z=2y^2$ and $z=8-2y^2$ we get $y^2=2$ and $y=\pm\sqrt{2}$. The solid V can be viewed as a type-1(3D) region: the corresponding elementary (2D) region D is the rectangle $[0,1]\times[-\sqrt{2},\sqrt{2}]$; the bottom surface is $z=2y^2$ and the top surface is $z=8-2y^2$. It follows that

$$\begin{aligned}\iiint_V (2z-5)\,dV &= \iint_D\left(\int_{2y^2}^{8-2y^2} (2z-5)dz\right)dA\\ &= \int_{-\sqrt{2}}^{\sqrt{2}}\left(\int_0^1\left(\int_{2y^2}^{8-2y^2} (2z-5)dz\right)dx\right)dy\\ &= \int_{-\sqrt{2}}^{\sqrt{2}}\left(\int_0^1 (z^2-5z)\Big|_{2y^2}^{8-2y^2} dx\right)dy\\ &= \int_{-\sqrt{2}}^{\sqrt{2}}\left(\int_0^1 (24-12y^2)dx\right)dy\end{aligned}$$

$$= \int_{-\sqrt{2}}^{\sqrt{2}} (24 - 12y^2)x\Big|_0^1 \, dy$$

$$= \int_{-\sqrt{2}}^{\sqrt{2}} (24 - 12y^2)\, dy = (24y - 4y^3)\Big|_{-\sqrt{2}}^{\sqrt{2}} = 32\sqrt{2}.$$

7. Combining $z = 12 - x^2 - y^2$ and $z = 2x^2 + 2y^2$ we get $x^2 + y^2 = 4$. It follows that the solid V can be viewed as a type-1(3D) region: the corresponding elementary (2D) region D is the disk $x^2 + y^2 \le 4$; the bottom surface is $z = 2x^2 + 2y^2$ and the top surface is $12 - x^2 - y^2$. Therefore

$$\iiint_V 4y\, dV = \iint_D \left(\int_{2x^2+2y^2}^{12-x^2-y^2} 4ydz \right) dA.$$

Passing to cylindrical coordinates, we get

$$\iiint_V 4y\, dV = \int_0^{2\pi} \left(\int_0^2 \left(\int_{2r^2}^{12-r^2} (4r\sin\theta) rdz \right) dr \right) d\theta$$

$$= 4\left(\int_0^{2\pi} \sin\theta\, d\theta \right) \left(\int_0^2 \left(\int_{2r^2}^{12-r^2} 4r^2 dz \right) dr \right) = 0,$$

since $\int_0^{2\pi} \sin\theta d\theta = 0$.

9. The solid V can be viewed as a type-1(3D) region: the corresponding elementary (2D) region D is the disk $x^2 + y^2 \le 1$; the bottom surface is $z = -\sqrt{2 - x^2 - y^2}$ and the top surface is $z = \sqrt{2 - x^2 - y^2}$. Therefore

$$\iiint_V xyz\, dV = \iint_D \left(\int_{-\sqrt{2-x^2-y^2}}^{\sqrt{2-x^2-y^2}} xyzdz \right) dA.$$

Passing to cylindrical coordinates, we get

$$\iiint_V xyz\, dV = \int_0^{2\pi} \left(\int_0^1 \left(\int_{-\sqrt{2-r^2}}^{\sqrt{2-r^2}} (r^2 z\cos\theta\sin\theta) rdz \right) dr \right) d\theta.$$

The product $\cos\theta\sin\theta$ can be factored out of the inner two integrations, so that

$$\iiint_V xyz\, dV = \left(\int_0^{2\pi} \cos\theta\sin\theta\, d\theta \right) \left(\int_0^1 \left(\int_{-\sqrt{2-r^2}}^{\sqrt{2-r^2}} r^3 z\cos\theta\sin\theta dz \right) dr \right).$$

Since

$$\int_0^{2\pi} \cos\theta\sin\theta\, d\theta = \tfrac{1}{2} \int_0^{2\pi} \sin 2\theta\, d\theta = 0,$$

it follows that $\iiint_V xyz\, dV = 0$.

11. The solid V that is the region of integration is defined by the inequalities $0 \le y \le 1$, $0 \le x \le 2 - 2y$ and $0 \le z \le 4 - 2x - 4y$. The corresponding elementary (2D) region D in the xy-plane is the triangular region bounded by $y = 0$, $x = 0$ and $x = 2 - 2y$ (or $y = 1 - x/2$). V is bounded from above by the plane $z = 4 - 2x - 4y$, that crosses the xy-plane when

$4-2x-4y=0$; i.e., when $x=2-2y$. Consequently, the region of integration is the three-dimensional region in the first octant bounded from above by the plane $z=4-2x-4y$.

The iterated integral is computed to be

$$\begin{aligned}\int_0^1\left(\int_0^{2-2y}\left(\int_0^{4-2x-4y}3dz\right)dx\right)dy &= \int_0^1\left(\int_0^{2-2y}(3z)\Big|_0^{4-2x-4y}dx\right)dy\\ &= \int_0^1\left(\int_0^{2-2y}(12-6x-12y)dx\right)dy\\ &= \int_0^1 \left(12x-3x^2-12xy\right)\Big|_0^{2-2y}dy\\ &= \int_0^1 (12-24y+12y^2)\,dy\\ &= (12y-12y^2+4y^3)\Big|_0^1 = 4.\end{aligned}$$

13. The limits $0\le\theta\le 2\pi$ and $0\le r\le 2$ define the disk in the xy-plane of radius 2 centered at the origin. The equation $z=r^2$ represents the paraboloid $z=x^2+y^2$. It follows that the region of integration is the three-dimensional solid inside the cylinder $x^2+y^2=4$, above the xy-plane and below the paraboloid $z=x^2+y^2$. The iterated integral is computed to be

$$\begin{aligned}\int_0^{2\pi}\left(\int_0^2\left(\int_0^{r^2}r^2zdz\right)dr\right)d\theta &= \left(\int_0^{2\pi}d\theta\right)\left(\int_0^2 \tfrac12 r^2z^2\Big|_0^{r^2}dr\right)\\ &= 2\pi\tfrac12\int_0^2 r^6\,dr = \tfrac{128}{7}\pi.\end{aligned}$$

15. The inequalities $-\sqrt8\le x\le\sqrt8$ and $-\sqrt{8-x^2}\le y\le\sqrt{8-x^2}$ define the disk $x^2+y^2=8$ in the xy-plane. The region of integration is the three-dimensional solid inside the cylinder, bounded from below by the plane $z=-3$ and from above by the paraboloid $z=8-x^2-y^2$. Passing to cylindrical coordinates, we get

$$\begin{aligned}\int_{-\sqrt8}^{\sqrt8}\left(\int_{-\sqrt{8-x^2}}^{\sqrt{8-x^2}}\left(\int_{-3}^{8-x^2-y^2}2dz\right)dy\right)dx &= \int_0^{2\pi}\left(\int_0^{\sqrt8}\left(\int_{-3}^{8-r^2}2rdz\right)dr\right)d\theta\\ &= 2\left(\int_0^{2\pi}d\theta\right)\left(\int_0^{\sqrt8} rz\Big|_{-3}^{8-r^2}dr\right)\\ &= 4\pi\int_0^{\sqrt8}(11r-r^3)\,dr\\ &= 4\pi\left(\tfrac{11}{2}r^2-\tfrac14 r^4\right)\Big|_0^{\sqrt8} = 112\pi.\end{aligned}$$

17. The tetrahedron in question (call it V) is the solid in the first octant bounded by the plane that goes through the points $(1,0,0)$, $(0,1,0)$, and $(0,0,1)$. We need the equation of that plane — and to get it, we need a normal. The vectors $\mathbf{v}=(-1,1,0)$ from $(1,0,0)$ to

$(0,1,0)$ and $\mathbf{w} = (-1,0,1)$ from $(1,0,0)$ to $(0,0,1)$ lie in the plane. So their cross product

$$\mathbf{v} \times \mathbf{w} = \begin{vmatrix} \mathbf{i} & \mathbf{j} & \mathbf{k} \\ -1 & 1 & 0 \\ -1 & 0 & 1 \end{vmatrix} = \mathbf{i} + \mathbf{j} + \mathbf{k},$$

is normal to the plane; its equation is (we need one point — take $(1,0,0)$):

$$1(x-1) + 1(y-0) + 1(z-0) = 0;$$

i.e., $x + y + z = 1$. The region D in the xy-plane is the triangular region bounded by the coordinate axes and the line $x + y = 1$. So the volume of the tetrahedron is

$$\begin{aligned} V = \iiint_V 1\,dV &= \iint_D \left(\int_0^{1-x-y} 1dz \right) dA \\ &= \int_0^1 \left(\int_0^{1-x} \left(\int_0^{1-x-y} dz \right) dy \right) dx \\ &= \int_0^1 \left(\int_0^{1-x} (1-x-y)dy \right) dx \\ &= \int_0^1 \left(\tfrac{1}{2} - x + \tfrac{1}{2}x^2 \right) dx = \left(\tfrac{1}{2}x - \tfrac{1}{2}x^2 + \tfrac{1}{6}x^3 \right) \Big|_0^1 = \tfrac{1}{6}. \end{aligned}$$

19. The sphere has radius 2. The semi-axes of the ellipsoid (rewrite the equation as $x^2 + y^2 + z^2/4 = 1$) are 1 (in the x-direction), 1 (in the y-direction) and 2 (in the z-direction). So the ellipsoid is inside the sphere, touching it at $(0,0,\pm 2)$. Therefore, to get the volume of the solid in question, we will subtract the volume of the ellipsoid from $32\pi/3$ (which is the volume of the sphere).

When $z = 0$, the equation of the ellipsoid implies $x^2 + y^2 = 1$; hence, the region D in the xy-plane is the disk $x^2 + y^2 \le 1$. The solid is bounded from below by $z = -\sqrt{4 - 4x^2 - 4y^2} = -2\sqrt{1 - x^2 - y^2}$ and from above by $z = \sqrt{4 - 4x^2 - 4y^2} = 2\sqrt{1 - x^2 - y^2}$. Its volume is (we use cylindrical coordinates)

$$\begin{aligned} v = \iiint_V 1\,dV &= \iint_D \left(\int_{-2\sqrt{1-x^2-y^2}}^{2\sqrt{1-x^2-y^2}} dz \right) dA \\ &= \int_0^{2\pi} \left(\int_0^1 \left(\int_{-2\sqrt{1-r^2}}^{2\sqrt{1-r^2}} rdz \right) dr \right) d\theta \\ &= \left(\int_0^{2\pi} d\theta \right) \left(\int_0^1 rz \Big|_{-2\sqrt{1-r^2}}^{2\sqrt{1-r^2}} dr \right) \\ &= 2\pi \int_0^1 4r\sqrt{1-r^2}\,dr \\ &= 8\pi \left(-\tfrac{1}{3} \right) (1 - r^2)^{3/2} \Big|_0^1 = \tfrac{8}{3}\pi. \end{aligned}$$

So the volume of the solid in question is $32\pi/3 - 8\pi/3 = 8\pi$.

21. The solid in question (denote it by V) is the region below the cone $z^2 = x^2 + y^2$ and above the disk centered at $(1,0)$ of radius 1 in the xy-plane. Its volume is

$$v = \iiint_V dV = \iint_D \left(\int_0^{\sqrt{x^2+y^2}} dz \right) dA = \iint_D \sqrt{x^2+y^2}\, dA.$$

In polar coordinates, the equation $(x-1)^2 + y^2 = 1$ can be expressed as $r^2 - 2r\cos\theta = 0$; i.e., $r = 2\cos\theta$. Notice that $-\pi/2 \le \theta \le \pi/2$. Hence

$$\begin{aligned}\iint_D \sqrt{x^2+y^2}\, dA &= \int_{-\pi/2}^{\pi/2} \left(\int_0^{2\cos\theta} r^2 dr \right) d\theta \\ &= \frac{1}{3} \int_{-\pi/2}^{\pi/2} r^3 \Big|_0^{2\cos\theta} d\theta \\ &= \frac{8}{3} \int_{-\pi/2}^{\pi/2} \cos^3\theta\, d\theta\end{aligned}$$

(use a formula from a table of integrals)

$$= \frac{8}{3}\frac{1}{3}(2 + \cos^2\theta)\sin\theta \Big|_{-\pi/2}^{\pi/2} = \frac{32}{9}.$$

23. The paraboloid intersects the xy-plane along the circle $x^2 + y^2 = 16$ (substitute $z = 0$ into $z = 16 - x^2 - y^2$). The solid in question (denote it by V) is the three-dimensional region below $z = 16 - x^2 - y^2$ and above the smaller of the two two-dimensional regions in the xy-plane defined by $x + y = 4$ and $x^2 + y^2 = 16$; see Figure 6.5.23 (call that region D). The volume of V is

$$\begin{aligned}v(V) = \iiint_V dV &= \iint_D \left(\int_0^{16-x^2-y^2} dz \right) dA \\ &= \iint_D (16 - x^2 - y^2)\, dA \\ &= \int_0^4 \left(\int_{4-x}^{\sqrt{16-x^2}} (16 - x^2 - y^2) dy \right) dx.\end{aligned}$$

In cylindrical coordinates (from $x + y = 4$ it follows that $r\sin\theta + r\cos\theta = 4$ and $r = 4/(\sin\theta + \cos\theta)$),

$$v(V) = \iint_D (16 - x^2 - y^2)\, dA = \int_0^{\pi/2} \left(\int_{4/(\sin\theta+\cos\theta)}^4 (16 - r^2) r dr \right) d\theta.$$

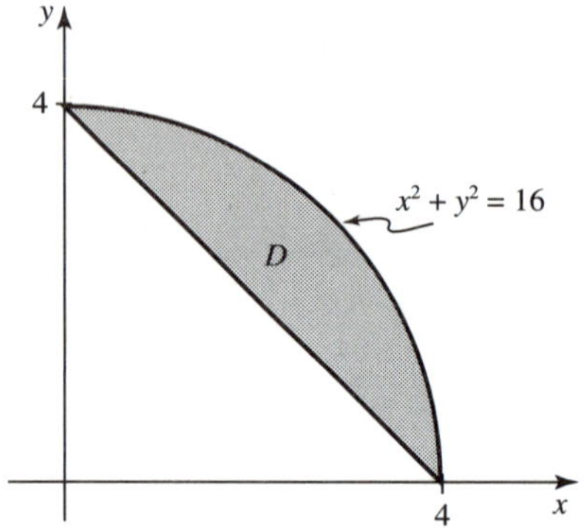

Figure 6.5.23

25. The solid V is bounded on the sides by the cylinder and from above and from below by the sphere. The region of integration D in the xy-plane is the disk $x^2 + y^2 \le b^2$. The volume of V is

$$v(V) = \iiint_V dV = \iint_D \left(\int_{-\sqrt{a^2-x^2-y^2}}^{\sqrt{a^2-x^2-y^2}} dz \right) dA = 2 \iint_D \sqrt{a^2 - x^2 - y^2}\, dA.$$

Passing to cylindrical coordinates, we get

$$\begin{aligned} v(V) &= 2\int_0^{2\pi} \left(\int_0^b r\sqrt{a^2 - r^2} dr \right) d\theta \\ &= 2\left(\int_0^{2\pi} d\theta \right) \left(-\tfrac{1}{3}(a^2 - r^2)^{3/2} \right) \Big|_0^b \\ &= -\tfrac{4}{3}\pi((a^2 - b^2)^{3/2} - a^3) = \tfrac{4}{3}\pi(a^3 - (a^2 - b^2)^{3/2}). \end{aligned}$$

27. The region of integration is defined by $-\sqrt{2} \le x \le \sqrt{2}$, $-\sqrt{2 - x^2} \le y \le \sqrt{2 - x^2}$ and $x^2 + y^2 - 2 \le z \le 2 - x^2 - y^2$. In words, it is the solid bounded by the paraboloids $z = x^2 + y^2 - 2$ and $z = 2 - x^2 - y^2$. The corresponding region in the xy-plane is the circle $x^2 + y^2 = 2$. Hence, in cylindrical coordinates,

$$\int_{-\sqrt{2}}^{\sqrt{2}} \int_{-\sqrt{2-x^2}}^{\sqrt{2-x^2}} \int_{x^2+y^2-2}^{2-x^2-y^2} (x^2 + y^2 - 2) dz dy dx = \int_0^{2\pi} \left(\int_0^{\sqrt{2}} \left(\int_{r^2-2}^{2-r^2} (r^2 - 2) r dz \right) dr \right) d\theta.$$

29. We use the change of variables $u = x + y$, $v = y + 2z$ and $w = 2x + y + z$. The new region of integration V^* is defined by $1 \le u \le 3$, $-2 \le v \le 4$ and $-1 \le w \le 1$. To find the Jacobian, we have to solve for x, y and z.

Subtracting the equation for v from the equation for u, we get $u - v = x - 2z$; eliminating y from equations for v and w, we get $v - w = z - 2x$; now solve this new system for x and z: $x =$

$-u/3-v/3+2w/3$ and $z=-2u/3+v/3+w/3$; consequently, $y=u-x=4u/3+v/3-2w/3$. It follows that the change of variables can be represented as the map

$$T(u,v,w)=\left(-\tfrac{1}{3}u-\tfrac{1}{3}v+\tfrac{2}{3}w, \tfrac{4}{3}u+\tfrac{1}{3}v-\tfrac{2}{3}w, -\tfrac{2}{3}u+\tfrac{1}{3}v+\tfrac{1}{3}w\right).$$

Then

$$DT=\begin{bmatrix}-1/3 & -1/3 & 2/3\\ 4/3 & 1/3 & -2/3\\ -2/3 & 1/3 & 1/3\end{bmatrix}$$

and

$$\det(DT)=\frac{1}{27}\begin{vmatrix}-1 & -1 & 2\\ 4 & 1 & -2\\ -2 & 1 & 1\end{vmatrix}=\frac{1}{27}(-3+0+12)=\frac{1}{3}.$$

So

$$\begin{aligned} v=\iiint_V dV &= \iiint_{V^*}\tfrac{1}{3}\,dV^* \\ &= \int_1^3\left(\int_{-2}^4\left(\int_{-1}^1 \tfrac{1}{3}dw\right)dv\right)du \\ &= \tfrac{1}{3}\left(\int_1^3 du\right)\left(\int_{-2}^4 dv\right)\left(\int_{-1}^1 dw\right) \\ &= \tfrac{1}{3}(2)(6)(2)=8. \end{aligned}$$

31. Combining $z=1$ and $z=x^2+y^2$ we get $x^2+y^2=1$. Similarly, combining $z=1$ and $z=2(x^2+y^2)$ we get $x^2+y^2=1/2$. So the solid V is bounded from above by $z=2(x^2+y^2)$ and from below by $z=x^2+y^2$, if (x,y) belongs to the disk $x^2+y^2\le 1/2$. If (x,y) belongs to the region $1/2\le x^2+y^2\le 1$, then V is bounded from below by $z=x^2+y^2$ and from above by $z=1$; see Figure 6.5.31. Consequently,

$$\begin{aligned}\iiint_V 2x\,dV &= \iint_{\{x^2+y^2\le 1/2\}}\left(\int_{x^2+y^2}^{2(x^2+y^2)} 2xdz\right)dA \\ &\quad+\iint_{\{1/2\le x^2+y^2\le 1\}}\left(\int_{x^2+y^2}^{1} 2xdz\right)dA.\end{aligned}$$

Passing to cylindrical coordinates, we get

$$\begin{aligned}\iiint_V 2x\,dV &= \int_0^{2\pi}\left(\int_0^{\sqrt{1/2}}\left(\int_{r^2}^{2r^2} 2r^2\cos\theta dz\right)dr\right)d\theta \\ &\quad+\int_0^{2\pi}\left(\int_{\sqrt{1/2}}^{1}\left(\int_{r^2}^{1} 2r^2\cos\theta dz\right)dr\right)d\theta.\end{aligned}$$

Actually it is not hard to evaluate this integral. We can factor out $\cos\theta$ from the inner two integrations in both integrals in cylindrical coordinates, thus getting $\int_0^{2\pi}\cos\theta d\theta$ as a factor. But $\int_0^{2\pi}\cos\theta d\theta=0$, and so $\iiint_V 2x\,dV=0$.

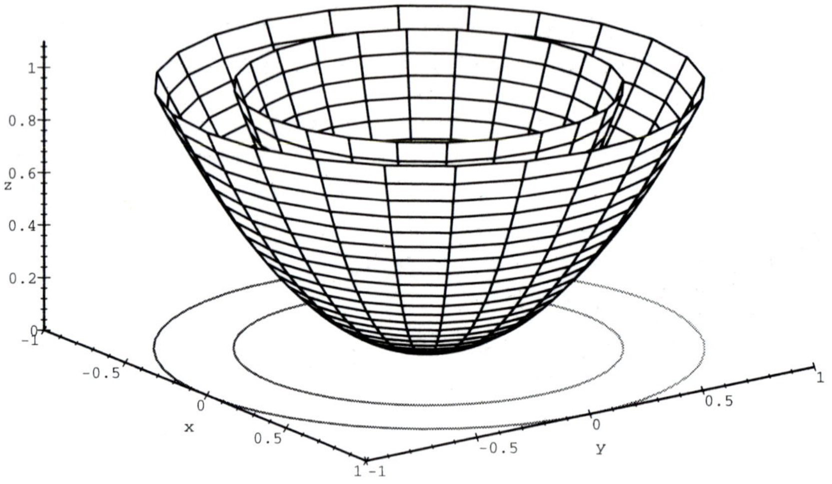

Figure 6.5.31

6.6 Chapter Review

Solutions to Computer Projects

1. (a) Since $0 \leq x, y \leq 1$, it follows that $0 \leq x^2 + y^2 \leq 2$. The largest difference between x and y (for $0 \leq x, y \leq 1$) is 1, and the smallest difference is -1; i.e., $-1 \leq x - y \leq 1$. Hence $e^{-1} \leq e^{x-y} \leq e$ and

$$0 \leq f(x,y) = e^{x-y}\sqrt{x^2+y^2} \leq e\sqrt{2}.$$

Consequently,

$$\iint_D 0\,dA \leq \iint_D f\,dA \leq \iint_D e\sqrt{2}\;dA,$$

or

$$0 \leq \iint_D f\,dA \leq e\sqrt{2} \approx 3.84,$$

since $\iint_D dA = \text{area}(D) = 1$.

(b) See Figure 6.6.1 (a); the maximum value of f is about 2.8. Our estimate for the minimum is the best possible, since $f(0,0) = 0$. So $0 \leq f(x,y) \leq 2.8$ on D, and so

$$0 \leq \iint_D f\,dA \leq \iint_D 2.8\,dA = 2.8.$$

(c) Write $\iint_D f\,dA$ as the iterated integral

$$\iint_D f\,dA = \int_0^1 \left(\int_0^1 e^{x-y}\sqrt{x^2+y^2}\;dx \right) dy$$

and use a computer to find its value — it is approximately 0.83748 (the computation requires a lot of memory and time). Since the area of D is 1, it follows that the average value $\overline{f}$ of f is 0.83748.

(d) We have to find all points in D where $f(x,y) = \overline{f} = 0.83748$. Using implicit plot command, we request the plot of $e^{x-y}\sqrt{x^2+y^2} = 0.83748$ for $-1 \le x, y \le 1$; see Figure 6.6.1 (b).

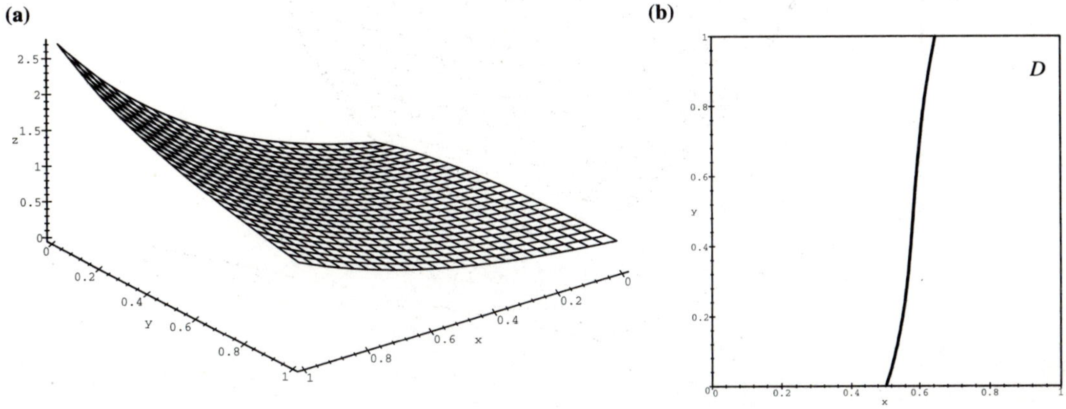

Figure 6.6.1

2. (a) If a is a side of a square, then

$$I(\text{square}) = \frac{4a}{(a^2)^{1/2}} = 4.$$

(b) The area of a disk of radius a is πa^2 and the length of its boundary is $2\pi a$; thus

$$I(\text{disk}) = \frac{2\pi a}{(\pi a^2)^{1/2}} = 2\sqrt{\pi} \approx 3.54491.$$

(c) The area of D_m is m and the length of its boundary is $2m+2$; so

$$I(D_m) = \frac{2m+2}{\sqrt{m}} = 2\sqrt{m} + \frac{2}{\sqrt{m}}.$$

It follows that, as $m \to \infty$ and as $m \to 0$, the isoperimetric ratio $I(D_m)$ approaches ∞.

From $I'(D_m) = m^{-1/2} - m^{-3/2} = 0$ it follows that $m^{1/2} = m^{3/2}$, or $m = 1$. By checking the sign of $I'(D_m)$ or by looking at the plot of $I(D_m)$ (see Figure 6.6.2 (a)) we conclude that $m = 1$ is a minimum point. So, of all rectangles considered, the square has the smallest isoperimetric ratio.

(d) Let us find formulas for the length and area of the objects involved. From $\mathbf{c}_n(t) = (\cos^n t, \sin^n t)$ it follows that

$$\mathbf{c}_n'(t) = (-n\cos^{n-1} t \sin t, n \sin^{n-1} t \cos t)$$

and

$$\|\mathbf{c}_n'(t)\| = n\sqrt{\cos^{2n-2} t \sin^2 t + \sin^{2n-2} t \cos^2 t}.$$

Hence

$$\ell(\mathbf{c}_n) = n \int_0^{\pi/2} \sqrt{\cos^{2n-2} t \sin^2 t + \sin^{2n-2} t \cos^2 t}\, dt,$$

and so the total length of the boundary of D_n is $4\ell(\mathbf{c}_n)$.

Since $\mathbf{c}_n$ can be expressed in Cartesian coordinates as $x^{2/n} + y^{2/n} = 1$, the region $\overline{D}_n$ can be descibed as the region in the first quadrant bounded by $x^{2/n} + y^{2/n} = 1$. So $y = (1 - x^{2/n})^{n/2}$, and the area of $\overline{D}_n$ is

$$A_n = \iint_{\overline{D}_n} dA = \int_0^1 (1 - x^{2/n})^{n/2}\, dx;$$

consequently, the area of D_n is $4A_n$. See Figure 6.6.2 (b).

When $n = 1$, we obtain a disk and its boundary circle — so $I(D_1) = 2\sqrt{\pi} \approx 3.54491$, by (b). When $n = 2$,

$$\ell(\mathbf{c}_2) = 2\int_0^{\pi/2} \sqrt{2\cos^2 t \sin^2 t}\, dt = 2\sqrt{2}\int_0^{\pi/2} \cos t \sin t\, dt = \sqrt{2}.$$

Since

$$A_2 = \int_0^1 (1 - x)\, dx = \tfrac{1}{2},$$

it follows that

$$I(D_2) = \frac{4\ell(\mathbf{c}_2)}{(4A_2)^{1/2}} = 4.$$

For $n = 3$,

$$\ell(\mathbf{c}_3) = 3\int_0^{\pi/2} \sqrt{\cos^4 t \sin^2 t + \sin^4 t \cos^2 t}\, dt = 3\int_0^{\pi/2} \cos t \sin t\, dt = \tfrac{3}{2}.$$

Since (using a computer)

$$A_3 = \int_0^1 (1 - x^{2/3})^{3/2}\, dx \approx 0.29452,$$

it follows that

$$I(D_3) = \frac{4(3/2)}{(4 \cdot 0.29452)^{1/2}} \approx 5.52795.$$

Finally, when $n = 1/2$, (using a computer)

$$\ell(\mathbf{c}_{1/2}) = \tfrac{1}{2}\int_0^{\pi/2} \sqrt{\cos^{-1} t \sin^2 t + \sin^{-1} t \cos^2 t}\, dt \approx 1.75442.$$

The area is

$$A_{1/2} = \int_0^1 (1 - x^4)^{1/4}\, dx \approx 0.92704,$$

and it follows that

$$I(D_{1/2}) = \frac{4(1.75442)}{(4 \cdot 0.92704)^{1/2}} \approx 3.64431.$$

(e) The disk has the smallest isoperimetric ratio of all region considered. As a matter of fact, the disk has the smallest isoperimetric ratio of *all* plane regions (the proof is beyond the scope of this book).

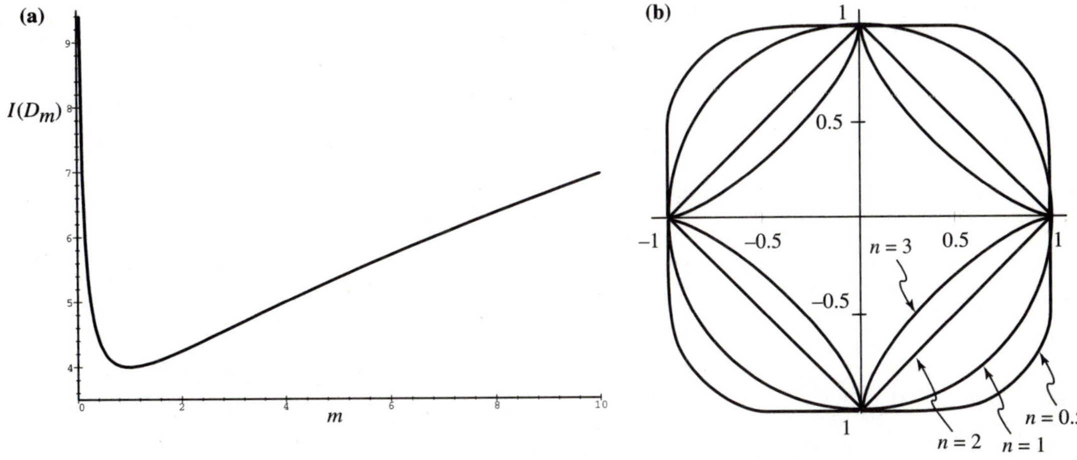

Figure 6.6.2

Solutions to Further Explorations

1. The region of integration (call it V) is defined by $0 \le \theta \le 2\pi$, $0 \le \phi \le \pi/4$ and $0 \le \rho \le 2+\cos\theta$. From the inequalities for θ and ϕ we conclude that V is inside the circular cone whose axis of symmetry is the z-axis and whose intersection with the yz-plane (or with any plane perpendicular to the xy-plane that contains thc z-axis) consists of two lines that make the angles of $\pm\pi/4$ with respect to the positive z-axis (if you rotate the part of the line $z = y$ satisfying $y \ge 0$ about the z-axis, you will obtain that cone). The surface $\rho = 2 + \cos\theta$ bounds V from above.

The integral is computed to be

$$\int_0^{2\pi}\left(\int_0^{2+\cos\theta}\left(\int_0^{\pi/4}\rho^2\sin\phi\,d\phi\right)d\rho\right)d\theta = \int_0^{2\pi}\left(\int_0^{2+\cos\theta}\left(-\rho^2\cos\phi\right)\Big|_0^{\pi/4}d\rho\right)d\theta$$

$$= \left(1-\tfrac{\sqrt{2}}{2}\right)\int_0^{2\pi}\left(\int_0^{2+\cos\theta}\rho^2 d\rho\right)d\theta$$

$$= \tfrac{1}{3}\left(1-\tfrac{\sqrt{2}}{2}\right)\int_0^{2\pi}(2+\cos\theta)^3\,d\theta$$

$$= \tfrac{1}{3}\left(1-\tfrac{\sqrt{2}}{2}\right)\int_0^{2\pi} 8 + 12\cos\theta + 6\cos^2\theta + cos^3\theta\,d\theta.$$

Now $\int_0^{2\pi} cos^3\theta\,d\theta = 0$; draw the graph of $\cos^3\theta$ and interpret the integral as "area above the x-axis – area below the x-axis," or use the formula $\cos^3\theta = (2+\cos^2\theta)\sin\theta/2$ and a substitution. Next,

$$\int_0^{2\pi}\cos^2\theta\,d\theta = \left(\tfrac{1}{2}\theta + \tfrac{1}{4}\sin 2\theta\right)\Big|_0^{2\pi} = \pi$$

and $\int_0^{2\pi}\cos\theta\,d\theta = 0$. It follows that the given integral is equal to

$$\tfrac{1}{3}\left(1-\tfrac{\sqrt{2}}{2}\right)(16\pi + 6\pi) = \tfrac{22}{3}\pi\left(1-\tfrac{\sqrt{2}}{2}\right).$$

3. The volume is given by

$$v = \iint_D \left(\int_0^{my+am} dz \right) dA,$$

where D is the disk $x^2 + y^2 \leq a^2$ in the xy-plane. Passing to cylindrical coordinates, we get

$$\begin{aligned}
v &= \int_0^{2\pi} \left(\int_0^a \left(\int_0^{mr\sin\theta+am} r dz \right) dr \right) d\theta \\
&= \int_0^{2\pi} \int_0^a rz\Big|_0^{mr\sin\theta+am} dr\, d\theta \\
&= \int_0^{2\pi} \left(\int_0^a (mr^2\sin\theta + amr) dr \right) d\theta = \int_0^{2\pi} \left(\tfrac{1}{3}mr^3\sin\theta + \tfrac{1}{2}amr^2\right)\Big|_0^a d\theta \\
&= \int_0^{2\pi} \left(\tfrac{1}{3}ma^3\sin\theta + \tfrac{1}{2}a^3 m\right) d\theta \\
&= \tfrac{1}{3}a^3 m \int_0^{2\pi} \sin\theta\, d\theta + \tfrac{1}{2}a^3 m \int_0^{2\pi} d\theta = a^3 m\pi,
\end{aligned}$$

since $\int_0^{2\pi} \sin\theta d\theta = 0$.

5. Assume that $\iint_R f\, dA = 0$ and that there is a point $(x_0, y_0) \in R$ such that $f(x_0, y_0) > 0$. We have to show that this leads to a contradiction (and thus, $f(x_0, y_0) = 0$ for all $(x_0, y_0) \in R$).

If $f(x_0, y_0) > 0$, then there must be a ball $B((x_0, y_0), \delta)$ of radius δ centered at (x_0, y_0), so that $f(x, y) > 0$ for all (x, y) in that ball. This statement follows directly from the definition of continuity. To be precise: choose $\epsilon > 0$ so that $f(x_0, y_0) - \epsilon > 0$; for that ϵ, by continuity, there must be a ball $B((x_0, y_0), \delta)$ of radius $\delta > 0$ so that, for all (x, y) in $B((x_0, y_0), \delta)$,

$$|f(x, y) - f(x_0, y_0)| < \epsilon;$$

i.e., $f(x, y)$ belongs to the interval $(f(x_0, y_0) - \epsilon, f(x_0, y_0) + \epsilon)$. However, from the way ϵ was chosen, it follows that all numbers in this interval are positive. Therefore, $f(x, y) > 0$ for all x in $B((x_0, y_0), \delta)$.

Now

$$\iint_R f\, dA = \iint_{B((x_0,y_0),\delta)} f dA + \iint_{R-B((x_0,y_0),\delta)} f dA,$$

where $R - B((x_0, y_0), \delta)$ is the part of the rectangle that does not include the ball $B((x_0, y_0), \delta)$. Since $f \geq 0$, it follows that

$$\iint_{R-B((x_0,y_0),\delta)} f dA \geq 0.$$

However, on $B((x_0, y_0), \delta)$, the function f satisfies $f(x, y) > 0$, and so

$$\iint_{B((x_0,y_0),\delta)} f dA > 0.$$

Therefore, $\iint_R f\, dA > 0$, and that contradicts the assumption.

7. Let us make a sketch of the region in question (call it D). The equation $r = 2$ represents the circle $x^2 + y^2 = 4$. The equation $r = 4\cos\theta$ implies (multiply both sides by r) $r^2 = r\cos\theta$; i.e., $x^2 + y^2 = 4x$ and (after completing the square) $(x - 2)^2 + y^2 = 4$. So D is the region common to both the circle centered at the origin of radius 2 and the circle centered at $(2,0)$ of radius 2; see Figure 6.6.7.

Let D_1 be the region inside $x^2+y^2=4x$ and outside $x^2+y^2=4$. We will compute the area of D_1 (since the integration is easier; see the comment at the end of this solution), and then get the area of D from $A(D)+A(D_1)=4\pi$.

Combining $r=2$ and $r=4\cos\theta$ we get $\cos\theta=1/2$; i.e., $\theta=\pm\pi/3$. Therefore, $-\pi/3\le\theta\le\pi/3$; since $2\le r\le 4\cos\theta$, it follows that the area of D_1 is

$$A(D_1)=\iint_{D_1} dA=\int_{-\pi/3}^{\pi/3}\left(\int_2^{4\cos\theta} r dr\right)d\theta=2\int_0^{\pi/3}\left(\int_2^{4\cos\theta} r dr\right)d\theta,$$

by symmetry. So

$$\begin{aligned}A(D_1)&=2\int_0^{\pi/3}\frac{1}{2}r^2\Big|_2^{4\cos\theta}d\theta\\&=\int_0^{\pi/3}(16\cos^2\theta-4)\,d\theta\\&=\left(16\left(\tfrac{1}{2}\theta+\tfrac{1}{4}\sin 2\theta\right)-4\theta\right)\Big|_0^{\pi/3}=\tfrac{4}{3}\pi+2\sqrt{3}.\end{aligned}$$

We used the identity $\cos^2\theta=\frac{1}{2}(1+\cos 2\theta)$ in computing the integral of $\cos^2\theta$. Therefore,

$$A(D)=4\pi-\left(\tfrac{4}{3}\pi+2\sqrt{3}\right)=\tfrac{8}{3}\pi-2\sqrt{3}.$$

Comment: the area of D can be computed directly, as

$$A(D)=\int_{-\pi/2}^{-\pi/3}\left(\int_0^{4\cos\theta} r dr\right)d\theta+\int_{-\pi/3}^{\pi/3}\left(\int_0^{2} r dr\right)d\theta+\int_{\pi/3}^{\pi/2}\left(\int_0^{4\cos\theta} r dr\right)d\theta.$$

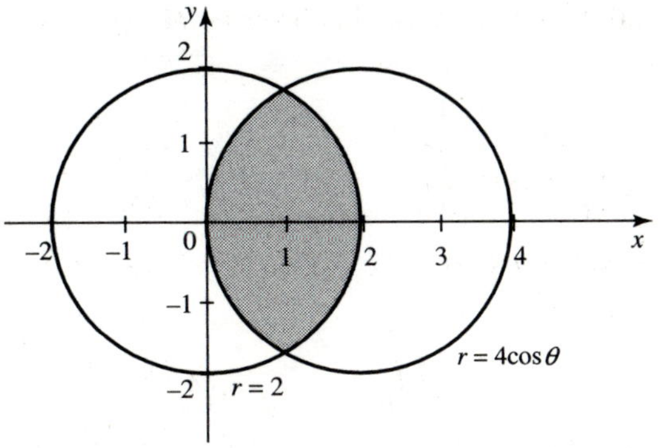

Figure 6.6.7

9. Enclose D into a rectangle R and divide R into n^2 equal subrectangles R_{ij}, $i,j=1,\dots,n$. Take only those that have a non-empty intersection with D. Since D is symmetric with respect to the x-axis, the subdivision is symmetric as well: every rectangle above the x-axis has its symmetric counterpart below the x-axis (if a rectangle crosses the x-axis, then it is symmetric to itself). Notice that from $f(x,-y) = -f(x,y)$ it follows that (read it with $y = 0$) $f(x,0) = -f(x,0)$; i.e., $f(x,0) = 0$, so the value of f along the x-axis is zero. Form a Riemann sum

$$\mathcal{R}_n = \sum_{i=1}^{n} \sum_{j=1}^{n} f(x_i^*, y_j^*)\Delta A_{ij}$$

in the following way: if a rectangle R_{ij} contains a point on the x-axis, then choose (x_i^*, y_j^*) to be that point (so, in this case, $f(x_i^*, y_j^*) = 0$). If a rectangle does not intersect the x-axis, take any point (x_i^*, y_j^*) in it (that is, of course, in D). For its symmetric counterpart, select the point $(x_i^*, -y_j^*)$. The contribution to $\mathcal{R}_n$ from this pair of rectangles is

$$f(x_i^*, y_j^*)\Delta A_{ij} + f(x_i^*, -y_j^*)\Delta A_{ij}^s$$

(where ΔA_{ij} is the area of R_{ij} and ΔA_{ij}^s is the area of the rectangle that is symmetric to R_{ij}; by construction, $\Delta A_{ij}^s = \Delta A_{ij}$). Since $f(x_i^*, -y_j^*) = -f(x_i^*, y_j^*)$, the above contribution is zero. It follows that the Riemann sum thus constructed is zero (for any n). Hence

$$\iint_R f\,dA = \lim_{n\to\infty} \mathcal{R}_n = \lim_{n\to\infty} 0 = 0.$$

7. INTEGRATION OVER SURFACES PROPERTIES AND APPLICATIONS OF INTEGRALS

7.1 Parametrized Surfaces

1. Let us consider the sphere first. The lines $u = 0$ and $u = 2\pi$ map to the same meridian (i.e., the semi-circle) $(a\cos v, 0, a\sin v)$, $-\pi/2 \le v \le \pi/2$, that is the intersection of the sphere and the xz-plane, with $x \ge 0$. The line $v = \pi/2$ maps to $(0, 0, a)$ and the line $v = -\pi/2$ maps to $(0, 0, -a)$. So the top and the bottom line segments map to the north and the south poles respectively, and the vertical boundary line segments map to the meridian that is the intersection of the sphere and the xz-plane with $x \ge 0$.

Consider the cylinder $\mathbf{r}(u, v) = (a\cos u, a\sin u, v)$, $0 \le v \le b$, $0 \le u \le 2\pi$. Both vertical lines $u = 0$ and $u = 2\pi$ map to the line segment $(a, 0, v)$, $0 \le v \le b$, that is the intersection of the cylinder and the xz-plane with $x \ge 0$. The horizontal lines $v = 0$ and $v = b$ map to the bottom and the top boundary circles $\mathbf{r}(u, 0) = (a\cos u, a\sin u, 0)$, $0 \le u \le 2\pi$ (in the xy-plane) and $\mathbf{r}(u, b) = (a\cos u, a\sin u, b)$, $0 \le u \le 2\pi$ (in the plane $y = b$).

3. We take a parametrization of the sphere $x^2 + y^2 + z^2 = a^2$ (for example, from Example 7.1) and adjust the range of the parameter(s) so that the z-coordinate is positive. So the required parametrization is

$$\mathbf{r}(u, v) = (a\cos v\cos u, a\cos v\sin u, a\sin v), \quad 0 \le u \le 2\pi, \quad 0 \le v \le \pi/2.$$

Notice that $0 \le v \le \pi/2$ implies that the z-coordinate $a\sin v$ is positive, which is exactly what is needed.

There are other parametrizations; for example,

$$\mathbf{r}(u, v) = (a\cos u\sin v, a\sin u\sin v, a\cos v),$$

where $0 \le u \le 2\pi$ and $0 \le v \le \pi/2$.

5. Parametrize the plane as the graph of $z = 2 + 3y - x$; hence $\mathbf{r}(u, v) = (u, v, 2 + 3v - u)$. We need only those points on the plane that are inside the cylinder; i.e., that satisfy $u^2 + v^2 \le 4$. The required parametrization $\mathbf{r}\colon D \subseteq \mathbb{R}^2 \to \mathbb{R}^3$ is given by

$$\mathbf{r}(u, v) = (u, v, 2 + 3v - u), \qquad (u, v) \in D,$$

where D is the disk $u^2 + v^2 \le 4$.

7. Parametrize the given plane as the graph of the function $z = 6 - x - 2y$; the domain $D \subseteq \mathbb{R}^2$ of the parametrization is defined by the requirement that only the part in the first octant is needed. So $x \ge 0$, $y \ge 0$ and $6 - x - 2y \ge 0$, or $x + 2y \le 6$. It follows that D is the triangular region bounded by the coordinate axes and the line $x + 2y = 6$; i.e., the boundary of D is the triangle with vertices at $(0, 0)$, $(6, 0)$ and $(0, 3)$. So the required parametrization is

$$\mathbf{r}(u, v) = (u, v, 6 - u - 2v), \qquad (u, v) \in D,$$

where D is the triangular region bounded by the coordinate lines and the line $u + 2v = 6$.

9. Parametrize the paraboloid as the graph of the surface $z = x^2 + y^2$; i.e., $\mathbf{r}(u,v) = (u, v, u^2 + v^2)$. To make sure that $\mathbf{r}(u,v)$ belongs to the first octant, we must take $u \geq 0$ and $v \geq 0$. Therefore

$$\mathbf{r}(u,v) = (u, v, u^2 + v^2), \qquad u \geq 0 \quad v \geq 0$$

is a required parametrization.

11. (a) From $\mathbf{r}(u,v) = (2u, u^2 + v, v^2)$ it follows that $\mathbf{T}_u = (2, 2u, 0)$, $\mathbf{T}_v = (0, 1, 2v)$ and

$$\mathbf{N} = \mathbf{T}_u \times \mathbf{T}_v = \begin{vmatrix} \mathbf{i} & \mathbf{j} & \mathbf{k} \\ 2 & 2u & 0 \\ 0 & 1 & 2v \end{vmatrix} = (4uv, -4v, 2).$$

(b) Since the z-component of $\mathbf{N}$ is 2, $\mathbf{N} \neq \mathbf{0}$, no matter what u and v are; so S is smooth for all $u, v \geq 0$.

13. (a) From $\mathbf{r}(u,v) = (\sin u \cos v, \sin u \sin v, 2\cos u)$ it follows that

$$\mathbf{T}_u = (\cos u \cos v, \cos u \sin v, -2\sin u),$$
$$\mathbf{T}_v = (-\sin u \sin v, \sin u \cos v, 0)$$

and

$$\begin{aligned} \mathbf{N} = \mathbf{T}_u \times \mathbf{T}_v &= \begin{vmatrix} \mathbf{i} & \mathbf{j} & \mathbf{k} \\ \cos u \cos v & \cos u \sin v & -2\sin u \\ -\sin u \sin v & \sin u \cos v & 0 \end{vmatrix} \\ &= (2\sin^2 u \cos v, 2\sin^2 u \sin v, \sin u \cos u \cos^2 v + \sin u \cos u \sin^2 v) \\ &= \sin u(2\sin u \cos v, 2\sin u \sin v, \cos u). \end{aligned}$$

(b) To find the points where S is not smooth, we compute the magnitude of the normal $\mathbf{N}$:

$$\begin{aligned} \|\mathbf{N}\| &= |\sin u|\sqrt{4\sin^2 u \cos^2 v + 4\sin^2 u \sin^2 v + \cos^2 u} \\ &= |\sin u|\sqrt{4\sin^2 u + \cos^2 u}. \end{aligned}$$

Now $\|\mathbf{N}\| = 0$ implies that $|\sin u| = 0$; i.e., $u = 0, \pi, 2\pi$. Notice that $4\sin^2 u + \cos^2 u > 0$ (since there is no u such that $\sin u = \cos u = 0$). It follows that S is not smooth at $\mathbf{r}(0,v) = (0,0,2)$, $\mathbf{r}(\pi, v) = (0,0,-2)$, and $\mathbf{r}(2\pi, v) = \mathbf{r}(0,v) = (0,0,2)$.

15. (a) From $\mathbf{r}(u,v) = ((1+\cos v)\cos u, (1+\cos v)\sin u, \sin v)$ we get

$$\mathbf{T}_u = (-(1+\cos v)\sin u, (1+\cos v)\cos u, 0),$$
$$\mathbf{T}_v = (-\sin v \cos u, -\sin v \sin u, \cos v)$$

and

$$\mathbf{N} = \mathbf{T}_u \times \mathbf{T}_v = \begin{vmatrix} \mathbf{i} & \mathbf{j} & \mathbf{k} \\ -(1+\cos v)\sin u & (1+\cos v)\cos u & 0 \\ -\sin v \cos u & -\sin v \sin u & \cos v \end{vmatrix}$$

$$= \big((1+\cos v)\cos v \cos u, (1+\cos v)\sin u \cos v,$$
$$(1+\cos v)\sin^2 u \sin v + (1+\cos v)\sin v \cos^2 u\big)$$
$$= (1+\cos v)(\cos u \cos v, \sin u \cos v, \sin v).$$

(b) To find the points where S is not smooth, we compute the magnitude of the surface normal $\mathbf{N}$:

$$\|\mathbf{N}\| = |1+\cos v|\sqrt{\cos^2 u \cos^2 v + \sin^2 u \cos^2 v + \sin^2 v}$$
$$= |1+\cos v|\sqrt{\cos^2 v + \sin^2 v} = |1+\cos v|.$$

Since $|1+\cos v| = 0$ only when $v = \pi$, it follows that S is smooth at all points except at $\mathbf{r}(u,\pi) = (0,0,0)$.

17. (a) From $\mathbf{r}(u,v) = (u, v, 1-(u^2+v^2))$ we get $\mathbf{T}_u = (1,0,-2u)$, $\mathbf{T}_v = (0,1,-2v)$ and

$$\mathbf{N} = \mathbf{T}_u \times \mathbf{T}_v = \begin{vmatrix} \mathbf{i} & \mathbf{j} & \mathbf{k} \\ 1 & 0 & -2u \\ 0 & 1 & -2v \end{vmatrix} = (2u, 2v, 1).$$

(b) Since the z-coordinate of $\mathbf{N}$ is 1, it follows that $\mathbf{N}(u,v) \neq 0$ for all (u,v); so S is smooth for all $0 \le u, v < \infty$.

19. A normal to the surface is given by

$$\mathbf{N} = \mathbf{T}_u \times \mathbf{T}_v = \begin{vmatrix} \mathbf{i} & \mathbf{j} & \mathbf{k} \\ e^u & 0 & v \\ 0 & e^v & u \end{vmatrix} = (-ve^v, -ue^u, e^u e^v).$$

The point $(1,1,0)$ is obtained for $u = 0$ and $v = 0$; so $\mathbf{N}(0,0) = (0,0,1)$ is a normal to the surface at $\mathbf{r}(0,0) = (1,1,0)$. Consequently, the tangent plane has the equation

$$0(x-1) + 0(y-1) + 1(z-0) = 0;$$

i.e., $z = 0$.

21. Parametrize the given surface as $\mathbf{r}(u,v) = (u, u^2+2uv, v)$. In that case, the point of tangency is $\mathbf{r}(1,1) = (1,3,1)$. A normal to the surface is given by

$$\mathbf{N}(u,v) = \mathbf{T}_u \times \mathbf{T}_v = \begin{vmatrix} \mathbf{i} & \mathbf{j} & \mathbf{k} \\ 1 & 2u+2v & 0 \\ 0 & 2u & 1 \end{vmatrix} = (2u+2v, -1, 2u),$$

and hence $\mathbf{N}(1,1) = (4,-1,2)$. The equation of the tangent plane is

$$4(x-1) - 1(y-3) + 2(z-1) = 0;$$

i.e., $4x - y + 2z - 3 = 0$.

23. Parametrize the cone as in Example 7.3 as

$$\mathbf{r}(u,v) = (v\cos u, v\sin u, v), \qquad 0 \le u \le 2\pi;$$

but this time, $v \in \mathbb{R}$. The normal vector $\mathbf{N}$ is

$$\mathbf{N} = \mathbf{T}_u \times \mathbf{T}_v = \begin{vmatrix} \mathbf{i} & \mathbf{j} & \mathbf{k} \\ -v\sin u & v\cos u & 0 \\ \cos u & \sin u & 1 \end{vmatrix} = (v\cos u, v\sin u, -v).$$

Since

$$\|\mathbf{N}\| = \sqrt{v^2\cos^2 u + v^2\sin^2 u + v^2} = |v|\sqrt{2},$$

it follows that the cone is not smooth only when $v = 0$; i.e., at the point $\mathbf{r}(u,0) = (0,0,0)$. At any other point $\mathbf{r}(u_0, v_0) = (v_0\cos u_0, v_0\sin u_0, v_0)$, $0 \le u_0 \le 2\pi$, $v \neq 0$, the tangent plane has the equation

$$v_0\cos u_0(x - v_0\cos u_0) + v_0\sin u_0(y - v_0\sin u_0) - v_0(z - v_0) = 0;$$

i.e.,

$$(v_0\cos u_0)x + (v_0\sin u_0)y - v_0 z = 0.$$

Clearly, this plane goes through the origin.

25. An ellipsoid can be obtained by stretching or compressing a sphere in the directions of the coordinate axes. So we take a parametrization of a sphere (say, from Example 7.1) and adjust it (i.e., change the "radii" in the coordinate directions); thus

$$\mathbf{r}(u,v) = (a\cos v\cos u, b\cos v\sin u, c\sin v), \qquad 0 \le u \le 2\pi, \quad -\pi/2 \le v \le \pi/2.$$

It follows that

$$\frac{x^2}{a^2} + \frac{y^2}{b^2} + \frac{z^2}{c^2} = \cos^2 v\cos^2 u + \cos^2 v\sin^2 u + \sin^2 v = 1.$$

Here is another parametrization (and there are many possibilities):

$$\mathbf{r}(u,v) = (a\cos u\sin v, b\sin u\sin v, c\cos u), \qquad 0 \le u \le 2\pi, \quad 0 \le v \le \pi.$$

27. We have to find all points on the given surface where the surface normal $\mathbf{N}$ is parallel to $(-1,1,1)$ (which is a normal vector to the plane $-x + y + z = 0$). Now

$$\mathbf{N} = \mathbf{T}_u \times \mathbf{T}_v = \begin{vmatrix} \mathbf{i} & \mathbf{j} & \mathbf{k} \\ 2uv & v^2 & 0 \\ u^2 & 2uv & 0 \end{vmatrix} = (0, 0, 3u^2v^2).$$

It follows that there are no points where $(0,0,3u^2v^2)$ is parallel to $(-1,1,1)$.

7.2 Surface Integrals of Real-Valued Functions

1. Parametrize S by

$$\mathbf{r}(u,v) = (u, v, u^2 + v^2), \qquad (u,v) \in D,$$

where D is the region in the uv-plane defined by $u^2 + v^2 \le 4$. Then $\mathbf{T}_u = (1, 0, 2u)$, $\mathbf{T}_v = (0, 1, 2v)$ and

$$\mathbf{N} = \mathbf{T}_u \times \mathbf{T}_v = \begin{vmatrix} \mathbf{i} & \mathbf{j} & \mathbf{k} \\ 1 & 0 & 2u \\ 0 & 1 & 2v \end{vmatrix} = (-2u, -2v, 1),$$

and so $\|\mathbf{N}\| = \sqrt{1 + 4(u^2 + v^2)}$. The values of f at points on the surface are $f(\mathbf{r}(u,v)) = f(u, v, u^2 + v^2) = uv$; it follows that

$$\iint_S xy\, dS = \iint_D uv\sqrt{1 + 4(u^2 + v^2)}\, dA.$$

Passing to polar coordinates, we get

$$\begin{aligned} \iint_S xy\, dS &= \int_0^{2\pi} \left(\int_0^2 r^2 \sin\theta \cos\theta \sqrt{1 + 4r^2}\, r dr \right) d\theta \\ &= \left(\int_0^{2\pi} \sin\theta \cos\theta\, d\theta \right) \left(\int_0^2 r^3 \sqrt{1 + 4r^2}\, dr \right). \end{aligned}$$

Since

$$\int_0^{2\pi} \sin\theta \cos\theta\, d\theta = \tfrac{1}{2} \int_0^{2\pi} \sin 2\theta\, d\theta = 0,$$

it follows that $\iint_S xy\, dS = 0$.

3. S is a piecewise smooth surface: it consists of four smooth surfaces, that are the faces of the tetrahedron. We will integrate f along each smooth surface and then add the results.

Let S_1 be the face containing the points $(0,0,0)$, $(2,0,0)$ and $(0,2,0)$. It is a part of the xy-plane, so parametrize it by $\mathbf{r}_1(u,v) = (u, v, 0)$, $(u,v) \in D_1$, where D_1 is the triangular region in the uv-plane defined by $u = 0$, $v = 0$ and $u + v = 2$ (that is the line joining $(2,0)$ and $(0,2)$). The normal is

$$\mathbf{N} = \begin{vmatrix} \mathbf{i} & \mathbf{j} & \mathbf{k} \\ 1 & 0 & 0 \\ 0 & 1 & 0 \end{vmatrix} = \mathbf{k}$$

(as expected) and so

$$\begin{aligned} \iint_{S_1} (y + x)\, dS &= \iint_{D_1} (u + v)\, 1\, dA = \int_0^2 \left(\int_0^{2-v} (u + v) du \right) dv \\ &= \int_0^2 \left(\tfrac{1}{2}u^2 + uv \right) \Big|_0^{2-v} dv = \int_0^2 \left(2 - \tfrac{1}{2}v^2 \right) dv = \left(2v - \tfrac{1}{6}v^3 \right) \Big|_0^2 = \tfrac{8}{3}. \end{aligned}$$

Let S_2 be the face containing the points $(0,0,0)$, $(0,2,0)$ and $(0,0,2)$. It is a part of the yz-plane, so parametrize it by $\mathbf{r}_2(u,v) = (0, u, v)$, $(u,v) \in D_2$; D_2 is the triangular region in the uv-plane (that, in this case, coincides with the yz-plane) defined by $u = 0$, $v = 0$ and $u + v = 2$. The normal is

$$\mathbf{N} = \begin{vmatrix} \mathbf{i} & \mathbf{j} & \mathbf{k} \\ 0 & 1 & 0 \\ 0 & 0 & 1 \end{vmatrix} = \mathbf{i}$$

and so

$$\iint_{S_2} (y + x)\, dS = \iint_{D_2} u dA = \int_0^2 \left(\int_0^{2-v} u du \right) dv$$
$$= \int_0^2 \left(\tfrac{1}{2}u^2\right)\Big|_0^{2-v} dv = \tfrac{1}{2} \int_0^2 (-2 + v)^2\, dv = \tfrac{1}{2}\tfrac{1}{3}(v - 2)^3 \Big|_0^2 = \tfrac{4}{3}.$$

Similarly, let S_3 be the face containing the points $(0, 0, 0)$, $(2, 0, 0)$ and $(0, 0, 2)$. It is a part of the xz-plane, so parametrize it by $\mathbf{r}_3(u, v) = (u, 0, v)$, $(u, v) \in D_3$; D_3 is the triangular region defined by $u = 0$, $v = 0$ and $u + v = 2$. The normal is

$$\mathbf{N} = \begin{vmatrix} \mathbf{i} & \mathbf{j} & \mathbf{k} \\ 1 & 0 & 0 \\ 0 & 0 & 1 \end{vmatrix} = -\mathbf{j}$$

and so

$$\iint_{S_3} (y + x)\, dS = \iint_{D_3} u dA = \int_0^2 \left(\int_0^{2-v} u du \right) dv = \tfrac{4}{3};$$

the result is borrowed from the integration over S_2.

Finally, parametrize the face S_4 containing the points $(2, 0, 0)$, $(0, 2, 0)$ and $(0, 0, 2)$ (the equation of the plane containing these three points is $x+y+z = 2$) as $\mathbf{r}_4(u, v) = (u, v, 2-u-v)$, $(u, v) \in D_4 = D_1$. The surface normal $\mathbf{N}$ is

$$\mathbf{N} = \begin{vmatrix} \mathbf{i} & \mathbf{j} & \mathbf{k} \\ 1 & 0 & -1 \\ 0 & 1 & -1 \end{vmatrix} = \mathbf{i} + \mathbf{j} + \mathbf{k};$$

so $\|\mathbf{N}\| = \sqrt{3}$, and thus

$$\iint_{S_4} (y + x)\, dS = \iint_{D_4} (u + v)\sqrt{3} dA = \tfrac{8}{3}\sqrt{3},$$

(borrowing the result of $\iint_{S_1} (y + x)\, dS$).

Hence

$$\iint_S (y + x)\, dS = \tfrac{8}{3} + \tfrac{4}{3} + \tfrac{4}{3} + \tfrac{8}{3}\sqrt{3} = \tfrac{16}{3} + \tfrac{8}{3}\sqrt{3}.$$

5. Imitating the parametrization used in Example 7.3 in Section 7.1, we get $\mathbf{r}(u, v) = (v, v \cos u, v \sin u)$. The requirement that $\mathbf{r}(u, v)$ be in the first octant implies $v \geq 0$ and $0 \leq u \leq \pi/2$. It is also given that $v \leq 1$. Therefore,

$$\mathbf{r}(u, v) = (v, v \cos u, v \sin u), \qquad 0 \leq u \leq \pi/2, \quad 0 \leq v \leq 1$$

is a desired parametrization. The normal vector $\mathbf{N}$ is

$$\mathbf{N} = \begin{vmatrix} \mathbf{i} & \mathbf{j} & \mathbf{k} \\ 0 & -v\sin u & v\cos u \\ 1 & \cos u & \sin u \end{vmatrix} = (-v, v\cos u, v\sin u),$$

and its magnitude is

$$\|\mathbf{N}\| = \sqrt{v^2 + v^2\cos^2 u + v^2\sin^2 u} = |v|\sqrt{2} = v\sqrt{2}.$$

The value of f at points on S is $f(\mathbf{r}(u,v)) = 2v\cos u - v$, and therefore

$$\iint_S (2y - x)\, dS = \iint_D (2v\cos u - v)v\sqrt{2}\, dA,$$

where D is defined by $0 \le u \le \pi/2$, $0 \le v \le 1$. Hence

$$\begin{aligned}\iint_S (2y - x)\, dS &= \sqrt{2}\int_0^{\pi/2}\left(\int_0^1 v^2(2\cos u - 1)dv\right)du \\ &= \sqrt{2}\int_0^{\pi/2} \tfrac{1}{3}(2\cos u - 1)\, du \\ &= \frac{\sqrt{2}}{3}(2\sin u - u)\Big|_0^{\pi/2} = \frac{\sqrt{2}}{3}\left(2 - \tfrac{1}{2}\pi\right) = \tfrac{1}{6}(4 - \pi)\sqrt{2}.\end{aligned}$$

7. The intersection of the paraboloid $z = 4 - x^2 - y^2$ and the xy-plane is the circle $x^2 + y^2 = 4$. Denote the disk $\{(x,y) \mid x^2 + y^2 \le 4\}$ by D, and parametrize S as

$$\mathbf{r}(u,v) = (u, v, 4 - u^2 - v^2), \qquad (u,v) \in D.$$

The values of f on S are $f(\mathbf{r}(u,v)) = (4u^2 + 4v^2 + 1)^{-1/2}$; the normal vector $\mathbf{N}$ is

$$\mathbf{N} = \begin{vmatrix} \mathbf{i} & \mathbf{j} & \mathbf{k} \\ 1 & 0 & -2u \\ 0 & 1 & -2v \end{vmatrix} = (2u, 2v, 1),$$

so $\|\mathbf{N}\| = \sqrt{4u^2 + 4v^2 + 1}$ and therefore

$$\begin{aligned}\iint_S (4x^2 + 4y^2 + 1)^{-1/2}\, dS &= \iint_D (4u^2 + 4v^2 + 1)^{-1/2}(4u^2 + 4v^2 + 1)^{1/2}\, dA \\ &= \iint_D 1\, dA = \text{area}(D) = 4\pi.\end{aligned}$$

9. From $\mathbf{r}(u,v) = (2u\cos v, 2u\sin v, v)$ it follows that $\mathbf{T}_u = (2\cos v, 2\sin v, 0)$ and $\mathbf{T}_v = (-2u\sin v, 2u\cos v, 1)$; so the surface normal $\mathbf{N}$ is

$$\mathbf{N} = \begin{vmatrix} \mathbf{i} & \mathbf{j} & \mathbf{k} \\ 2\cos v & 2\sin v & 0 \\ -2u\sin v & 2u\cos v & 1 \end{vmatrix} = (2\sin v, -2\cos v, 4u).$$

Hence $\|\mathbf{N}\| = \sqrt{4 + 16u^2} = 2\sqrt{1 + 4u^2}$ and the surface area is

$$\iint_S dS = \iint_D 2\sqrt{1 + 4u^2}\, dA,$$

where D is the region in the uv-plane defined by $0 \le u \le 2$ and $0 \le v \le \pi$. So

$$\iint_S dS = 2\int_0^{\pi}\left(\int_0^2 \sqrt{1+4u^2}du\right)dv = 2\left(\int_0^{\pi} dv\right)\left(\int_0^2 \sqrt{1+4u^2}\,du\right).$$

Using integration tables (or a trigonometric substitution) we get

$$\begin{aligned}\int_0^2 \sqrt{1+4u^2}\,du &= 2\int_0^2 \sqrt{\tfrac{1}{4}+u^2}\,du\\ &= 2\left(\tfrac{1}{2}u\sqrt{\tfrac{1}{4}+u^2}+\tfrac{1}{8}\ln\left|u+\sqrt{\tfrac{1}{4}+u^2}\right|\right)\Big|_0^2\\ &= 2\left(\tfrac{1}{2}\sqrt{17}+\tfrac{1}{8}\ln\left|2+\tfrac{1}{2}\sqrt{17}-\tfrac{1}{8}\ln\left(\tfrac{1}{2}\right)\right|\right).\end{aligned}$$

It follows that the surface area of the surface in question is

$$4\pi\left(\tfrac{1}{2}\sqrt{17}+\tfrac{1}{8}\ln\left|2+\tfrac{1}{2}\sqrt{17}+\tfrac{1}{8}\ln 2\right|\right) \approx 29.1966.$$

11. Recall the parametrization of the cone from Example 7.3 in Section 7.1:

$$\mathbf{r}(u,v) = (v\cos u, v\sin u, v), \qquad 0 \le u \le 2\pi, \quad 0 \le v \le h.$$

This is not what we are looking for: when $v = h$, $\mathbf{r}(u,h) = (h\cos u, h\sin u, h)$; i.e., we get a circle of radius h in the plane $z = h$. What we need is a circle of radius r — so, we adjust the above parametrization:

$$\mathbf{r}(u,v) = \left(\frac{rv}{h}\cos u, \frac{rv}{h}\sin u, v\right), \qquad 0 \le u \le 2\pi, \quad 0 \le v \le h.$$

This time, $\mathbf{r}(u,h) = (r\cos u, r\sin u, h)$, exactly as needed.

Now $\mathbf{T}_u = (-(rv/h)\sin u, (rv/h)\cos u, 0)$ and $\mathbf{T}_v = ((r/h)\cos u, (r/h)\sin u, 1)$; so the surface normal $\mathbf{N}$ is

$$\mathbf{N} = \begin{vmatrix} \mathbf{i} & \mathbf{j} & \mathbf{k} \\ -(rv/h)\sin u & (rv/h)\cos u & 0 \\ (r/h)\cos u & (r/h)\sin u & 1 \end{vmatrix} = \left(\frac{rv}{h}\cos u, \frac{rv}{h}\sin u, -\frac{r^2v}{h^2}\right).$$

Hence

$$\|\mathbf{N}\| = \sqrt{\frac{r^2v^2}{h^2}+\frac{r^4v^2}{h^4}} = |v|\sqrt{\frac{r^2}{h^2}\left(1+\frac{r^2}{h^2}\right)} = \frac{vr}{h}\sqrt{1+\frac{r^2}{h^2}}$$

and the surface area of the cone is

$$A(S) = \iint_S dS = \iint_D \|\mathbf{N}\|\,dA,$$

where D is the region in the uv-plane defined by $0 \le u \le 2\pi$ and $0 \le v \le h$. Hence

$$\begin{aligned}A(S) &= \int_0^{2\pi}\left(\int_0^h \frac{vr}{h}\sqrt{1+\frac{r^2}{h^2}}\,dv\right)du\\ &= \left(\int_0^{2\pi} du\right)\frac{r}{h}\sqrt{1+\frac{r^2}{h^2}}\left(\frac{v^2}{2}\Big|_0^h\right)\\ &= \frac{2\pi r}{h}\sqrt{1+\frac{r^2}{h^2}}\,\frac{h^2}{2} = \pi r h\sqrt{1+\frac{r^2}{h^2}} = \pi r\sqrt{h^2+r^2}.\end{aligned}$$

13. Recall that a surface integral $\iint_S f\,dS$ is the limit as $n \to \infty$ of Riemann sums of the form

$$\mathcal{R}_n = \sum (\text{value of } f) \cdot (\text{area of a small patch of } S),$$

where the summation goes over all small patches that cover S. Take a subdivision where all small patches have the same area. Then

$$\mathcal{R}_n = (\text{area of a small patch}) \sum (\text{value of } f).$$

At a point (x, y, z), the value of the function $f(x, y, z) = x$ is x. At a symmetric point $(-x, y, z)$ on S, the value is $-x$, and the two contributions to $\mathcal{R}_n$ will cancel each other. That will happen to every such pair, and consequently, $\mathcal{R}_n = 0$ for all n and so $\iint_S x\,dS = 0$. An analogous argument proves that $\iint_S x^3\,dS = 0$.

By symmetry,

$$\iint_S x^2\,dS = \iint_S y^2\,dS = \iint_S z^2\,dS,$$

and so

$$\begin{aligned} 3\iint_S x^2\,dS &= \iint_S x^2\,dS + \iint_S y^2\,dS + \iint_S z^2\,dS = \iint_S (x^2 + y^2 + z^2)\,dS \\ &= \iint_S a^2\,dS = 4\pi a^2\,a^2 = 4\pi a^4, \end{aligned}$$

since $\iint_S dS = \text{area}(S) = 4\pi a^2$. It follows that $\iint_S x^2\,dS = 4\pi a^4/3$.

15. See Figure 7.2.15. One side of S is of length a, and the other one is the hypotenuse of the right triangle with sides b and bm (the slope of $z = my$ is m). So the area of S is

$$\text{area}(S) = a\sqrt{b^2 + b^2m^2} = ab\sqrt{1 + m^2} = \text{area}(R)\sqrt{1 + m^2}.$$

Notice that the angle between $\mathbf{k}$ and the upward normal $\mathbf{N}$ is equal to α. Since

$$\cos\alpha = \frac{b}{\sqrt{b^2 + b^2m^2}} = \frac{1}{\sqrt{1 + m^2}},$$

it follows that $\sec\alpha = \sqrt{1 + m^2}$ and (area of S) $= \sec\alpha$ (area of R).

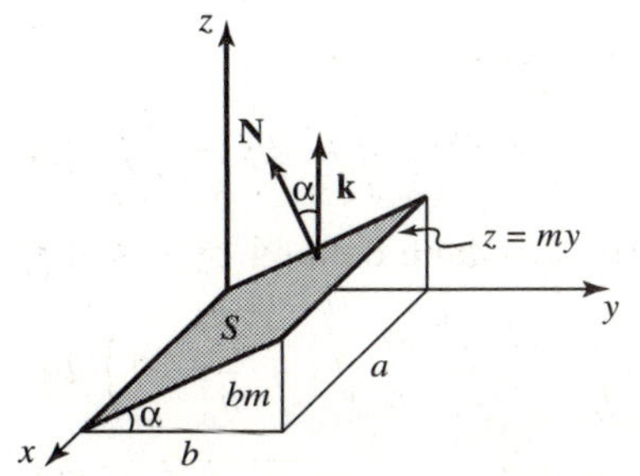

Figure 7.2.15

17. Parametrize S by $\mathbf{r}(u,v) = (u, v, 1-v^2)$, $(u,v) \in D$, where D is defined by $0 \le u \le 2$ and $0 \le v \le 1$ (obtained from $z = 1 - y^2 = 0$ and the fact that S is in the first octant). The values of f on S are $f(\mathbf{r}(u,v)) = 4uv$; the normal is

$$\mathbf{N} = \begin{vmatrix} \mathbf{i} & \mathbf{j} & \mathbf{k} \\ 1 & 0 & 0 \\ 0 & 1 & -2v \end{vmatrix} = 2v\mathbf{j} + \mathbf{k};$$

hence $\|\mathbf{N}\| = \sqrt{1+4v^2}$ and therefore

$$\begin{aligned}\iint_S 4xy\,dS &= \iint_D 4uv\sqrt{1+4v^2}\,dA \\ &= \int_0^2 \left(\int_0^1 4uv\sqrt{1+4v^2}\,dv\right) du \\ &= \left(\int_0^2 u\,du\right)\left(2\tfrac{1}{3}(1+4v^2)^{3/2}\right)\Big|_0^1 \\ &= \tfrac{4}{3}(5^{3/2}-1).\end{aligned}$$

Let us recompute this integral using a projection onto the xy-plane; that projection is the rectangle (call it D) $0 \le x, y \le 1$. Since the unit normal is $\mathbf{n} = (2y\mathbf{j}+\mathbf{k})/\sqrt{(1+4y^2)}$, it follows that

$$\iint_S f\,dS = \iint_D 4xy\,\frac{dA}{|\mathbf{n}\cdot\mathbf{k}|} = \iint_D 4xy\sqrt{1+4y^2}dA.$$

We proceed as above.

19. Let us project onto the xy-plane. In that case (take $z = 0$) $x + y^2 = 4$, and the projection D is defined by $x = 0$, $y = 0$ and $x + y^2 = 4$. A normal vector to S is (view it as the level surface of $g(x,y,z) = x + y^2 + z$ of value 4) $\nabla g = (1, 2y, 1)$, and a unit normal is $\mathbf{n} = \nabla g/\|\nabla g\| = (1, 2y, 1)/\sqrt{2+4y^2}$. Thus

$$\begin{aligned}\iint_S y\,dS &= \iint_D y\,\frac{dA}{|\mathbf{n}\cdot\mathbf{k}|} \\ &= \iint_D y\sqrt{2+4y^2}\,dA \\ &= \int_0^4 \left(\int_0^{\sqrt{4-x}} y\sqrt{2+4y^2}\,dy\right) dx \\ &= \int_0^4 \left(\tfrac{1}{12}(2+4y^2)^{3/2}\right)\Big|_0^{\sqrt{4-x}} dx \\ &= \tfrac{1}{12}\int_0^4 \left((18-4x)^{3/2} - 2^{3/2}\right) dx \\ &= \tfrac{1}{12}\left(-\tfrac{2}{20}(18-4x)^{5/2} - 2^{3/2}x\right)\Big|_0^4 \\ &= \tfrac{1}{12}\left(-\tfrac{1}{10}2^{5/2} - 2^{3/2}\,4 + \tfrac{1}{10}18^{5/2}\right) \approx 10.4652.\end{aligned}$$

21. Let $g(x,y,z) = x^2 + y^2 + z^2$; then S is the level surface of g of value a^2, and so

$$\mathbf{n} = \frac{\nabla g}{\|\nabla g\|} = \frac{(2x, 2y, 2z)}{\sqrt{4x^2 + 4y^2 + 4z^2}} = \frac{(x,y,z)}{a}$$

is its unit normal. Project S onto the xz-plane; the projection D is the disk $x^2 + z^2 \leq a^2$. It follows that the surface area is

$$\begin{aligned} A(S) &= \iint_D \frac{dA}{|\mathbf{n}\cdot\mathbf{j}|} \\ &= \iint_D \frac{dA}{y/a} = a \iint_D (a^2 - x^2 - z^2)^{-1/2}. \end{aligned}$$

Passing to polar coordinates $x = r\cos\theta$, $z = r\sin\theta$, we get

$$\begin{aligned} A(S) &= a \int_0^{2\pi} \left(\int_0^a (a^2 - r^2)^{-1/2}\, r dr \right) d\theta \\ &= a \left(\int_0^{2\pi} d\theta \right) \left(-(a^2 - r^2)^{1/2} \Big|_0^a \right) \\ &= 2\pi a (0 - (-(a^2)^{1/2})) = 2\pi a^2. \end{aligned}$$

7.3 Surface Integrals of Vector Fields

1. Parametrize S by $\mathbf{r}(u,v) = (u, v, 3u + v - 1)$, where $u^2 + v^2 \leq 4$ (so that only the part of the plane inside the cylinder is considered). Then $\mathbf{T}_u = (1,0,3)$, $\mathbf{T}_v = (0,1,1)$ and

$$\mathbf{N} = \mathbf{T}_u \times \mathbf{T}_v = \begin{vmatrix} \mathbf{i} & \mathbf{j} & \mathbf{k} \\ 1 & 0 & 3 \\ 0 & 1 & 1 \end{vmatrix} = -3\mathbf{i} - \mathbf{j} + \mathbf{k}$$

(since the $\mathbf{k}$ component of $\mathbf{N}$ is positive, $\mathbf{N}$ is an upward normal). It follows that

$$\begin{aligned} \iint_S (4y\mathbf{i} + (3x-1)\mathbf{j} + z\mathbf{k}) \cdot d\mathbf{S} &= \iint_D (4y\mathbf{i} + (3x-1)\mathbf{j} + z\mathbf{k}) \cdot (-3\mathbf{i} - \mathbf{j} + \mathbf{k})\, dA \\ &= \iint_D (-12y - 3x + 1 + z)\, dA \end{aligned}$$

where D is the region in the uv-plane defined by $u^2 + v^2 \leq 4$. Passing to polar coordinates $u = r\cos\theta$, $v = r\sin\theta$, we get

$$\begin{aligned} \iint_S (4y\mathbf{i} + (3x-1)\mathbf{j} + z\mathbf{k}) \cdot d\mathbf{S} &= \iint_D (-12v - 3u + 1 + 3u + v - 1)\, dA \\ &= \iint_D (-11v)\, dA \\ &= -11 \int_0^{2\pi} \left(\int_0^2 r^2 \sin\theta dr \right) d\theta \\ &= -11 \left(\int_0^{2\pi} \sin\theta\, d\theta \right) \left(\int_0^2 \tfrac{1}{3} r^3\, dr \right) = 0, \end{aligned}$$

since $\int_0^{2\pi} \sin\theta\, d\theta = 0$.

3. Parametrize S by

$$\mathbf{r}(u,v) = 3\cos v\cos u\,\mathbf{i} + 3\cos v\sin u\,\mathbf{j} + 3\sin v\,\mathbf{k}, \qquad 0 \le u \le 2\pi, \quad 0 \le v \le \pi/2$$

(if $0 \le v \le \pi/2$, then $3\sin v \ge 0$, as needed). Let D be the rectangle $D = [0, 2\pi] \times [0, \pi/2]$ in the uv-plane. The parametrization we use has been introduced in Example 7.1; the surface normal $\mathbf{N}$ was computed in Example 7.4 to be $\mathbf{N} = 3\cos v\,\mathbf{r}(u,v)$ (and this is an outward normal, since it points away from the origin). Therefore,

$$\begin{aligned}
&\iint_S (x^2\mathbf{i} + 2z\mathbf{k}) \cdot d\mathbf{S} \\
&= \iint_D (9\cos^2 v\cos^2 u\,\mathbf{i} + 6\sin v\,\mathbf{k}) \cdot 3\cos v\,(3\cos v\cos u\,\mathbf{i} + 3\cos v\sin u\,\mathbf{j} + 3\sin v\,\mathbf{k})\, dA \\
&= \iint_D (81\cos^4 v\cos^3 u + 54\sin^2 v\cos v)\, dA \\
&= 81\int_0^{2\pi}\left(\int_0^{\pi/2} \cos^4 v\cos^3 u\, dv\right) du + 54\int_0^{2\pi}\left(\int_0^{\pi/2} \sin^2 v\cos v\, dv\right) du \\
&= 81\left(\int_0^{2\pi} \cos^3 u\, du\right)\left(\int_0^{\pi/2} \cos^4 v\, dv\right) + 54\left(\int_0^{2\pi} du\right)\left(\tfrac{1}{3}\sin^3 v\Big|_0^{\pi/2}\right).
\end{aligned}$$

Since $\int_0^{2\pi} \cos^3 u\, du = 0$ (no computation is needed — sketch the graph of $\cos^3 u$ and think of a definite integral as a "signed area"; i.e., "area above the x-axis" – "area below the x-axis"), it follows that

$$\iint_S (x^2\mathbf{i} + 2z\mathbf{k}) \cdot d\mathbf{S} = 54(2\pi)\tfrac{1}{3} = 36\pi.$$

5. Combining $z = x^2 + y^2$ with $z = 1$ and with $z = 2$, we get $1 \le x^2 + y^2 \le 2$. So we can parametrize S by $\mathbf{r}(u,v) = (u, v, u^2 + v^2)$, where $(u,v) \in D$, and D is the annulus $1 \le u^2 + v^2 \le 2$ in the uv-plane. The surface normal $\mathbf{N}$ is computed to be

$$\mathbf{N} = \mathbf{T}_u \times \mathbf{T}_v = \begin{vmatrix} \mathbf{i} & \mathbf{j} & \mathbf{k} \\ 1 & 0 & 2u \\ 0 & 1 & 2v \end{vmatrix} = -2u\mathbf{i} - 2v\mathbf{j} + \mathbf{k}.$$

Since the $\mathbf{k}$ component of $\mathbf{N}$ is positive, $\mathbf{N}$ points inward, into the paraboloid. We can do one of two things: use $\mathbf{N}$ to compute the surface integral and then change the sign of the result, or change the sign of the normal $\mathbf{N}$ right away. We will change it right away:

$$\begin{aligned}
\iint_S z\mathbf{k} \cdot d\mathbf{S} &= \iint_D z\mathbf{k} \cdot (-\mathbf{N})\, dA \\
&= \iint_D (u^2 + v^2)\mathbf{k} \cdot (2u\mathbf{i} + 2v\mathbf{j} - \mathbf{k})\, dA \\
&= -\iint_D (u^2 + v^2)\, dA
\end{aligned}$$

(passing to polar coordinates $u = r\cos\theta$, $v = r\sin\theta$)

$$= -\int_0^{2\pi}\left(\int_1^{\sqrt{2}} r^3 dr\right) d\theta$$
$$= -\left(\int_0^{2\pi} d\theta\right)\left(\tfrac{1}{4}r^4\Big|_1^{\sqrt{2}}\right) = -2\pi\left(1-\tfrac{1}{4}\right) = -\tfrac{3}{2}\pi.$$

7. From $\mathbf{T}_u = (e^u\cos v, e^u\sin v, 0)$ and $\mathbf{T}_v = (-e^u\sin v, e^u\cos v, 1)$ it follows that

$$\mathbf{N} = \begin{vmatrix} \mathbf{i} & \mathbf{j} & \mathbf{k} \\ e^u\cos v & e^u\sin v & 0 \\ -e^u\sin v & e^u\cos v & 1 \end{vmatrix} = (e^u\sin v, -e^u\cos v, e^{2u})$$

(this is an upward pointing normal, as needed). The value of $\mathbf{F}$ along S is $\mathbf{F}(\mathbf{r}(u,v)) = (e^u\cos v, e^u\sin v, v)$; so

$$\mathbf{F}\cdot\mathbf{N} = (e^u\cos v, e^u\sin v, v)\cdot(e^u\sin v, -e^u\cos v, e^{2u}) = e^{2u}v,$$

and (let D be the rectangle $[0,\ln 2]\times[0,\pi]$ in the uv-plane)

$$\iint_S (x\mathbf{i}+y\mathbf{j}+z\mathbf{k})\cdot d\mathbf{S} = \iint_D e^{2u}v\,dA$$
$$= \int_0^{\ln 2}\left(\int_0^{\pi} e^{2u}v dv\right) du$$
$$= \left(\int_0^{\ln 2} e^{2u}\,du\right)\left(\int_0^{\pi} v\,dv\right)$$
$$= \left(\tfrac{1}{2}e^{2u}\Big|_0^{\ln 2}\right)\left(\tfrac{1}{2}v^2\Big|_0^{\pi}\right) = \tfrac{3}{4}\pi^2.$$

9. From $\mathbf{F} = -\nabla T$ we get $\mathbf{F}(x,y,z) = -(2x, 2y, 6z)$. Parametrize the cylinder by $\mathbf{r}(u,v) = (\cos u, \sin u, v)$, where $(u,v)\in D = [0,2\pi]\times[-1,1]$. The surface normal is

$$\mathbf{N} = \mathbf{T}_u\times\mathbf{T}_v = \begin{vmatrix} \mathbf{i} & \mathbf{j} & \mathbf{k} \\ -\sin u & \cos u & 0 \\ 0 & 0 & 1 \end{vmatrix} = (\cos u, \sin u, 0).$$

Since $\mathbf{T}_u\times\mathbf{T}_v$ points away from the origin (hence away from the surface of the cylinder), it follows that $\mathbf{T}_u\times\mathbf{T}_v$ is an outward normal. The flux of $\mathbf{F}$ is

$$-\iint_S (2x, 2y, 6z)\cdot d\mathbf{S} = -\iint_D (2\cos u, 2\sin u, 6v)\cdot(\cos u, \sin u, 0)\,dA$$
$$= -\iint_D 2\,dA = -2\text{area}(D) = -8\pi.$$

11. (a) View S as the level surface of $f(x,y,z) = x^2+y^2+z^2$ of value a^2. Then

$$\frac{\nabla f}{\|\nabla f\|} = \frac{(2x,2y,2z)}{\sqrt{4x^2+4y^2+4z^2}} = \frac{(x,y,z)}{a}$$

is a normal vector. Since (x,y,z) points away from the origin, $\nabla f/\|\nabla f\|$ is an outward unit normal. So

$$\iint_S \mathbf{F}\cdot\mathbf{n}\,dS = \iint_S c\mathbf{k}\cdot\frac{1}{a}(x\mathbf{i}+y\mathbf{j}+z\mathbf{k})\,dS = \iint_S \frac{c}{a}z\,dS = \frac{c}{a}\iint_S z\,dS.$$

The projection D of S onto the xy-plane is the disk $x^2 + y^2 \le a^2$. Therefore, with $\mathbf{n} = \nabla f / \|\nabla f\|$,

$$\iint_S z\,dS = \iint_D z \frac{dA}{|\mathbf{n} \cdot \mathbf{k}|} = \iint_D z \frac{dA}{z/a} = a \iint_D dA = a(\pi a^2) = a^3 \pi.$$

It follows that

$$\iint_S \mathbf{F} \cdot \mathbf{n}\,dS = \frac{c}{a} a^3 \pi = a^2 c \pi.$$

(b) Parametrize the disk $x^2 + y^2 \le a^2$ in the xy-plane as (think of polar coordinates!)

$$\mathbf{r}(u, v) = (u \cos v, u \sin v, 0), \qquad (u, v) \in D = [0, a] \times [0, 2\pi].$$

Then

$$\mathbf{N} = \begin{vmatrix} \mathbf{i} & \mathbf{j} & \mathbf{k} \\ \cos v & \sin v & 0 \\ -u \sin v & u \cos v & 0 \end{vmatrix} = u\mathbf{k},$$

and therefore (recall that $\mathbf{F} = c\mathbf{k}$)

$$\begin{aligned} \iint_S \mathbf{F} \cdot \mathbf{n}\,dS &= \iint_D c\mathbf{k} \cdot u\mathbf{k}\,dA = \iint_D cu\,dA \\ &= c \int_0^{2\pi} \left(\int_0^a u du \right) dv \\ &= c(2\pi)\tfrac{1}{2}a^2 = a^2 c \pi. \end{aligned}$$

(c) The flow is constant and occurs in the vertical direction. The volume of fluid that flows across the surface of the disk is the same as the volume of fluid that flows across the hemisphere, since their projections onto the xy-plane are the same.

13. Let S_1 denote the paraboloid $z = x^2 + y^2$ and let S_2 denote the paraboloid $z = 12 - x^2 - y^2$. Combining the two equations we get $x^2 + y^2 = 6$; it follows that the region of integration D is the disk $x^2 + y^2 \le 6$ (for both surfaces). Parametrize S_1 by $\mathbf{r}_1(u, v) = (u, v, u^2 + v^2)$. Its surface normal

$$\mathbf{N}_1 = \begin{vmatrix} \mathbf{i} & \mathbf{j} & \mathbf{k} \\ 1 & 0 & 2u \\ 0 & 1 & 2v \end{vmatrix} = (-2u, -2v, 1)$$

points inwards, into the paraboloid. We will compute the surface integral anyway, and will change the sign of the result.

$$\begin{aligned} \iint_{S_1} \mathbf{F} \cdot d\mathbf{S} &= \iint_D \mathbf{F}(\mathbf{r}_1(u, v)) \cdot \mathbf{N}_1\,dA \\ &= \iint_D u\mathbf{i} \cdot (-2u\mathbf{i} - 2v\mathbf{j} + \mathbf{k})\,dA \\ &= -2 \iint_D u^2\,dA \end{aligned}$$

(change to polar coordinates $u = r \cos \theta$, $v = r \sin \theta$)

$$= -2\int_0^{2\pi}\left(\int_0^{\sqrt{6}} r^3\cos^2\theta dr\right)d\theta$$
$$= -2\left(\int_0^{2\pi}\cos^2\theta\, d\theta\right)\left(\int_0^{\sqrt{6}} r^3\, dr\right)$$
$$= (-2)\left(\tfrac{1}{2}\theta + \tfrac{1}{4}\sin(2\theta)\right)\Big|_0^{2\pi}\left(\tfrac{1}{4}r^4\Big|_0^{\sqrt{6}}\right) = (-2)(\pi)(9) = -18\pi.$$

It follows that the outward flow is 18π. Now parametrize S_2 by $\mathbf{r}_2(u,v) = (u, v, 12 - u^2 - v^2)$. Its surface normal

$$\mathbf{N}_2 = \begin{vmatrix} \mathbf{i} & \mathbf{j} & \mathbf{k} \\ 1 & 0 & -2u \\ 0 & 1 & -2v \end{vmatrix} = (2u, 2v, 1)$$

points outward, and

$$\iint_{S_2} \mathbf{F}\cdot d\mathbf{S} = \iint_D u\mathbf{i}\cdot(2u\mathbf{i} + 2v\mathbf{j} + \mathbf{k})\, dA = 2\iint_D u^2\, dA = 18\pi,$$

as above. So the total flux across S is $18\pi + 18\pi = 36\pi$.

15. The given plane intersects the xy-plane along the line $2x+y = 16$. Therefore, its projection onto the xy-plane is the triangular region D bounded by $x = 0$, $y = 0$ and $2x + y = 16$. Parametrize the given plane by $\mathbf{r}(u,v) = (u, v, 16 - 2u - v)$, $(u,v) \in D$. Then

$$\mathbf{N} = \mathbf{T}_u \times \mathbf{T}_v = \begin{vmatrix} \mathbf{i} & \mathbf{j} & \mathbf{k} \\ 1 & 0 & -2 \\ 0 & 1 & -1 \end{vmatrix} = 2\mathbf{i} + \mathbf{j} + \mathbf{k}$$

points outward; i.e., away from the origin. The required flux is

$$\iint_S y^3(\mathbf{j} - \mathbf{k})\cdot d\mathbf{S} = \iint_D v^3(\mathbf{j} - \mathbf{k})\cdot(2\mathbf{i} + \mathbf{j} + \mathbf{k})\, dA$$
$$= \iint_D 0\, dA = 0.$$

17. Parametrize the given surface by $\mathbf{r}(u,v) = (u, v, 2u^2v)$, $0 \le u \le 1$, $0 \le v \le 2$. The surface normal

$$\mathbf{N} = \mathbf{T}_u \times \mathbf{T}_v = \begin{vmatrix} \mathbf{i} & \mathbf{j} & \mathbf{k} \\ 1 & 0 & 4uv \\ 0 & 1 & 2u^2 \end{vmatrix} = -4uv\mathbf{i} - 2u^2\mathbf{j} + \mathbf{k}$$

points upward, as required. Since $\mathbf{F}(\mathbf{r}(u,v)) = u^2v\mathbf{i} + uv^3\mathbf{j} + 4u^3v^2\mathbf{k}$ and $\mathbf{F}\cdot\mathbf{N} = -2u^3v^3$, it follows that

$$\iint_S (x^2y\mathbf{i} + xy^3\mathbf{j} + 2xyz\mathbf{k})\cdot d\mathbf{S} = -2\iint_{D=[0,1]\times[0,2]} u^3v^3 dA$$
$$= -2\int_0^1\left(\int_0^2 u^3v^3 dv\right)du$$
$$= -2\left(\int_0^1 u^3\, du\right)\left(\int_0^2 v^3\, dv\right)$$
$$= -2\left(\tfrac{1}{4}u^4\Big|_0^1\right)\left(\tfrac{1}{4}v^4\Big|_0^2\right) = -2.$$

19. We use the parametrization $\mathbf{r}(\theta,\phi) = a\cos\theta\sin\phi\,\mathbf{i} + a\sin\theta\sin\phi\,\mathbf{j} + a\cos\phi\,\mathbf{k}$, where $0 \le \theta \le 2\pi$ and $0 \le \phi \le \pi$ (we usually use u and v; however, the parameters in this case come from spherical coordinates, and are denoted accordingly). The surface normal $\mathbf{N}$ is computed to be (see Example 7.13 in Section 7.1) $\mathbf{N}(\theta,\phi) = -a\sin\phi\,\mathbf{r}(\theta,\phi)$. Since $0 \le \phi \le \pi$ implies that $\sin\phi \ge 0$, the vector $\mathbf{N}$ points inward, Therefore, we use $-\mathbf{N} = a\sin\phi\,\mathbf{r}(\theta,\phi)$. Recall that (see Section 2.7) $\mathbf{r}(\theta,\phi) = a\mathbf{e}_\rho$; therefore, $-\mathbf{N} = a^2\sin\phi\,\mathbf{e}_\rho$. It follows that

$$\begin{aligned}\iint_S \mathbf{F}\cdot d\mathbf{S} &= \iint_{D=[0,2\pi]\times[0,\pi]} \mathbf{F}\cdot(-\mathbf{N})dA \\ &= \int_0^{2\pi}\left(\int_0^{\pi} a^2\sin\phi\mathbf{F}\cdot\mathbf{e}_\rho d\phi\right)d\theta \\ &= a^2\int_0^{2\pi}\left(\int_0^{\pi} F_\rho\sin\phi\, d\phi\right)d\theta,\end{aligned}$$

where $F_\rho = \mathbf{F}\cdot\mathbf{e}_\rho$ is the component of $\mathbf{F}$ in the direction of $\mathbf{e}_\rho$.

21. The projection of the plane $x + 2y + 8z = 8$ onto the first quadrant in the xy-plane is the triangular region D bounded by $x = 0$, $y = 0$ and $x + 2y = 8$. A unit normal is given by $\overline{\mathbf{n}} = \pm\nabla f/\|\nabla f\|$, where $f = x+2y+8z$; so $\overline{\mathbf{n}} = \pm(1,2,8)/\sqrt{69}$. We need a downward-pointing normal, so we must take $\mathbf{n} = -(1,2,8)/\sqrt{69}$. Thus

$$\begin{aligned}\iint_S \mathbf{F}\cdot d\mathbf{S} &= \iint_D \mathbf{F}\cdot\mathbf{n}\frac{dA}{|\mathbf{n}\cdot\mathbf{k}|} \\ &= -\iint_D ((x+y)\mathbf{i}+\mathbf{j}+z\mathbf{k})\cdot\frac{\mathbf{i}+2\mathbf{j}+8\mathbf{k}}{\sqrt{69}}\,\frac{dA}{8/\sqrt{69}} \\ &= -\tfrac{1}{8}\iint_D (x+y+2+8z)\,dA = -\tfrac{1}{8}\iint_D (10-y)\,dA \\ &= -\tfrac{1}{8}\int_0^4\left(\int_0^{8-2y}(10-y)dx\right)dy \\ &= -\tfrac{1}{8}\int_0^4 (10-y)(8-2y)\,dy \\ &= -\tfrac{1}{8}\int_0^4 (80-28y+2y^2)\,dy \\ &= -\tfrac{1}{8}\left(80y-14y^2+\tfrac{2}{3}y^3\right)\Big|_0^4 = -\tfrac{1}{8}\left(320-224+\tfrac{128}{3}\right) = -\tfrac{52}{3}.\end{aligned}$$

23. The projection onto the xy-plane is the disk D given by $x^2+y^2 \le 9$. The hemisphere can be viewed as the level surface of $g(x,y,z) = x^2+y^2+z^2$ of value 9; so $\nabla g/\|\nabla g\| = (2x,2y,2z)/6 = (x,y,z)/3$ gives a unit normal direction. Since the vector (x,y,z) points away from the origin (and hence away from the sphere), it follows that $\nabla g/\|\nabla g\|$ is the outward unit normal to S. Hence (let $\mathbf{n} = (x\mathbf{i}+y\mathbf{j}+z\mathbf{k})/3$)

$$\begin{aligned}\iint_S (x^2\mathbf{i}+2z\mathbf{k})\cdot d\mathbf{S} &= \iint_D (x^2\mathbf{i}+2z\mathbf{k})\cdot\mathbf{n}\frac{dA}{|\mathbf{n}\cdot\mathbf{k}|} \\ &= \iint_D (x^2\mathbf{i}+2z\mathbf{k})\cdot\tfrac{1}{3}(x\mathbf{i}+y\mathbf{j}+z\mathbf{k})\,\frac{dA}{\tfrac{1}{3}|(x\mathbf{i}+y\mathbf{j}+z\mathbf{k})\cdot\mathbf{k}|}\end{aligned}$$

$$= \iint_D \frac{x^3 + 2z^2}{|z|} \, dA$$
$$= \iint_D \left(\frac{x^3}{z} + 2z \right) dA,$$

since $z \geq 0$ (by assumption) and so $|z| = z$. Eliminate z and change to polar coordinates:

$$\iint_D \left(\frac{x^3}{z} + 2z \right) dA = \iint_D \left(\frac{x^3}{\sqrt{9 - x^2 - y^2}} + 2\sqrt{9 - x^2 - y^2} \right) dA$$
$$= \int_0^{2\pi} \left(\int_0^3 \left(\frac{r^3 \cos^3 \theta}{\sqrt{9 - r^2}} + 2\sqrt{9 - r^2} \right) r dr \right) d\theta$$
$$= \left(\int_0^{2\pi} \cos^3 \theta \, d\theta \right) \left(\int_0^3 \frac{r^4}{\sqrt{9 - r^2}} dr \right)$$
$$+ \left(\int_0^{2\pi} d\theta \right) \left(\int_0^3 2r\sqrt{9 - r^2} \, dr \right).$$

This is not as bad as it looks. The first summand is zero, since $\int_0^{2\pi} \cos^3 \theta \, d\theta = 0$ (think of the graph of $\cos^3 \theta$ — it looks somewhat like $\cos \theta$ (the precise shape of the graph is not needed here); what is important is the fact that the area between $\cos^3 \theta$ and the x-axis above the x-axis is the same as the area between $\cos^3 \theta$ and the x-axis below the x-axis (due to the symmetry of $\cos^3 \theta$). Therefore, (integral="area above" − "area below") the integral of $\cos^3 \theta$ over $[0, 2\pi]$ is zero. Since

$$\int_0^3 2r\sqrt{9 - r^2} \, dr = -\tfrac{2}{3}(9 - r^2)^{3/2} \Big|_0^3 = \tfrac{2}{3} 9^{3/2} = 18,$$

it follows that $\iint_S (x^2 \mathbf{i} + 2z\mathbf{k}) \cdot d\mathbf{S} = 2\pi(18) = 36\pi$.

25. Parametrize S by $\mathbf{r}(u, v) = (\cos u, \sin u, v)$, $(u, v) \in D$, where D is the rectangle in the uv-plane given by $0 \leq u \leq 2\pi$, $0 \leq v \leq 3$. By definition,

$$\int_S (xdxdy + ydydz + zdzdx) = \iint_D \left(x\frac{\partial(x, y)}{\partial(u, v)} + y\frac{\partial(y, z)}{\partial(u, v)} z\frac{\partial(z, x)}{\partial(u, v)} \right) dA,$$

where

$$\frac{\partial(x, y)}{\partial(u, v)} = \begin{vmatrix} -\sin u & 0 \\ \cos u & 0 \end{vmatrix} = 0,$$

$$\frac{\partial(y, z)}{\partial(u, v)} = \begin{vmatrix} \cos u & 0 \\ 0 & 1 \end{vmatrix} = \cos u$$

and

$$\frac{\partial(z, x)}{\partial(u, v)} = \begin{vmatrix} 0 & 1 \\ -\sin u & 0 \end{vmatrix} = \sin u;$$

It follows that

$$\int_S xdxdy + ydydz + zdzdx = \iint_D (\sin u \cos u + v \sin u) \, dA$$
$$= \int_0^{2\pi} \left(\int_0^3 \tfrac{1}{2} \sin 2u \, dv \right) du + \int_0^{2\pi} \left(\int_0^3 v \sin u dv \right) du = 0,$$

since $\int_0^{2\pi} \sin 2u\, du = 0$ and $\int_0^{2\pi} \sin u\, du = 0$.

27. Parametrize S by $\mathbf{r}(u,v) = (u, v, u^2+v^2)$, $(u,v) \in D$, where D is the disk $x^2+y^2 \le 4$. Then

$$\int_S (dxdy + zdzdx) = \iint_D \left(\frac{\partial(x,y)}{\partial(u,v)} + z\frac{\partial(z,x)}{\partial(u,v)} \right) dA\,,$$

where

$$\frac{\partial(x,y)}{\partial(u,v)} = \begin{vmatrix} 1 & 0 \\ 0 & 1 \end{vmatrix} = 1$$

and

$$\frac{\partial(z,x)}{\partial(u,v)} = \begin{vmatrix} 2u & 2v \\ 1 & 0 \end{vmatrix} = -2v.$$

So

$$\begin{aligned}\int_S (dxdy + zdzdx) &= \iint_D (1 + (u^2+v^2)(-2v))\, dA \\ &= \iint_D 1\, dA - 2\iint_D (u^2v + v^3)\, dA.\end{aligned}$$

Changing to polar coordinates $u = r\cos\theta$, $v = r\sin\theta$, we get

$$\begin{aligned}\iint_D (u^2v+v^3)\,dA &= \int_0^{2\pi}\left(\int_0^2 (r^3\cos^2\theta\sin\theta + r^3\sin^3\theta) r dr\right) d\theta \\ &= \left(\int_0^{2\pi}\cos^2\theta\sin\theta\, d\theta\right)\left(\int_0^2 r^4\, dr\right) + \left(\int_0^{2\pi}\sin^3\theta\, d\theta\right)\left(\int_0^2 r^4\, dr\right) \\ &= \left(-\tfrac{1}{3}\cos^3\theta\Big|_0^{2\pi}\right)\left(\tfrac{1}{5}r^5\Big|_0^2\right) + 0 \\ &= \left(-\tfrac{1}{3} - (-\tfrac{1}{3})\right)\tfrac{32}{5} = 0,\end{aligned}$$

since $\int_0^{2\pi}\sin^3\theta\, d\theta = 0$ (use integration tables or symmetry). It follows that

$$\int_S (dxdy + zdzdx) = \iint_D 1\, dA = 4\pi.$$

7.4 Integrals: Properties and Applications

1. Since the surface area of a hemisphere of radius a is $2\pi a^2$, it follows that

$$\overline{p} = \frac{1}{2\pi a^2}\iint_S (1-z^2)\, dS,$$

where S is given by $x^2+y^2+z^2 = a^2$, $z \ge 0$. Parametrize it by

$$\mathbf{r}(u,v) = (a\cos v\cos u, a\cos v\sin u, a\sin v), \qquad (u,v) \in D,$$

where D is the rectangle $[0,2\pi]\times[0,\pi/2]$ in the uv-plane (this parametrization was introduced in Example 7.1). From Example 7.4, $\mathbf{N} = a\cos v\,\mathbf{r}(u,v)$, and so (since $\|\mathbf{r}(u,v)\| = a$)

$$\|\mathbf{N}\| = a|\cos v|\,\|\mathbf{r}(u,v)\| = a^2\cos v.$$

It follows that

$$\iint_S (1 - z^2)\, dS = \iint_D (1 - a^2 \sin^2 v) a^2 \cos v\, dA$$
$$= a^2 \int_0^{2\pi} \left(\int_0^{\pi/2} \cos v dv \right) du - a^4 \int_0^{2\pi} \left(\int_0^{\pi/2} \sin^2 v \cos v dv \right) du.$$

Now $\int_0^{\pi/2} \cos v\, dv = 1$ and

$$\int_0^{\pi/2} \sin^2 v \cos v\, dv = \tfrac{1}{3} \sin^3 v \Big|_0^{\pi/2} = \tfrac{1}{3}$$

and therefore

$$\overline{p} = \frac{1}{2\pi a^2} \left[a^2 (2\pi) - a^4 (2\pi) \tfrac{1}{3} \right] = 1 - \tfrac{1}{3} a^2.$$

3. Parametrize the surface S by (see Example 7.3)

$$\mathbf{r}(u, v) = (v \cos u, v \sin u, v), \qquad (u, v) \in D,$$

where D is the rectangle in the uv-plane defined by $0 \le u \le 2\pi$ and $1 \le v \le 4$. The surface normal $\mathbf{N}$ is (see Example 7.5)

$$\mathbf{N} = \mathbf{T}_u \times \mathbf{T}_v = \begin{vmatrix} \mathbf{i} & \mathbf{j} & \mathbf{k} \\ -v \sin u & v \cos u & 0 \\ \cos u & \sin u & 1 \end{vmatrix} = v \cos u \mathbf{i} + v \sin u \mathbf{j} - v \mathbf{k}.$$

So $\|\mathbf{N}\| = |v|\sqrt{2} = v\sqrt{2}$. The surface area of S is

$$A(S) = \iint_S dS = \iint_D v\sqrt{2}\, dA$$
$$= \int_0^{2\pi} \left(\int_1^4 v\sqrt{2}\, dv \right) du = \left(\int_0^{2\pi} du \right) \sqrt{2}\, \tfrac{1}{2} v^2 \Big|_1^4 = 15\pi\sqrt{2}.$$

The distance from (x, y, z) to the z-axis (i.e., to the point $(0, 0, z)$) is given by $d(x, y, z) = \sqrt{x^2 + y^2}$. So

$$\iint_S d(x, y, z)\, dS = \iint_S \sqrt{x^2 + y^2}\, dS = \iint_D v\, (v\sqrt{2})\, dA$$
$$= \sqrt{2} \int_0^{2\pi} \left(\int_1^4 v^2 dv \right) du$$
$$= \sqrt{2}(2\pi)\, \tfrac{1}{3} v^3 \Big|_1^4 = 42\pi\sqrt{2}.$$

The average distance is

$$\overline{d} = \frac{1}{A(S)} \iint_S d(x, y, z)\, dS = \frac{42\pi\sqrt{2}}{15\pi\sqrt{2}} = \frac{42}{15}.$$

5. The total charge is given by the triple integral

$$\iiint_V \rho(x, y, z)\, dV = \iiint_V (2x^2 + y)\, dV,$$

where V is the given tetrahedron. The limits of integration are given by $0 \le x \le 1$, $0 \le y \le 3 - 3x$ (the line through $(1, 0)$ and $(0, 3)$ in the xy-plane has the equation $3x + y = 3$), and $0 \le z \le 4 - 4x - 4y/3$ (the plane through $(1,0,0)$, $(0,3,0)$ and $(0,0,4)$ has the equation $x/1 + y/3 + z/4 = 1$). It follows that

$$\iiint_V (2x^2 + y)\, dV = \int_0^1 \left(\int_0^{3-3x} \left(\int_0^{4-4x-4y/3} (2x^2+y) dz \right) dy \right) dx$$

$$= \int_0^1 \left(\int_0^{3-3x} (2x^2+y) z \Big|_0^{4-4x-4y/3} dy \right) dx$$

$$= \int_0^1 \left(\int_0^{3-3x} -\tfrac{4}{3}(2x^2+y)(-3+3x+y) dy \right) dx$$

$$= \int_0^1 \left(-\tfrac{4}{9}y^3 - \tfrac{2}{3}(2x^2-3+3x)y^2 - \tfrac{8}{3}x^2(-3+3x)y \right) \Big|_0^{3-3x} dx$$

$$= \int_0^1 (6 - 18x + 30x^2 - 30x^3 + 12x^4)\, dx = \tfrac{19}{10}.$$

7. By symmetry, $\overline{x} = 0$; so we need $\overline{y}$ only. The mass m of the wire is (density) $\cdot$ (length) $=$ $\rho\, a\pi$. The moment about the y-axis is

$$M_x = \int_{\mathbf{c}} y\rho ds = \rho \int_{\mathbf{c}} y ds,$$

where $\mathbf{c}$ is a parametrization of the wire. Take $\mathbf{c}(t) = (a\cos t, a\sin t)$, $0 \le t \le \pi$. Then $\mathbf{c}'(t) = (-a\sin t, a\cos t)$ and $\|\mathbf{c}'(t)\| = a$; so

$$M_x = \rho \int_0^{\pi} a\sin t\; a\, dt = 2a^2\rho.$$

Therefore,

$$\overline{y} = \frac{M_x}{m} = \frac{2a^2\rho}{a\pi\rho} = \frac{2a}{\pi}.$$

Hence the coordinates of the center of mass are $(0, 2a/\pi)$.

9. Parametrize the wire as $\mathbf{c}(t) = (a\cos t, a\sin t)$, $t \in [0, \pi/2]$; then $\mathbf{c}'(t) = (-a\sin t, a\cos t)$ and $\|\mathbf{c}'(t)\| = a$. The mass of the wire is

$$m = \int_{\mathbf{c}} \rho\, ds = \int_{\mathbf{c}} bx\, ds = \int_0^{\pi/2} ab\cos t\; a\, dt = a^2 b\, \sin t \Big|_0^{\pi/2} = a^2 b.$$

The moments are

$$M_x = \int_{\mathbf{c}} y\rho\, ds = \int_0^{\pi/2} a^2 b \sin t \cos t\; a\, dt$$

$$= \tfrac{1}{2}a^3 b \int_0^{\pi/2} \sin 2t\, dt = \tfrac{1}{2}a^3 b \left(-\tfrac{1}{2}\cos 2t \Big|_0^{\pi/2} \right) = \tfrac{1}{2}a^3 b$$

and

$$M_y = \int_{\mathbf{c}} x\rho\, ds = \int_0^{\pi/2} a^2 b \cos^2 t\; a\, dt$$

$$= a^3 b \int_0^{\pi/2} \cos^2 t\, dt$$

$$= a^3 b \left(\tfrac{1}{2}t - \tfrac{1}{4}\sin 2t \right) \Big|_0^{\pi/2} = \tfrac{1}{4}a^3 b\pi.$$

Therefore,

$$\overline{x} = \frac{M_y}{m} = \frac{a\pi}{4} \quad \text{and} \quad \overline{y} = \frac{M_x}{m} = \frac{a}{2}$$

are the coordinates of the center of mass.

11. Parametrize the paraboloid as $\mathbf{r}(u, v) = (u, v, 2u^2 + 2v^2)$, where $(u, v) \in D$, and D is the disk $x^2 + y^2 \leq 4$ (obtained by substituting $z = 8$ into $z = 2(x^2 + y^2)$). Then

$$\mathbf{N} = \mathbf{T}_u \times \mathbf{T}_v = \begin{vmatrix} \mathbf{i} & \mathbf{j} & \mathbf{k} \\ 1 & 0 & 4u \\ 0 & 1 & 4v \end{vmatrix} = -4u\mathbf{i} - 4v\mathbf{j} + \mathbf{k},$$

and $\|\mathbf{N}\| = \sqrt{1 + 4u^2 + 4v^2}$. In the evaluation of the integral we will use polar coordinates $u = r\cos\theta$, $v = r\sin\theta$. The mass is

$$\begin{aligned} m &= \iint_S \rho\, dS = \rho \iint_S dS \\ &= \rho \iint_D \sqrt{1 + 4u^2 + 4v^2}\, dA = \rho \int_0^{2\pi} \left(\int_0^2 \sqrt{1 + 4r^2}\, r dr \right) d\theta \\ &= \rho \left(\theta \Big|_0^{2\pi} \right) \left(\tfrac{1}{12}(1 + 4r^2)^{3/2} \Big|_0^2 \right) \\ &= \rho\, 2\pi\, \tfrac{1}{12}(17^{3/2} - 1) = \tfrac{1}{6}\pi\rho\, (17^{3/2} - 1). \end{aligned}$$

By symmetry, $\overline{x} = 0$ and $\overline{y} = 0$. The moment with respect to the xy-plane is

$$\begin{aligned} M_{xy} &= \iint_S z\rho\, dS = \rho \iint_S z\, dS \\ &= \rho \iint_D (2u^2 + 2v^2)\sqrt{1 + 4u^2 + 4v^2}\, dA \\ &= 2\rho \int_0^{2\pi} \left(\int_0^2 r^3 \sqrt{1 + 4r^2}\, dr \right) d\theta \end{aligned}$$

(proceed using formulas from a table of integrals)

$$\begin{aligned} &= 2\rho(2\pi) \left(\tfrac{1}{20} r^2 (1 + 4r^2)^{3/2} - \tfrac{1}{120}(1 + 4r^2)^{3/2} \right) \Big|_0^2 \\ &= 4\pi\rho \left(\left(\tfrac{4}{20} 17^{3/2} - \tfrac{1}{120} 17^{3/2} \right) - \left(-\tfrac{1}{120} \right) \right) \\ &= \tfrac{1}{30}\pi\rho\, (23(17^{3/2}) + 1). \end{aligned}$$

Therefore,

$$\overline{z} = \frac{M_{xy}}{m} = \frac{\frac{1}{30}\pi\rho(23(17^{3/2}) + 1)}{\frac{1}{6}\pi\rho(17^{3/2} - 1)} \approx 4.6695.$$

The moment of inertia about the z-axis is

$$\begin{aligned} I_z &= \iint_S (x^2 + y^2)\rho\, dS = \rho \iint_S (x^2 + y^2)\, dS \\ &= \rho \iint_D (u^2 + v^2)\sqrt{1 + 4u^2 + 4v^2}\, dA \\ &= \rho \int_0^{2\pi} \left(\int_0^2 r^3 \sqrt{1 + 4r^2}\, dr \right) d\theta \end{aligned}$$

(proceed as in computing M_{xy})

$$= \tfrac{1}{60}\pi\rho\left(23(17^{3/2})+1\right) \approx 84.4635\rho.$$

13. The mass of V is ($v(V)$ denotes the volume of V)

$$m = \iiint_V \rho\, dV = bv(V) = b\tfrac{2}{3}\pi a^3 = \tfrac{2}{3}\pi a^3 b.$$

By symmetry, $\overline{x} = \overline{y} = 0$. The moment with respect to the xy-plane is $M_{xy} = \iiint_V bz\, dV$. Passing to spherical coordinates $x = \rho\sin\phi\cos\theta$, $y = \rho\sin\phi\sin\theta$ and $z = \rho\cos\phi$, we get

$$\begin{aligned}
M_{xy} &= \int_0^{2\pi}\left(\int_0^{\pi/2}\left(\int_0^a b\rho\cos\phi\ \rho^2\sin\phi\ d\rho\right) d\phi\right) d\theta \\
&= b\int_0^{2\pi}\left(\int_0^{\pi/2} \sin\phi\cos\phi\ \tfrac{1}{4}\rho^4\Big|_0^a\, d\phi\right)\ d\theta \\
&= \tfrac{1}{4}a^4 b\int_0^{2\pi}\left(\int_0^{\pi/2} \tfrac{1}{2}\sin 2\phi d\phi\right) d\theta \\
&= \tfrac{1}{4}a^4 b\int_0^{2\pi} \left(-\tfrac{1}{4}\cos 2\phi\right)\Big|_0^{\pi/2}\ d\theta \\
&= \tfrac{1}{4}a^4 b\,(2\pi)\,\tfrac{1}{2} = \tfrac{1}{4}a^4 b\pi.
\end{aligned}$$

Therefore,

$$\overline{z} = \frac{M_{xy}}{m} = \frac{a^4 b\pi/4}{2\pi a^3 b/3} = \tfrac{3}{8}a.$$

15. Parametrize the hemisphere by $\mathbf{r}(u,v) = (a\cos u\sin v, a\sin u\sin v, a\cos v)$, where $0 \le u \le 2\pi$ and $0 \le v \le \pi/2$. Then (see Example 7.13) $\mathbf{N} = a(-\sin v)\cdot\mathbf{r}(u,v)$, and since $\|\mathbf{r}(u,v)\| = a$, it follows that $\|\mathbf{N}\| = a^2\sin v$. The highest point is $z_0 = a$, and so

$$\begin{aligned}
\iint_S \mu(a-z)\, dS &= \mu\int_0^{2\pi}\left(\int_0^{\pi/2}(a - a\cos v)a^2\sin v dv\right) du \\
&= a^3\mu\left(\int_0^{2\pi} du\right)\left(\int_0^{\pi/2}(1-\cos v)\sin v\, dv\right) \\
&= 2\pi\mu a^3\int_0^{\pi/2}(\sin v - \sin v\cos v)\, dv \\
&= 2\pi\mu a^3\left(-\cos v + \tfrac{1}{4}\cos 2v\right)\Big|_0^{\pi/2} = \pi\mu a^3.
\end{aligned}$$

17. Parametrize the cylinder by $\mathbf{r}(u,v) = (2\cos u, 2\sin u, v)$, where $(u,v) \in D$, and D is the rectangle $[0,2\pi]\times[0,10]$ in the uv-plane. Then

$$\mathbf{N} = \begin{vmatrix} \mathbf{i} & \mathbf{j} & \mathbf{k} \\ -2\sin u & 2\cos u & 0 \\ 0 & 0 & 1 \end{vmatrix} = (2\cos u, 2\sin u, 0)$$

and $\|\mathbf{N}\| = 2$. The highest point has its z-coordinate equal to 10, and so

$$\iint_S \mu(10-z)\, dS = \mu\int_0^{2\pi}\left(\int_0^{10}(10-v)dv\right) du$$

$$= \mu\left(\int_0^{2\pi} du\right)\left(10v - \tfrac{1}{2}v^2\right)\Big|_0^{10}$$
$$= \mu(2\pi)(50) = 100\pi\mu.$$

19. For example, let $f(x,y,z) = z$ and $\mathbf{c}(t) = (0,0,t)$, where $-1 \le t \le 1$. Then $\|\mathbf{c}'(t)\| = 1$ and

$$\left|\int_{\mathbf{c}} f\,ds\right| = \left|\int_{-1}^{1} t\,dt\right| = 0,$$

whereas

$$\int_{\mathbf{c}} |f| = \int_{-1}^{1} |t|\,dt = 1.$$

21. For points on S, $0 \le z \le 1$; it follows that $1 \le z^2 + 1 \le 2$, and hence

$$\iint_S 1\,dS \le \iint_S (z^2+1)\,dS \le \iint_S 2\,dS$$

and therefore

$$2\pi \le \iint_S (z^2+1)\,dS \le 4\pi$$

(since the surface area of a hemisphere of radius 1 is 2π).

23. Since $-1 \le \sin a \le 1$ for any real number a, it follows that $e^{-1} \le e^{\sin(x+y)} \le e$ and

$$e^{-1}\ell(\mathbf{c}) \le \int_{\mathbf{c}} e^{\sin(x+y)}\,ds \le e\ell(\mathbf{c}),$$

where $\ell(\mathbf{c})$ is the length of $\mathbf{c}$. From $\mathbf{c}'(t) = (-2\sin t, 2\cos t, 1)$ and $\|\mathbf{c}'(t)\| = \sqrt{5}$, it follows that

$$\ell(\mathbf{c}) = \int_{\pi}^{4\pi} \sqrt{5}\,dt = 3\pi\sqrt{5},$$

and therefore

$$\frac{3\pi\sqrt{5}}{e} \le \int_{\mathbf{c}} e^{\sin(x+y)}\,ds \le 3\pi e\sqrt{5}.$$

7.5 Chapter Review

Solutions to Computer Projects

1. (a) The volume of a cube of side a is a^3, and its surface area is $6a^2$; hence $I(\text{cube}) = 6a^2/(a^3)^{2/3} = 6$.

(b) Let $I(m)$ denote the isoperimetric ratio for the box with dimensions a, a and ma. Then

$$I(m) = \frac{2a^2 + 4ma^2}{(a^3m)^{2/3}} = \frac{2+4m}{m^{2/3}}.$$

From

$$I'(m) = \frac{4m^{2/3} - (2+4m)(2/3)m^{-1/3}}{m^{4/3}} = 0$$

it follows that $4m - (2/3)(2+4m) = 0$; i.e., $m = 1$. Either by checking the values of $I'(m)$ or from the plot of $I(m)$ (see Figure 7.5.1 (a)) we conclude that the smallest value of $I(m)$ is 6, and is attained for $m = 1$. So a cube of side a has the smallest isoperimetric ratio of all rectangular boxes considered. As $m \to 0$, $I(m) \to \infty$; as $m \to \infty$, $I(m) \to \infty$ as well. In words, the higher (or the lower) the box, the larger the isoperimetric ratio.

(c) The volume of a ball of radius a is $4\pi a^3/3$, and is surface area is $4\pi a^2$; so

$$I(\text{ball}) = \frac{4\pi a^2}{(4\pi a^3/3)^{2/3}} = (4\pi)^{1/3} 3^{2/3} \approx 4.83598.$$

(d) The volume of an ellipsoid with semi-axes 1, 1 and n is $4\pi n/3$ (recall that the volume of an ellipsoid with semi-axes a, b and c is $4abc\pi/3$). Thus

$$I(n) = \left(\tfrac{4}{3}\pi n\right)^{-2/3} 2\pi \left(1 + \frac{n^2}{\sqrt{n^2-1}} \arcsin\left(\sqrt{n^2-1}/n\right)\right),$$

and since (notice that $4^{-2/3}\,2 = 4^{1/3}/2$)

$$\left(\tfrac{4}{3}\pi\right)^{-2/3} 2\pi = 3^{2/3}\pi^{1/3}4^{-2/3}2 = 3^{2/3}\pi^{1/3}4^{1/3}/2$$

it follows that

$$I(n) = 3^{2/3}(4\pi)^{1/3}\tfrac{1}{2}n^{-2/3}\left(1 + \frac{n^2}{\sqrt{n^2-1}} \arcsin\left(\sqrt{n^2-1}/n\right)\right).$$

The plot of $I(n)$ (see Figure 7.5.1 (b)) shows that $I(n)$ is increasing for $n > 1$. We can write

$$\begin{aligned} I(n) &= Cn^{-2/3} + Cn^{-2/3}\frac{n^2}{\sqrt{n^2-1}} \arcsin\left(\sqrt{n^2-1}/n\right) \\ &= Cn^{-2/3} + C\frac{n^{4/3}}{\sqrt{n^2-1}} \arcsin\left(\sqrt{n^2-1}/n\right), \end{aligned}$$

where $C = 3^{2/3}(4\pi)^{1/3}/2$. As $n \to \infty$, the first term goes to zero. Since $\sqrt{n^2-1}/n \to 1$ as $n \to \infty$, it follows that $\arcsin(\sqrt{n^2-1}/n) \to \pi/2$, and the second term is of the form $C\pi n^{1/3}/2$. In other words, as $n \to \infty$, $I(n)$ behaves as $C\pi n^{1/3}/2$; thus $I(n)$ approaches ∞ as $n \to \infty$.

To compute $\lim_{n\to 1+} I(n)$ we can use an appropriate command or L'Hôpital's rule. We choose the latter: from

$$\begin{aligned} &\lim_{n\to 1+} \frac{1}{\sqrt{n^2-1}} \arcsin\left(\sqrt{n^2-1}/n\right) \\ &\quad = \lim_{n\to 1+} \frac{1}{n/\sqrt{n^2-1}} \frac{1}{\sqrt{1-(n^2-1)/n^2}} \frac{n^2/\sqrt{n^2-1} - \sqrt{n^2-1}}{n^2} \\ &\quad = \lim_{n\to 1+} \frac{\sqrt{n^2-1}}{n} n \frac{n^2-(n^2-1)}{n^2\sqrt{n^2-1}} \\ &\quad = \lim_{n\to 1+} \frac{1}{n^2} = 1 \end{aligned}$$

it follows that

$$\lim_{n\to 1+} I(n) = \lim_{n\to 1+} Cn^{-2/3} + C\frac{n^{4/3}}{\sqrt{n^2-1}} \arcsin\left(\sqrt{n^2-1}/n\right)$$
$$= 2C = 3^{2/3}(4\pi)^{1/3}.$$

In words, as $n \to 1^+$, ellipsoids $x^2 + y^2 + z^2/n^2 = 1$ get deformed into a sphere, and their isoperimetric ratios converge to the isoperimetric ratio of a sphere. Since $I(n)$ is increasing, it follows that the sphere has the smallest isoperimetric ratio of all ellipsoids considered. (As a matter of fact, the sphere has the smallest isoperimetric ratio of *all* (piecewise smooth) surfaces — we state this fact without a proof.) Comparing this result with (c), we see that the more "ideal" the object (square, ball — as compared to a rectangular box, ellipsoid, respectively), the smaller the isoperimetric ratio.

(e) The sphere has the smallest isoperimetric ratio.

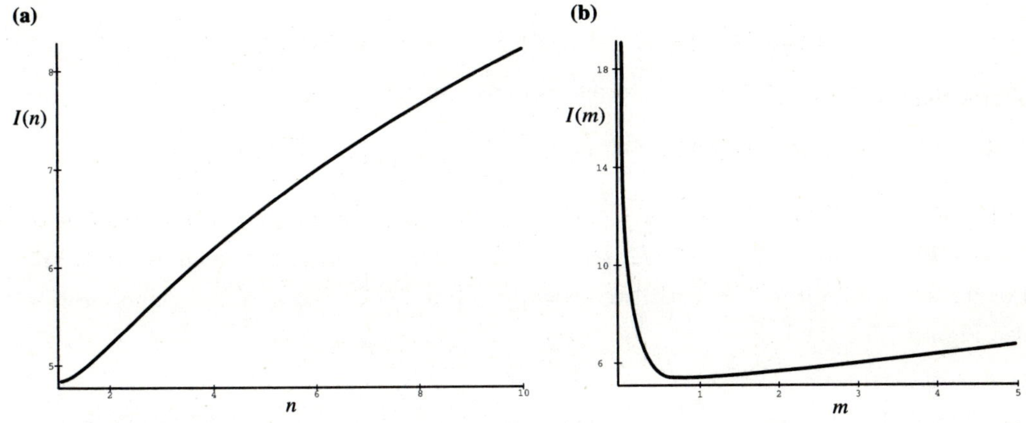

Figure 7.5.1

2. (a) Pick a point A on the circle of radius 2; its coordinates are $(2\cos t, 2\sin t, 0)$ for some fixed value of t; so the vector $\overrightarrow{OA}$ can be written as $\overrightarrow{OA} = (2\cos t, 2\sin t, 0)$. Now $\overrightarrow{AP} = \overrightarrow{AQ} + \overrightarrow{QP}$, see Figure 7.5.2 (a). Since $\overrightarrow{AQ}$ is parallel to $\overrightarrow{OA}$ and $\|\overrightarrow{AQ}\| = |\cos u|$ (this is due to $\|\overrightarrow{AP}\| = 1$; we need $|\cos u|$ to ensure that $\|\overrightarrow{AQ}\| \geq 0$), it follows that

$$\overrightarrow{AQ} = \cos u \,\frac{\overrightarrow{OA}}{\|\overrightarrow{OA}\|} = \cos u \,\frac{(2\cos t, 2\sin t, 0)}{2} = (\cos u\cos t, \cos u \sin t, 0).$$

Since $\overrightarrow{QP}$ is perpendicular to the xy-plane and $\|\overrightarrow{QP}\| = |\sin u|$, it follows that $\overrightarrow{QP} = \sin u\,\mathbf{k}$. Hence

$$\overrightarrow{OP} = \overrightarrow{OA} + \overrightarrow{AQ} + \overrightarrow{QP}$$
$$= (2\cos t + \cos u\cos t, 2\sin t + \cos u \sin t, \sin u)$$
$$= (\cos t(\cos u + 2), \sin t(\cos u + 2), \sin u),$$

where $0 \leq t, u \leq 2\pi$, represents a parametrization of the torus in question. Its plot is shown in Figure 7.5.2 (b). Repeating the above construction with the values R for the radius of the circle in the xy-plane (instead of 2) and r for the radius of the smaller circle (instead of 1), we get

$$\mathbf{r}(t,u) = (\cos t(r\cos u + R), \sin t(r\cos u + R), r\sin u),$$

where $0 \leq t, u \leq 2\pi$. Figure 7.5.2 (c) shows a torus with $r = 1$ and $R = 5$.

(b) Pick a point on the circle $\mathbf{c}(t)$; then $\overrightarrow{OA} = (\cos t, \sin t, 0)$, for some fixed value of t. Now $\overrightarrow{AP} = \overrightarrow{AQ} + \overrightarrow{QP}$; see Figure 7.5.2 (d). Since $\overrightarrow{AQ}$ is parallel to $\overrightarrow{OA}$ and $\|\overrightarrow{AQ}\| = u|\cos\theta|$, it follows that

$$\overrightarrow{AQ} = u\cos\theta\frac{\overrightarrow{OA}}{\|\overrightarrow{OA}\|} = u\cos(t/2)(\cos t, \sin t, 0).$$

Since $\overrightarrow{QP}$ is perpendicular to the xy-plane and $\|\overrightarrow{QP}\| = u|\sin\theta|$, it follows that $\overrightarrow{QP} = u\sin\theta\,\mathbf{k} = (0, 0, u\sin(t/2))$. Hence

$$\begin{aligned}\overrightarrow{OP} &= \overrightarrow{OA} + \overrightarrow{AQ} + \overrightarrow{QP}\\ &= (\cos t, \sin t, 0) + u\cos(t/2)(\cos t, \sin t, 0) + (0, 0, u\sin(t/2)),\end{aligned}$$

and the Möbius strip can be parametrized as

$$\mathbf{r}(t,u) = (\cos t(1 + u\cos(t/2)), \sin t(1 + u\cos(t/2)), u\sin(t/2)),$$

where $0 \leq t \leq 2\pi$ and $-1/2 \leq u \leq 1/2$; see Figure 7.5.2 (e). If the line segment rotates through 360 degrees, we obtain a strip with two twists (replace $t/2$ in the above parametrization by t); see Figure 7.5.2 (f).

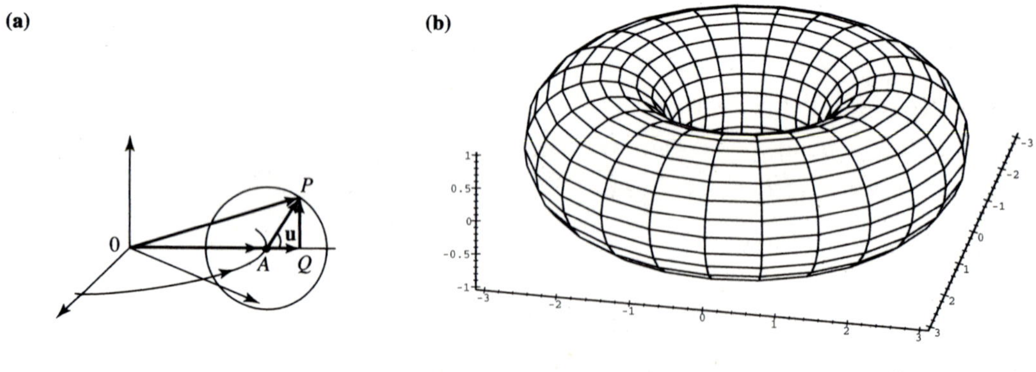

Figure 7.5.2

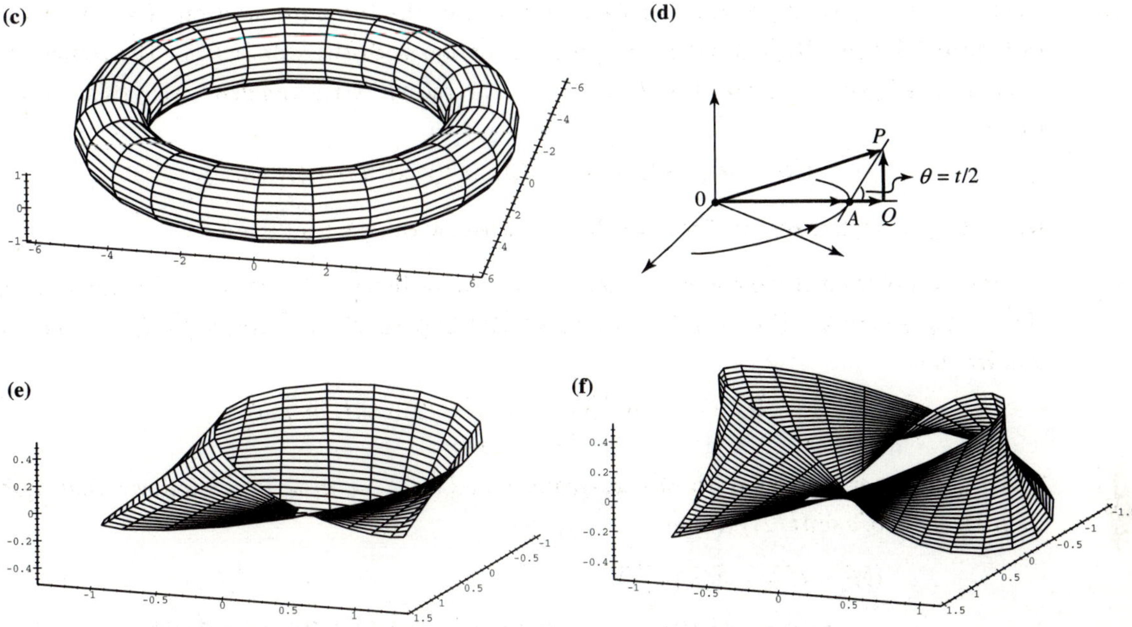

Figure 7.5.2

Solutions to Further Explorations

1. For example, let $f(x, y, z) = z$ and let S be the part of the plane $z = x$ defined by $-1 \leq x \leq 1$ and $0 \leq y \leq 1$. Parametrize the plane by $\mathbf{r}(u, v) = (u, v, u)$, $(u, v) \in D = [-1, 1] \times [0, 1]$; then

$$\mathbf{N} = \mathbf{T}_u \times \mathbf{T}_v = \begin{vmatrix} \mathbf{i} & \mathbf{j} & \mathbf{k} \\ 1 & 0 & 1 \\ 0 & 1 & 0 \end{vmatrix} = -\mathbf{i} + \mathbf{k}$$

and $\|\mathbf{N}\| = \sqrt{2}$. It follows that

$$\left| \iint_S f\, dS \right| = \left| \iint_D u\sqrt{2}\, dA \right| = \sqrt{2} \left| \int_0^1 \left(\int_{-1}^1 u du \right) dv \right| = 0,$$

since $\int_{-1}^1 u\, du = 0$. However,

$$\iint_S |f|\, dS = \iint_D |u|\sqrt{2}\, dA = \sqrt{2} \int_0^1 \left(\int_{-1}^1 |u| du \right) dv$$
$$= \sqrt{2} \int_0^1 \left(-\int_{-1}^0 u\, du + \int_0^1 u\, du \right) dv = \sqrt{2} \int_0^1 dv = \sqrt{2}.$$

3. Using the projection onto the xy-plane, we get ($A(S)$ denotes the surface area)

$$A(S) = \iint_S dS = \iint_D \frac{dA}{|\mathbf{n} \cdot \mathbf{k}|},$$

where $\mathbf{n}$ is a unit normal to S. Since the gradient of f is a vector field perpendicular to S, it follows that

$$\mathbf{n} = \pm\frac{\nabla f}{\|\nabla f\|} = \pm\frac{f_x\mathbf{i} + f_y\mathbf{j} + f_z\mathbf{k}}{\sqrt{f_x^2 + f_y^2 + f_z^2}}.$$

In either case,

$$|\mathbf{n}\cdot\mathbf{k}| = \frac{|f_z|}{\sqrt{f_x^2 + f_y^2 + f_z^2}}$$

and

$$A(S) = \iint_D \frac{\sqrt{f_x^2 + f_y^2 + f_z^2}}{|f_z|}\,dA,$$

where dA refers to integration with respect to x and y.

In the case $z = g(x,y)$, $(x,y) \in D$, we let $f(x,y,z) = g(x,y) - z$ and use the formula for $A(S)$ we have just derived; thus $f_x = g_x$, $f_y = g_y$ and $f_z = -1$ and

$$A(S) = \iint_D \frac{\sqrt{g_x^2 + g_y^2 + 1}}{1}\,dA = \iint_D \sqrt{g_x^2 + g_y^2 + 1}\,dA.$$

5. By definition,

$$\begin{aligned}&\iint_S P(x,y,z)dydz + Q(x,y,z)dzdx + R(x,y,z)dxdy\\ &= \iint_D \left(P(u,v,f(u,v))\frac{\partial(y,z)}{\partial(u,v)} + Q(u,v,f(u,v))\frac{\partial(z,x)}{\partial(u,v)} + R(u,v,f(u,v))\frac{\partial(x,y)}{\partial(u,v)}\right) dA.\end{aligned}$$

The Jacobian determinants are computed to be

$$\frac{\partial(y,z)}{\partial(u,v)} = \begin{vmatrix} 0 & 1 \\ f_u(u,v) & f_v(u,v)\end{vmatrix} = -f_u(u,v),$$

$$\frac{\partial(z,x)}{\partial(u,v)} = \begin{vmatrix} f_u(u,v) & f_v(u,v) \\ 1 & 0\end{vmatrix} = -f_v(u,v)$$

and

$$\frac{\partial(x,y)}{\partial(u,v)} = \begin{vmatrix} 1 & 0 \\ 0 & 1\end{vmatrix} = 1.$$

7. By definition,

$$\iint_S dS = \iint_D \|\mathbf{N}\|\,dA,$$

where (use subscripts for partial derivatives)

$$\mathbf{N} = \mathbf{T}_u \times \mathbf{T}_v = (y_u z_v - y_v z_u, z_u x_v - z_v x_u, x_u y_v - x_v y_u)\,;$$

see formula (7.2) in Section 7.1. Hence

$$\begin{aligned}\|\mathbf{N}\| &= (y_u z_v - y_v z_u)^2 + (z_u x_v - z_v x_u)^2 + (x_u y_v - x_v y_u)^2\\ &= y_u^2 z_v^2 + y_v^2 z_u^2 - 2y_u y_v z_u z_v + z_u^2 x_v^2 + z_v^2 x_u^2 - 2x_u x_v z_u z_v + x_u^2 y_v^2 + x_v^2 y_u^2 - 2x_u x_v y_u y_v.\end{aligned}$$

On the other hand,

$$\begin{aligned}EG - F^2 &= \|\mathbf{r}_u\|^2 \, \|\mathbf{r}_v\|^2 - (\mathbf{r}_u \cdot \mathbf{r}_v)^2\\ &= (x_u^2 + y_u^2 + z_u^2)(x_v^2 + y_v^2 + z_v^2) - (x_u x_v + y_u y_v + z_u z_v)^2\\ &= x_u^2 x_v^2 + x_u^2 y_v^2 + x_u^2 z_v^2 + y_u^2 x_v^2 + y_u^2 y_v^2 + y_u^2 z_v^2 + z_u^2 x_v^2 + z_u^2 y_v^2 + z_u^2 z_v^2\\ &\quad - x_u^2 x_v^2 - y_u^2 y_v^2 - z_u^2 z_v^2 - 2x_u x_v y_u y_v - 2x_u x_v z_u z_v - 2y_u y_v z_u z_v\\ &= \|\mathbf{N}\|.\end{aligned}$$

Therefore,

$$\iint_S dS = \iint_D \|\mathbf{N}\| \, dA = \iint_D (EG - F^2) \, dA.$$

8. CLASSICAL INTEGRATION THEOREMS OF VECTOR CALCULUS

8.1 Green's Theorem

1. By Green's Theorem,

$$\int_{\mathbf{c}} (-2y\mathbf{i} + x\mathbf{j}) \cdot d\mathbf{s} = \iint_D \left(\frac{\partial}{\partial x}(x) - \frac{\partial}{\partial y}(-2y)\right) dA = \iint_D 3\, dA,$$

where D is the region enclosed by $\mathbf{c}$; it is an ellipse with semi-axes of length 2 (in the x-direction) and 1 (in the y-direction). It follows that

$$\iint_D 3\, dA = 3\,\text{area}(D) = 6\pi.$$

(Recall that the area of an ellipse with semi-axes a and b is $ab\pi$.)

3. Let $\mathbf{F} = e^{x+y}\mathbf{j} - e^{x-y}\mathbf{i}$; then

$$curl\,\mathbf{F} = \begin{vmatrix} \mathbf{i} & \mathbf{j} & \mathbf{k} \\ \partial/\partial x & \partial/\partial y & \partial/\partial z \\ -e^{x-y} & e^{x+y} & 0 \end{vmatrix} = (e^{x+y} - e^{x-y})\mathbf{k},$$

and, by Green's Theorem,

$$\int_{\mathbf{c}} (e^{x+y}\mathbf{j} - e^{x-y}\mathbf{i}) \cdot d\mathbf{s} = \iint_D (e^{x+y} - e^{x-y})\mathbf{k} \cdot \mathbf{k}\, dA = \iint_D (e^{x+y} - e^{x-y})\, dA$$

(D is the region in the xy-plane bounded by the lines $x = 0$, $y = 0$ and $y = x$). It follows that

$$\begin{aligned}
\int_{\mathbf{c}} (e^{x+y}\mathbf{j} - e^{x-y}\mathbf{i}) \cdot d\mathbf{s} &= \iint_D (e^{x+y} - e^{x-y})\, dA \\
&= \int_0^1 \left(\int_0^x (e^{x+y} - e^{x-y})dy\right) dx \\
&= \int_0^1 (e^{x+y} + e^{x-y}) \Big|_0^x\, dx \\
&= \int_0^1 (e^{2x} + 1 - 2e^x)\, dx \\
&= \left(\tfrac{1}{2}e^{2x} + x - 2e^x\right)\Big|_0^1 = \tfrac{1}{2}e^2 - 2e + \tfrac{5}{2}.
\end{aligned}$$

5. Let D be the region in the xy-plane bounded by the graphs of $y = 2x^3$ and $y = 2x$ on $[0, 1]$. By Green's Theorem,

$$\begin{aligned}
\int_{\mathbf{c}} (2x^2y^2\mathbf{i} - x\mathbf{j}) \cdot d\mathbf{s} &= \iint_D \left(\frac{\partial}{\partial x}(-x) - \frac{\partial}{\partial y}(2x^2y^2)\right) dA \\
&= \iint_D (-1 - 4x^2y)\, dA \\
&= \int_0^1 \left(\int_{2x^3}^{2x} (-1 - 4x^2y)dy\right) dx \\
&= -\int_0^1 (y + 2x^2y^2)\Big|_{2x^3}^{2x}\, dx
\end{aligned}$$

$$= -\int_0^1 (2x + 8x^4 - 2x^3 - 8x^8)\,dx$$
$$= -\left(x^2 + \tfrac{8}{5}x^5 - \tfrac{1}{2}x^4 - \tfrac{8}{9}x^9\right)\Big|_0^1$$
$$= -\left(1 + \tfrac{8}{5} - \tfrac{1}{2} - \tfrac{8}{9}\right) = -\tfrac{109}{90}.$$

7. Using Green's Theorem,

$$\int_{\mathbf{c}} (x^2y^2\mathbf{i} + y^4\mathbf{j}) \cdot d\mathbf{s} = -2\iint_D yx^2\,dA,$$

where D is the disk $x^2 + y^2 \leq 1$. So

$$\iint_D yx^2\,dA = \int_{-1}^1 \left(\int_{-\sqrt{1-x^2}}^{\sqrt{1-x^2}} yx^2 dy\right) dx = \int_{-1}^1 x^2 \left(\int_{-\sqrt{1-x^2}}^{\sqrt{1-x^2}} y\,dy\right) dx = 0,$$

since $\int_{-\sqrt{1-x^2}}^{\sqrt{1-x^2}} y\,dy = 0$; recall that an integral of an odd function over a symmetric interval (with respect to 0) is zero.

9. By Green's Theorem,

$$\int_{\mathbf{c}} (\cosh y\mathbf{i} + x\sinh y\mathbf{j}) \cdot d\mathbf{s} = \iint_D (\sinh y - \sinh y)\,dA = 0;$$

D is the region bounded by $y = 2x$, $y = 4x$ and $x = 1$.

11. The curve $\mathbf{c}$ encloses the disk D of radius $\sqrt{2}$ centered at the origin. By Green's Theorem,

$$\int_{\mathbf{c}} (\arctan(y/x)\mathbf{i} + \arctan(x/y)\mathbf{j}) \cdot d\mathbf{s} = \iint_D \left(\frac{\partial}{\partial x}(\arctan(x/y)) - \frac{\partial}{\partial y}(\arctan(y/x))\right) dA.$$

The integrand on the right side is equal to

$$\frac{1}{1 + x^2/y^2}\frac{1}{y} - \frac{1}{1 + y^2/x^2}\frac{1}{x} = \frac{y}{x^2 + y^2} - \frac{x}{x^2 + y^2} = \frac{y - x}{x^2 + y^2},$$

and hence (changing to polar coordinates)

$$\int_{\mathbf{c}} (\arctan(y/x)\mathbf{i} + \arctan(x/y)\mathbf{j}) \cdot d\mathbf{s} = \iint_D \frac{y - x}{x^2 + y^2}\,dA$$
$$= \int_0^{2\pi} \left(\int_0^{\sqrt{2}} \frac{r\sin\theta - r\cos\theta}{r^2}\,rdr\right) d\theta$$
$$= \int_0^{2\pi} \left(\int_0^{\sqrt{2}} (\sin\theta - \cos\theta)dr\right) d\theta$$
$$= \sqrt{2}\int_0^{2\pi} (\sin\theta - \cos\theta)\,d\theta = 0,$$

since $\int_0^{2\pi} \sin\theta\,d\theta = \int_0^{2\pi} \cos\theta\,d\theta = 0$.

13. Combining $y = 4x$ and $y = 2x^2$, we get $x^2 = 2x$; i.e., $x = 0$ and $x = 2$. We will use the formula $\int_{\mathbf{c}} xdy$ to compute the area of D, where $\mathbf{c}$ consists of the parabolic path $y = 2x^2$ (call it $\mathbf{c}_1$) from $(0,0)$ to $(2,8)$ and the straight line segment $y=4x$ (call it $\mathbf{c}_2$) from $(2,8)$ back to $(0,0)$. Parametrize $\mathbf{c}_1$ by $\mathbf{c}_1(t)=(t,2t^2)$, $0\le t\le 2$, so that

$$\int_{\mathbf{c}_1} x\,dy = \int_0^2 (t)(4t)\,dt = \int_0^2 4t^2\,dt = \tfrac{4}{3}t^3\Big|_0^2 = \tfrac{32}{3}.$$

Parametrize $\mathbf{c}_2$ by $\mathbf{c}_2(t)=(2-2t,8-8t)$, $0\le t\le 1$; it follows that

$$\int_{\mathbf{c}_2} x\,dy = \int_0^1 (2-2t)(-8)\,dt = -16\int_0^1 (1-t)\,dt = -16\left(t-\tfrac{1}{2}t^2\right)\Big|_0^1 = -8.$$

Consequently, the area of D is

$$A(D) = \int_{\mathbf{c}} x\,dy = \int_{\mathbf{c}_1} x\,dy + \int_{\mathbf{c}_2} x\,dy = \tfrac{32}{3} - 8 = \tfrac{8}{3}.$$

15. The boundary of D (oriented counterclockwise) consists of the following curves: the straight line segment from $(0,1)$ to $(0,0)$ (call it $\mathbf{c}_1$), straight line segment from $(0,0)$ to $(1,0)$ (call it $\mathbf{c}_2$) and the graph of $x^{2/3}+y^{2/3}=1$ from $(1,0)$ back to $(0,1)$ (call it $\mathbf{c}_3$).

Parametrize $\mathbf{c}_1$ by $\mathbf{c}_1(t)=(0,1-t)$, $t\in[0,1]$; then

$$\int_{\mathbf{c}_1} x\,dy = \int_0^1 0\,dt = 0.$$

Parametrize $\mathbf{c}_2$ by $\mathbf{c}_2(t)=(t,0)$, $t\in[0,1]$; then

$$\int_{\mathbf{c}_2} x\,dy = \int_0^1 0\,dt = 0.$$

Parametrize $\mathbf{c}_3$ by $\mathbf{c}_3(t)=(\cos^3 t,\sin^3 t)$, $t\in[0,\pi/2]$; then

$$\int_{\mathbf{c}_3} x\,dy = \int_0^{\pi/2} (\cos^3 t)(3\sin^2 t\cos t)\,dt = 3\int_0^{\pi/2} \cos^4 t\sin^2 t\,dt.$$

Using integration tables, or doing it by hand (expressing $\sin^2 t$ and $\cos^2 t$ using double-angle formulas) we get

$$\begin{aligned}3\int_0^{\pi/2} \cos^4 t\sin^2 t\,dt &= 3\left(-\tfrac{1}{6}\sin t\cos^5 t + \tfrac{1}{24}\cos^3 t\sin t + \tfrac{1}{16}\sin t\cos t + \tfrac{1}{16}t\right)\Big|_0^{\pi/2} \\ &= \tfrac{3}{32}\pi.\end{aligned}$$

Consequently, the area of D is

$$A(D) = \int_{\mathbf{c}} x\,dy = \int_{\mathbf{c}_1} x\,dy + \int_{\mathbf{c}_2} x\,dy + \int_{\mathbf{c}_3} x\,dy = \tfrac{3}{32}\pi.$$

17. Consider the triangular region D determined by the given points; it is bounded by the lines $y=0$, $y=2x$ and $y=3-x$. By Green's Theorem, the work of $\mathbf{F}$ is

$$W = \int_{\mathbf{c}} (x\mathbf{i} + (x^2+3y^2)\mathbf{j})\cdot d\mathbf{s} = \iint_D 2x\,dA$$

(combining $y=2x$ and $y=3-x$ we get $x=1$, and thus $y=2$)

$$= \int_0^2 \left(\int_{y/2}^{3-y} 2x dx \right) dy = \int_0^2 x^2 \Big|_{y/2}^{3-y} dy$$

$$= \int_0^2 \left(9 - 6y + y^2 - \tfrac{1}{4}y^2\right) dy = \left(\tfrac{1}{4}y^3 - 3y^2 + 9y\right)\Big|_0^2 = 8.$$

19. By Green's Theorem for differential forms,

$$\int_{\partial D} (f_x dy - f_y dx) = \iint_D d(f_x dy - f_y dx)$$

$$= \iint_D (f_{xx} dx + f_{xy} dy) \wedge dy - (f_{yx} dx + f_{yy} dy) \wedge dx$$

$$= \iint_D (f_{xx} + f_{yy}) dx dy = 0,$$

since $dydx = -dxdy$, and since, by assumption, $f_{xx} + f_{yy} = 0$.

21. The moments of inertia are given by

$$I_x = \iint_D y^2 \rho \, dA = \rho \iint_D y^2 \, dA$$

and

$$I_y = \iint_D x^2 \rho \, dA = \rho \iint_D x^2 \, dA,$$

since ρ is constant.

We have to find the components of a vector field $\mathbf{F} = P\mathbf{i} + Q\mathbf{j}$ such that

$$\int_{\partial D} \mathbf{F} \cdot d\mathbf{s} = \iint_D \left(\frac{\partial Q}{\partial x} - \frac{\partial P}{\partial y} \right) dA = \rho \iint_D y^2 \, dA.$$

There are many possibilities; for example, let $\partial P/\partial y = 0$, and $\partial Q/\partial x = y^2$; in that case, $P = C_1(x)$ and $Q = xy^2 + C_2(y)$. Therefore,

$$I_x = \rho \int_{\mathbf{c}} (C_1(x)\mathbf{i} + (xy^2 + C_2(y))\mathbf{j}) \cdot d\mathbf{s},$$

where $C_1(x)$ is any function of x only and $C_2(y)$ is any function of y only. Similarly,

$$I_y = \rho \int_{\mathbf{c}} (D_1(x)\mathbf{i} + (\tfrac{1}{3}x^3 + D_2(y))\mathbf{j}) \cdot d\mathbf{s},$$

where $D_1(x)$ and $D_2(y)$ are functions of the indicated variables.

23. From $f = e^x \cos y$ we get $f_x = f_{xx} = e^x \cos y$, $f_y = -e^x \sin y$ and $f_{yy} = -e^x \cos y$. It follows that $\Delta f = f_{xx} + f_{yy} = 0$, and therefore $\iint_D \Delta f \, dA = 0$.

Now we compute the right side. The gradient of f is given by $\nabla f = e^x \cos y \, \mathbf{i} - e^x \sin y \, \mathbf{j}$. Parametrize the line segment from $(0,0)$ to $(1,0)$ by $\mathbf{c}_1(t) = (t, 0)$, $t \in [0,1]$. The outward unit normal is $\mathbf{n}_1 = -\mathbf{j}$, and so

$$D_{\mathbf{n}_1} f = \nabla f \cdot (-\mathbf{j}) = (e^x \cos y \, \mathbf{i} - e^x \sin y \, \mathbf{j}) \cdot (-\mathbf{j}) = e^x \sin y$$

and

$$\int_{\mathbf{c}_1} D_{\mathbf{n}_1} f \, ds = \int_{\mathbf{c}_1} e^x \sin y \, ds = \int_0^1 0 \, dt = 0.$$

Parametrize the line segment from $(1,0)$ to $(1,2)$ by $\mathbf{c}_2(t) = (1,t)$, $t \in [0,2]$. The outward unit normal is $\mathbf{n}_2 = \mathbf{i}$, and so

$$D_{\mathbf{n}_2} f = \nabla f \cdot \mathbf{i} = (e^x \cos y\, \mathbf{i} - e^x \sin y\, \mathbf{j}) \cdot \mathbf{i} = e^x \cos y$$

and

$$\int_{\mathbf{c}_2} D_{\mathbf{n}_2} f\, ds = \int_{\mathbf{c}_2} e^x \cos y\, ds = \int_0^2 e \cos t\, dt = e \sin t \Big|_0^2 = e \sin 2.$$

Parametrize the line segment from $(1,2)$ to $(0,2)$ by $\mathbf{c}_3(t) = (1-t, 2)$, $t \in [0,1]$. The outward unit normal is $\mathbf{n}_3 = \mathbf{j}$, and so $D_{\mathbf{n}_3} f = -e^x \sin y$ and

$$\int_{\mathbf{c}_3} D_{\mathbf{n}_3} f\, ds = -\int_{\mathbf{c}_3} e^x \sin y\, ds = -\int_0^1 e^{1-t} \sin 2\; dt = (\sin 2) e^{1-t}\Big|_0^1 = (1-e)\sin 2.$$

Finally, parametrize the line segment from $(0,2)$ to $(0,0)$ by $\mathbf{c}_4(t) = (0, 2-t)$, $t \in [0,2]$. The outward unit normal is $\mathbf{n}_4 = -\mathbf{i}$, and so $D_{\mathbf{n}_4} f = -e^x \cos y$ and

$$\int_{\mathbf{c}_4} D_{\mathbf{n}_4} f\, ds = -\int_{\mathbf{c}_4} e^x \cos y\, ds = -\int_0^2 \cos(2-t)\, dt = \sin(2-t)\Big|_0^2 = -\sin 2.$$

It follows that

$$\begin{aligned}\int_{\mathbf{c}} D_{\mathbf{n}} f\, ds &= \int_{\mathbf{c}_1} D_{\mathbf{n}_1} f\, ds + \int_{\mathbf{c}_2} D_{\mathbf{n}_2} f\, ds + \int_{\mathbf{c}_3} D_{\mathbf{n}_3} f\, ds + \int_{\mathbf{c}_4} D_{\mathbf{n}_4} f\, ds \\ &= e \sin 2 + (1-e)\sin 2 - \sin 2 = 0.\end{aligned}$$

25. Let D be the triangular region in the xy-plane bounded by the lines $y = x$, $x = 0$ and $y = 1$ (in that case, $\mathbf{c}$ is its positively oriented boundary). By Green's Theorem,

$$\begin{aligned}\int_{\mathbf{c}} y^2 dx + xy dy &= \int_D d(y^2 dx + xy dy) \\ &= \int_D (2y dy dx + y dx dy) = -\int_D y\, dx dy\end{aligned}$$

(this is an integral of a 2-form; interpret it as an iterated integral)

$$\begin{aligned}&= -\iint_D y\, dA = -\int_0^1 \left(\int_x^1 y dy\right) dx \\ &= -\tfrac{1}{2}\int_0^1 (1-x^2)\, dx = -\tfrac{1}{2}\left(x - \tfrac{1}{3}x^3\right)\Big|_0^1 = -\tfrac{1}{3}.\end{aligned}$$

27. The curves $y = x^2$ and $y = x$ intersect at $(0,0)$ and $(1,1)$. Let D be the region between $y = x^2$ and $y = x$; then $\mathbf{c}$ is its positively oriented boundary. By Green's Theorem,

$$\begin{aligned}\int_{\mathbf{c}} (x-y)^2 dx + 2x dy &= \int_D d((x-y)^2 dx + 2x dy) \\ &= \int_D (-2(x-y) dy dx + 2 dx dy) = \int_D (2 + 2x - 2y) dx dy\end{aligned}$$

(this is an integral of a 2-form; interpret it as an iterated integral)

$$= \int_0^1 \left(\int_{x^2}^{x} (2 + 2x - 2y) dy \right) dx$$
$$= \int_0^1 (2y + 2xy - y^2) \Big|_{x^2}^{x} dx$$
$$= \int_0^1 (2x - x^2 - 2x^3 + x^4)\, dx$$
$$= (x^2 - \tfrac{1}{3}x^3 - \tfrac{1}{2}x^4 + \tfrac{1}{5}x^5) \Big|_0^1 = \tfrac{11}{30}.$$

29. By Green's Theorem, (D is the disk $x^2 + y^2 \leq 1$)

$$\int_{\mathbf{c}} 2xy^2 dx + 2x^2 y dy = \int_D d(2xy^2 dx + 2x^2 y dy)$$
$$= \int_D (4xy dy dx + 4xy dx dy) = \int_D 0 = 0.$$

31. Let D be the annulus $1 \leq x^2 + y^2 \leq 2$. By Green's Theorem,

$$\int_{\mathbf{c}} x dx + (x^2 + y^2) dy = \int_D d(x dx + (x^2 + y^2) dy)$$
$$= \int_D 2x\, dx dy = 2 \iint_D x\, dA$$

(change to polar coordinates)

$$= 2 \int_0^{2\pi} \left(\int_1^{\sqrt{2}} r^2 \cos\theta dr \right) d\theta$$
$$= 2 \left(\int_0^{2\pi} \cos\theta\, d\theta \right) \left(\int_1^{\sqrt{2}} r^2\, dr \right) = 0,$$

since $\int_0^{2\pi} \cos\theta\, d\theta = 0$.

8.2 The Divergence Theorem

1. Since

$$div\, ((y^2 + \sin z)\mathbf{i} + (e^{\sin z} + 2)\mathbf{j} + (xy + \ln x)\mathbf{k}) = 0,$$

the Divergence Theorem implies that

$$\iint_S ((y^2 + \sin z)\mathbf{i} + (e^{\sin z} + 2)\mathbf{j} + (xy + \ln x)\mathbf{k}) \cdot d\mathbf{S} = \iiint_V 0\, dV = 0$$

(V is the three-dimensional solid enclosed by S).

3. By the Divergence Theorem,

$$\iint_S ((x + y^2 + 1)\mathbf{i} + (y + xz)\mathbf{j}) \cdot d\mathbf{S} = \iiint_V div\, ((x + y^2 + 1)\mathbf{i} + (y + xz)\mathbf{j})\, dV$$
$$= \iiint_V 2\, dV,$$

where V is the part of the cone $z = x^2 + y^2$ between the planes $z = 1$ and $z = 2$. Instead of a somewhat messy integration, we proceed by noticing that

$$\iiint_V 2\,dV = 2\,\text{volume}(V).$$

Since volume(V) = (volume of the cone of radius 2 and height 2) − (volume of the cone of radius 1 and height 1) $= 8\pi/3 - \pi/3 = 7\pi/3$, it follows that $\iiint_V 2\,dV = 14\pi/3$.

5. Since

$$div\,(-e^x \cos y\mathbf{i} + e^x \sin y\mathbf{j} + \mathbf{k}) = 0,$$

the Divergence Theorem implies that

$$\iint_S (-e^x \cos y\mathbf{i} + e^x \sin y\mathbf{j} + \mathbf{k}) \cdot d\mathbf{S} = \iiint_V 0\,dV = 0$$

(V is the three-dimensional solid enclosed by S).

7. By the Divergence Theorem,

$$\iint_S \mathbf{F} \cdot d\mathbf{S} = \iiint_V div\,\mathbf{F}\,dV = 4\iiint_V x\,dV,$$

where V is the three-dimensional solid inside the hemisphere and above the xy-plane. Changing to polar coordinates, we get

$$\begin{aligned}\iint_S \mathbf{F} \cdot d\mathbf{S} &= 4\int_0^{2\pi}\left(\int_0^{\pi/2}\left(\int_0^1 \rho\sin\phi\cos\theta\,\rho^2\sin\phi d\rho\right)d\phi\right)d\theta \\ &= 4\left(\int_0^{2\pi}\cos\theta\,d\theta\right)\left(\int_0^{\pi/2}\left(\int_0^1 \rho^3\sin^2\phi d\rho\right)d\phi\right) = 0,\end{aligned}$$

since $\int_0^{2\pi}\cos\theta\,d\theta = 0$.

9. By the Divergence Theorem,

$$\iint_S (ye^z\mathbf{i} + yz\mathbf{k}) \cdot d\mathbf{S} = \iiint_V div\,(ye^z\mathbf{i} + yz\mathbf{k})\,dV = \iiint_V y\,dV,$$

where V is the tetrahedron bounded by the coordinate planes and the plane $x + 2y + z = 4$. It is defined by $0 \le y \le 2$, $0 \le x \le 4 - 2y$ (to get this, substitute $z = 0$ into $x + 2y + z = 4$) and $0 \le z \le 4 - x - 2y$. Thus

$$\begin{aligned}\iiint_V y\,dV &= \int_0^2\left(\int_0^{4-2y}\left(\int_0^{4-x-2y} ydz\right)dx\right)dy \\ &= \int_0^2\left(\int_0^{4-2y}(4 - x - 2y)y\,dx\right)dy \\ &= \int_0^2 \left(4xy - \tfrac{1}{2}x^2y - 2y^2x\right)\Big|_0^{4-2y} dy \\ &= \int_0^2 (2y^3 - 8y^2 + 8y)\,dy \\ &= \left(\tfrac{1}{2}y^4 - \tfrac{8}{3}y^3 + 4y^2\right)\Big|_0^2 = \tfrac{8}{3}.\end{aligned}$$

11. Write $\iint_S xyz\,dS$ as

$$\iint_S xyz\,dS = \iint_S xyz\mathbf{n}\cdot\mathbf{n}\,dS,$$

where $\mathbf{n}$ is the unit outward normal to S. Since S is a sphere centered at the origin, $\mathbf{n}$ has the same direction as the position vector of a point; i.e., $\mathbf{n} = (x\mathbf{i}+y\mathbf{j}+z\mathbf{k})/\sqrt{x^2+y^2+z^2} = x\mathbf{i}+y\mathbf{j}+z\mathbf{k}$ (since $x^2+y^2+z^2=1$). The divergence of $xyz\mathbf{n}$ is computed to be

$$div\,(xyz\mathbf{n}) = div\,xyz(x\mathbf{i}+y\mathbf{j}+z\mathbf{k}) = 6xyz,$$

and, consequently

$$\iint_S xyz\mathbf{n}\cdot\mathbf{n}\,dS = \iint_S xyz\mathbf{n}\cdot d\mathbf{S} = \iiint_V div\,(xyz\mathbf{n})\,dV = 6\iiint_V xyz\,dV,$$

where V is the ball $0 \le x^2+y^2+z^2 \le 1$. Using spherical coordinates to evaluate the integral, we get

$$6\int_0^{2\pi}\left(\int_0^{\pi}\left(\int_0^1 \rho^3\sin^2\phi\cos\phi\ \sin\theta\cos\theta\ \rho^2\sin\phi d\rho\right)d\phi\right)d\theta = 0.$$

The result was obtained without too much integration — factor $\sin\theta\cos\theta$ all the way to the outermost integration and use the fact that $\int_0^{2\pi}\sin\theta\cos\theta\,d\theta = \frac{1}{2}\int_0^{2\pi}\sin 2\theta\,d\theta = 0$.

13. If we show that

$$div\,(f\nabla g) = f\Delta g + \nabla f\cdot\nabla g,$$

then we are done — the statement follows from the Divergence Theorem.

Using a product rule for the divergence (see Section 4.3), we get

$$div\,(f\nabla g) = f div\,(\nabla g) + \nabla f\cdot\nabla g = f\Delta g + \nabla f\cdot\nabla g,$$

since, by definition, $div\,(\nabla g) = \Delta g$.

15. By definition, $D_{\mathbf{n}}f = \nabla f\cdot\mathbf{n}$, where $\mathbf{n}$ is a unit normal to the surface. So

$$\iint_S D_{\mathbf{n}}f\,dS = \iint_S \nabla f\cdot\mathbf{n}\,dS = \iint_S \nabla f\,d\mathbf{S}.$$

Now proceed using the Divergence Theorem and the definition of the Laplace operator:

$$\iint_S \nabla f\,d\mathbf{S} = \iiint_V div\,(\nabla f)\,dV = \iiint_V \Delta f\,dV.$$

17. Since

$$div\left(\frac{y\mathbf{i}}{x^2+y^2} - \frac{x\mathbf{j}}{x^2+y^2}\right) = \frac{-2xy}{(x^2+y^2)^2} - \frac{-2xy}{(x^2+y^2)^2} = 0,$$

it follows that

$$\iint_{\mathbf{c}} \mathbf{F}\cdot\mathbf{n}\,ds = \iint_R div\,\mathbf{F}\,dA = 0.$$

19. The outward flux of $\mathbf{F}$ is given by

$$\iint_{\mathbf{c}} \mathbf{F}\cdot\mathbf{n}\,ds = \iint_D div\,\mathbf{F}\,dA = \iint_D x\,dA$$

where D is the triangle bounded by the lines $y = x$, $y = 2x$ and $x = 1$. Hence

$$\iint_{\mathbf{c}} \mathbf{F}\cdot\mathbf{n}ds = \int_0^1 \left(\int_x^{2x} xdy\right) dx = \int_0^1 xy\Big|_x^{2x} dx = \int_0^1 x^2\, dx = \tfrac{1}{3}.$$

The counterclockwise circulation of $\mathbf{F}$ is computed using Green's Theorem:

$$\begin{aligned}\iint_{\mathbf{c}} \mathbf{F}\cdot d\mathbf{s} &= \iint_D curl\,\mathbf{F}\cdot\mathbf{k}\,dA = \iint_D (-3y)\,dA \\ &= -3\int_0^1 \left(\int_x^{2x} ydy\right) dx = -\tfrac{3}{2}\int_0^1 (y^2)\Big|_x^{2x} dx = -\tfrac{3}{2}\int_0^1 3x^2\, dx = -\tfrac{3}{2}.\end{aligned}$$

21. Since $div\,\mathbf{F} = e^x e^y + 2e^y = e^y(e^x + 2)$, the outward flux of $\mathbf{F}$ is given by

$$\iint_{\mathbf{c}} \mathbf{F}\cdot\mathbf{n}\;ds = \iint_R div\,\mathbf{F}\,dA = \iint_R e^y(e^x+2)\,dA.$$

From Fubini's Theorem, it follows that

$$\begin{aligned}\iint_R e^y(e^x+2)\,dA &= \int_0^3 \left(\int_0^4 e^y(e^x+2)dy\right) dx \\ &= (e^4-1)\int_0^3 (e^x+2)\,dx = (e^4-1)\,(e^x+2x)\Big|_0^3 = (e^4-1)(e^3+5).\end{aligned}$$

Since

$$curl\,\mathbf{F} = \begin{vmatrix} \mathbf{i} & \mathbf{j} & \mathbf{k} \\ \partial/\partial x & \partial/\partial y & \partial/\partial z \\ e^x e^y & 2e^y & 0 \end{vmatrix} = -e^x e^y \mathbf{k},$$

it follows that the counterclockwise circulation of $\mathbf{F}$ is (by Green's Theorem)

$$\begin{aligned}\iint_{\mathbf{c}} \mathbf{F}\cdot d\mathbf{s} &= \iint_R curl\,\mathbf{F}\cdot\mathbf{k}\,dA = -\iint_R e^x e^y\,dA \\ &= -\int_0^4 \left(\int_0^3 e^x e^y dx\right) dy \\ &= -(e^3-1)\int_0^4 e^y\,dy = -(e^3-1)(e^4-1).\end{aligned}$$

23. Since

$$d\alpha = d(ydydz + xzdzdx + zdxdy) = dxdydz,$$

it follows that

$$\int_S ydydz + xzdzdx + zdxdy = \int_V dxdydz = \iiint_V dV,$$

where V is the three-dimensional solid region bounded by S. Using cylindrical coordinates, we get

$$\begin{aligned}\iiint_V dV &= \int_0^{2\pi}\left(\int_0^1 \left(\int_0^{r^2} rdz\right) dr\right) d\theta \\ &= \int_0^{2\pi}\left(\int_0^1 r^3 dr\right) d\theta = \tfrac{1}{4}2\pi = \tfrac{1}{2}\pi.\end{aligned}$$

25. Since

$$d\alpha = d(xdydz + ydzdx + zdxdy) = 3dxdydz,$$

the Divergence Theorem implies that (V is the solid region between the spheres of radius 1 and 2)

$$\int_S xdydz + ydzdx + zdxdy = 3\int_V dxdydz$$
$$= 3\ \text{volume}(V) = 3\left(\tfrac{4}{3}\pi\, 2^3 - \tfrac{4}{3}\pi\, 1^3\right) = 28\pi.$$

8.3 Stokes' Theorem

1. Complete the square in $x^2 + 2x + y^2 = 3$ to get $(x+1)^2 + y^2 = 4$; parametrize the intersection $\mathbf{c}$ of the cylinder $x^2 + 2x + y^2 = 3$ and the plane $z = x$ as

$$\mathbf{c}(t) = (2\cos t - 1, 2\sin t, 2\cos t - 1), \qquad t \in [0, 2\pi].$$

Then $\mathbf{c}'(t) = (-2\sin t, 2\cos t, -2\sin t)$ and

$$\iint_{\mathbf{c}} \mathbf{F}\cdot d\mathbf{s} = \int_0^{2\pi} (4\cos^2 t, 0, -(2\cos t - 1)^2)\cdot(-2\sin t, 2\cos t, -2\sin t)\, dt$$
$$= \int_0^{2\pi} (-8\cos t\sin t + 2\sin t)\, dt$$
$$= \int_0^{2\pi} (-4\sin 2t + 2\sin t)\, dt = 0,$$

since $\int_0^{2\pi} \sin t\, dt = 0$ and $\int_0^{2\pi} \sin 2t\, dt = 0$.

Parametrize the part of the plane inside the cylinder (call it S) by $\mathbf{r}(u,v) = (u,v,u)$, where $(u+1)^2 + v^2 \le 4$. The surface normal vector is $\mathbf{N} = (-1,0,1)$ and, since $curl\,\mathbf{F} = 2x\mathbf{j}$, it follows that

$$\iint_S curl\,\mathbf{F}\cdot\mathbf{N}\, dS = \iint_S (0,2u,0)\cdot(-1,0,1)\, dS = \iint_S 0\, dS = 0.$$

3. The *curl* of $\mathbf{F}$ is

$$curl\,\mathbf{F} = \begin{vmatrix} \mathbf{i} & \mathbf{j} & \mathbf{k} \\ \partial/\partial x & \partial/\partial y & \partial/\partial z \\ 2x+y & 0 & -3x+y+x^2z \end{vmatrix} = \mathbf{i} + 3\mathbf{j} - \mathbf{k};$$

so, by Stokes' Theorem,

$$\int_{\mathbf{c}} \mathbf{F}\cdot d\mathbf{s} = \iint_S curl\,\mathbf{F}\cdot d\mathbf{S} = \iint_S (\mathbf{i} + 3\mathbf{j} - \mathbf{k})\cdot d\mathbf{S},$$

where S is the part of the plane $x + 4y + 3z = 1$ in the first octant, oriented by a normal that points away from the side that does not face the origin; i.e., that normal has positive $\mathbf{k}$ component. Parametrize the surface of the plane by

$$\mathbf{r}(u,v) = (u, v, (1 - u - 4v)/3), \qquad (u,v) \in D,$$

where D is the region in the uv-plane defined by $u = 0$, $v = 0$ and $u + 4v = 1$. Then

$$\mathbf{N} = \begin{vmatrix} \mathbf{i} & \mathbf{j} & \mathbf{k} \\ 1 & 0 & -1/3 \\ 0 & 1 & -4/3 \end{vmatrix} = \tfrac{1}{3}\mathbf{i} + \tfrac{4}{3}\mathbf{j} + \mathbf{k}$$

is a normal with the desired direction. It follows that

$$\iint_S (\mathbf{i} + 3\mathbf{j} - \mathbf{k}) \cdot d\mathbf{S} = \iint_D (\mathbf{i} + 3\mathbf{j} - \mathbf{k}) \cdot \left(\tfrac{1}{3}\mathbf{i} + \tfrac{4}{3}\mathbf{j} + \mathbf{k}\right) dA$$
$$= \iint_D \tfrac{10}{3}\, dA = \tfrac{10}{3}\,\tfrac{1}{8} = \tfrac{5}{12},$$

since $\iint_D dA = \text{area}(D) = 1/8$ (D is a right triangle with sides 1 and $1/4$).

5. The *curl* of $\mathbf{F}$ is computed to be

$$curl\, \mathbf{F} = \begin{vmatrix} \mathbf{i} & \mathbf{j} & \mathbf{k} \\ \partial/\partial x & \partial/\partial y & \partial/\partial z \\ x^2 + z^2 & y^2 z^2 & 0 \end{vmatrix} = -2y^2 z\mathbf{i} + 2z\mathbf{j};$$

by Stokes' Theorem,

$$\int_{\mathbf{c}} \mathbf{F} \cdot d\mathbf{s} = \iint_S curl\, \mathbf{F} \cdot d\mathbf{S} = \iint_S (-2y^2 z\mathbf{i} + 2z\mathbf{j}) \cdot d\mathbf{S},$$

where S is the part of the plane $z = y$ defined by $1 \le x \le 2$ and $0 \le y \le 4$. Parametrize S by

$$\mathbf{r}(u, v) = (u, v, v), \qquad (u, v) \in D = [1, 2] \times [0, 4].$$

The surface normal $\mathbf{N}$ is

$$\mathbf{N} = \begin{vmatrix} \mathbf{i} & \mathbf{j} & \mathbf{k} \\ 1 & 0 & 0 \\ 0 & 1 & 1 \end{vmatrix} = -\mathbf{j} + \mathbf{k}$$

(notice that the orientation requirement is satisfied); therefore,

$$\iint_S (-2y^2 z\mathbf{i} + 2z\mathbf{j}) \cdot d\mathbf{S} = \iint_D (-2v^3\mathbf{i} + 2v\mathbf{j}) \cdot (-\mathbf{j} + \mathbf{k})\, dA$$
$$= -\iint_D 2v\, dA = -\int_1^2 \left(\int_0^4 2v dv \right) du$$
$$= -\int_1^2 v^2 \Big|_0^4 du = -\int_1^2 16\, du = -16.$$

7. Since

$$curl\, \mathbf{F} = \begin{vmatrix} \mathbf{i} & \mathbf{j} & \mathbf{k} \\ \partial/\partial x & \partial/\partial y & \partial/\partial z \\ y^2 & y^2 & y^2 \end{vmatrix} = 2y\mathbf{i} - 2y\mathbf{k},$$

Stokes' Theorem implies that

$$\iint_{\mathbf{c}} y^2(\mathbf{i} + \mathbf{j} + \mathbf{k}) \cdot d\mathbf{s} = \iint_{\overline{S}} (2y\mathbf{i} - 2y\mathbf{k}) \cdot d\mathbf{S},$$

where $\overline{S}$ is a surface whose boundary is $\mathbf{c}$. Instead of taking $\overline{S}$ to be the part of the sphere (and thus going through a messy parametrization and evaluation of an integral), we use the plane instead. In other words, parametrize the part of the plane $z = 1/2$ (call it S) whose boundary is $\mathbf{c}$ by $\mathbf{r}(u,v) = (u, v, 1/2)$, where $0 \le u^2 + v^2 \le \frac{3}{4}$. The surface normal is $\mathbf{N} = \mathbf{k}$; that forces the boundary curve to be oriented counterclockwise as seen from above (which is precisely what the assumption states). Hence

$$\iint_S (2y\mathbf{i} - 2y\mathbf{k}) \cdot d\mathbf{S} = \iint_{\{0 \le u^2+v^2 \le 3/4\}} (2y\mathbf{i} - 2y\mathbf{k}) \cdot \mathbf{k}\, dS$$
$$= -2\int_0^{2\pi} \left(\int_0^{\sqrt{3}/2} r \sin\theta \, r dr \right) d\theta = 0,$$

since $\int_0^{2\pi} \sin\theta \, d\theta = 0$.

9. By Stokes' Theorem,

$$\int_{\mathbf{c}} (\mathbf{F} \times \mathbf{r}) \cdot d\mathbf{s} = \iint_S \mathit{curl}\,(\mathbf{F} \times \mathbf{r}) \cdot d\mathbf{S}.$$

Let $\mathbf{F} = (F_1, F_2, F_3)$ (since $\mathbf{F}$ is a constant vector field, its components are constants). So

$$\mathbf{F} \times \mathbf{r} = \begin{vmatrix} \mathbf{i} & \mathbf{j} & \mathbf{k} \\ F_1 & F_2 & F_3 \\ x & y & z \end{vmatrix} = (F_2 z - F_3 y)\mathbf{i} + (F_3 x - F_1 z)\mathbf{j} + (F_1 y - F_2 x)\mathbf{k}$$

and

$$\mathit{curl}\,(\mathbf{F} \times \mathbf{r}) = \begin{vmatrix} \mathbf{i} & \mathbf{j} & \mathbf{k} \\ \partial/\partial x & \partial/\partial y & \partial/\partial z \\ F_2 z - F_3 y & F_3 x - F_1 z & F_1 y - F_2 x \end{vmatrix}$$
$$= (F_1 + F_1)\mathbf{i} - (-F_2 - F_2)\mathbf{j} + (F_3 + F_3)\mathbf{k} = 2\mathbf{F}.$$

It follows that

$$\int_{\mathbf{c}} (\mathbf{F} \times \mathbf{r}) \cdot d\mathbf{s} = \iint_S \mathit{curl}\,(\mathbf{F} \times \mathbf{r}) \cdot d\mathbf{S} = 2\iint_S \mathbf{F} \cdot d\mathbf{S}.$$

11. The *curl* of $\mathbf{F}$ is

$$\mathit{curl}\,\mathbf{F} = \begin{vmatrix} \mathbf{i} & \mathbf{j} & \mathbf{k} \\ \partial/\partial x & \partial/\partial y & \partial/\partial z \\ 2x/(x^2+y) & 1/(x^2+y) & 0 \end{vmatrix}$$
$$= \left(-\frac{2x}{(x^2+y)^2} + \frac{2x}{(x^2+y)^2} \right)\mathbf{k} = \mathbf{0}.$$

The domain of $\mathbf{F}$ is $\mathbb{R}^2$, with the points on the parabola $y = -x^2$ removed — so it is not simply-connected. However, if we restrict its domain as shown in Figure 8.3.11, then it becomes simply-connected. By Theorem 5.8 in Section 5.4, the circulation is zero.

Alternatively, notice that $\mathbf{F} = \nabla \ln(x^2 + y)$; proceed by using the Fundamental Theorem of Calculus (see Theorem 5.5).

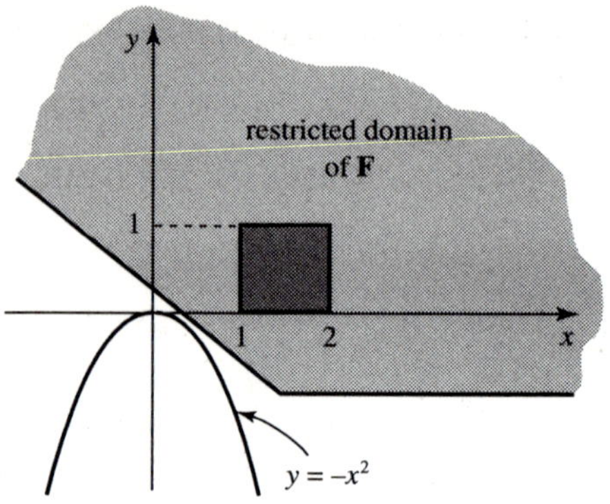

Figure 8.3.11

13. Since

$$curl\,\mathbf{F} = \begin{vmatrix} \mathbf{i} & \mathbf{j} & \mathbf{k} \\ \partial/\partial x & \partial/\partial y & \partial/\partial z \\ y & 2z & 3x \end{vmatrix} = -2\mathbf{i} - 3\mathbf{j} - \mathbf{k},$$

Stokes' Theorem implies that

$$\int_{\mathbf{c}} \mathbf{F} \cdot d\mathbf{s} = \iint_S (-2\mathbf{i} - 3\mathbf{j} - \mathbf{k}) \cdot d\mathbf{S},$$

where S is any surface (with the appropriate orientation) whose boundary is the intersection of $x^2 + y^2 = 1$ and $z = y$. Take the plane and use the parametrization $\mathbf{r}(u, v) = (u, v, v)$, where $(u, v) \in D$, and D is the disk $x^2 + y^2 \leq 1$. Then

$$\mathbf{N} = \begin{vmatrix} \mathbf{i} & \mathbf{j} & \mathbf{k} \\ 1 & 0 & 0 \\ 0 & 1 & 1 \end{vmatrix} = -\mathbf{j} + \mathbf{k},$$

and so

$$\begin{aligned} \iint_S \mathbf{F} \cdot d\mathbf{S} &= \iint_D (-2\mathbf{i} - 3\mathbf{j} - \mathbf{k}) \cdot (-\mathbf{j} + \mathbf{k})\, dA \\ &= \iint_D 2\, dA = 2\text{area}(D) = 2\pi. \end{aligned}$$

15. We will use Stokes' Theorem. Since

$$curl\,\mathbf{F} = \begin{vmatrix} \mathbf{i} & \mathbf{j} & \mathbf{k} \\ \partial/\partial x & \partial/\partial y & \partial/\partial z \\ x & -yz & 1 \end{vmatrix} = y\mathbf{i},$$

it follows that

$$\int_{\mathbf{c}} \mathbf{F} \cdot d\mathbf{s} = \iint_S y\mathbf{i} \cdot d\mathbf{S},$$

where S is any surface whose boundary is $\mathbf{c}$. Take S to be the part of the plane inside the paraboloid — parametrize it by $\mathbf{r}(u, v) = (u, v, 2v)$, where $(u, v) \in D$, and D is the disk

$u^2 + (v-1)^2 \le 1$ (this has been obtained by combining $z = x^2 + y^2$ and $z = 2y$). Then

$$\mathbf{N} = \begin{vmatrix} \mathbf{i} & \mathbf{j} & \mathbf{k} \\ 1 & 0 & 0 \\ 0 & 1 & 2 \end{vmatrix} = -2\mathbf{j} + \mathbf{k},$$

and so

$$\iint_S y\mathbf{i} \cdot d\mathbf{S} = \iint_D y\mathbf{i} \cdot (-2\mathbf{j} + \mathbf{k})\, dA = \iint_D 0\, dA = 0.$$

17. We are not able to use Stokes' Theorem, since in that case we would have to find a surface whose boundary is $\mathbf{c}$ (such surfaces exist, but their parametrizations are quite complicated).

We proceed by computing the path integrals directly (actually we do not have much choice, because $curl\,\mathbf{F} = -2\mathbf{k} \neq \mathbf{0}$ and we cannot use the Fundamental Theorem). From $\mathbf{c}_1(t) = (\cos t, \sin t, t)$ we get $\mathbf{c}_1'(t) = (-\sin t, \cos t, 1)$ and thus

$$\int_{\mathbf{c}_1} \mathbf{F} \cdot d\mathbf{s} = \int_0^{3\pi} (2\cos t + \sin t, 2\sin t - \cos t, 0) \cdot (-\sin t, \cos t, 1)\, dt = \int_0^{3\pi} (-1)\, dt = -3\pi.$$

Parametrize the line segment by

$$\mathbf{c}_2(t) = (-1, 0, 3\pi) + t(2, 0, -3\pi) = (-1 + 2t, 0, 3\pi - 3\pi t),$$

where $t \in [0,1]$. Then $\mathbf{c}_2'(t) = (2, 0, -3\pi)$ and

$$\int_{\mathbf{c}_2} \mathbf{F} \cdot d\mathbf{s} = \int_0^1 (-2 + 4t, 1 - 2t, 0) \cdot (2, 0, -3\pi)\, dt = \int_0^1 (-4 + 8t)\, dt = (-4t + 4t^2)\Big|_0^1 = 0.$$

It follows that

$$\int_{\mathbf{c}} \mathbf{F} \cdot d\mathbf{s} = \int_{\mathbf{c}_1} \mathbf{F} \cdot d\mathbf{s} + \int_{\mathbf{c}_2} \mathbf{F} \cdot d\mathbf{s} = -3\pi.$$

19. By Stokes' Theorem,

$$\int_{\mathbf{c}} f\nabla f \cdot d\mathbf{s} = \iint_S curl\,(f\nabla f) \cdot d\mathbf{S}.$$

Using a product rule (see Section 4.3) we get

$$curl\,(f\nabla f) = f curl\,(\nabla f) + \nabla f \times \nabla f = \mathbf{0},$$

since both summands are $\mathbf{0}$. It follows that $\int_{\mathbf{c}} f\nabla f \cdot d\mathbf{s} = 0$.

21. Parametrize the circle by $\mathbf{c}(t) = \cos t\,\mathbf{i} + \sin t\,\mathbf{j}$, $t \in [0, 2\pi]$. Then $\mathbf{F}(\mathbf{c}(t)) = -2\sin t\,\mathbf{i} + 2\cos t\,\mathbf{j}$, $\mathbf{c}'(t) = -\sin t\,\mathbf{i} + \cos t\,\mathbf{j}$ and

$$\int_{\mathbf{c}} \mathbf{F} \cdot d\mathbf{s} = \int_0^{2\pi} (-2\sin t\,\mathbf{i} + 2\cos t\,\mathbf{j}) \cdot (-\sin t\,\mathbf{i} + \cos t\,\mathbf{j})\, dt = \int_0^{2\pi} 2\, dt = 4\pi.$$

On the other hand, since $curl\,\mathbf{F} = \mathbf{0}$, it follows that

$$\iint_S curl\,\mathbf{F}\, d\mathbf{S} = \iint_S \mathbf{0}\, d\mathbf{S} = 0.$$

The vector field $\mathbf{F}$ is not defined at $(0,0)$, and therefore Stokes' Theorem does not apply.

8.4 Vector Calculus in Electromagnetism

1. There are infinitely many answers. For example,

$$\mathbf{A}_0 = (2xz^2 + f(x))\mathbf{i} + (xy + g(y))\mathbf{j} + (yz + h(z))\mathbf{k},$$

where f, g and h are differentiable functions of variables indicated.

In general, $curl\,\mathbf{A}_0 = curl\,\mathbf{A}$ implies $curl\,(\mathbf{A}_0 - \mathbf{A}) = \mathbf{0}$; i.e., $\mathbf{A}_0 - \mathbf{A} = grad\,f$ (this is due to Theorem 5.4 in Section 5.4; the domain of $\mathbf{A}$ and $\mathbf{A}_0$ is $\mathbb{R}^3$, which is simply-connected). It follows that $\mathbf{A}_0 = \mathbf{A} + grad\,f$, where f is any differentiable function.

3. Formula (8.15) states that

$$\mathbf{J} = \frac{1}{\mu_0} curl\,\mathbf{B} - \epsilon_0 \frac{\partial \mathbf{E}}{\partial t}.$$

Using the linearity of div and switching div and $\partial/\partial t$ we get

$$\begin{aligned} div\,\mathbf{J} &= div\left(\frac{1}{\mu_0} curl\,\mathbf{B} - \epsilon_0 \frac{\partial \mathbf{E}}{\partial t}\right) \\ &= \frac{1}{\mu_0} div\,(curl\,\mathbf{B}) - \epsilon_0 div\left(\frac{\partial \mathbf{E}}{\partial t}\right) = -\epsilon_0 \frac{\partial}{\partial t}(div\,\mathbf{E}), \end{aligned}$$

since $div\,(curl\,\mathbf{B}) = 0$. From (8.13) it follows that

$$div\,\mathbf{J} = -\epsilon_0 \frac{\partial}{\partial t}\left(\frac{\rho}{\epsilon_0}\right) = -\frac{\partial \rho}{\partial t}.$$

5. Since

$$div\,\mathbf{E} = div\,\left(e^t(x+y)\mathbf{i} + e^t(y+z)\mathbf{j} + e^t z\mathbf{k}\right) = 3e^t$$

and $3e^t = \rho/\epsilon_0$ (by assumption), equation (8.13) is satisfied. From

$$curl\,\mathbf{E} = e^t \begin{vmatrix} \mathbf{i} & \mathbf{j} & \mathbf{k} \\ \partial/\partial x & \partial/\partial y & \partial/\partial z \\ x+y & y+z & z \end{vmatrix} = e^t(-\mathbf{i} - \mathbf{k})$$

and

$$-\frac{\partial \mathbf{B}}{\partial t} = -(e^t\mathbf{i} + e^t\mathbf{k})$$

it follows that equation (8.14) is satisfied. Now

$$curl\,\mathbf{B} = \begin{vmatrix} \mathbf{i} & \mathbf{j} & \mathbf{k} \\ \partial/\partial x & \partial/\partial y & \partial/\partial z \\ e^t & x^2 - z^2 & e^t + 1 \end{vmatrix} = 2z\mathbf{i} + 2x\mathbf{k}$$

and

$$\begin{aligned} \mu_0\left(\mathbf{J} + \epsilon_0 \frac{\partial \mathbf{E}}{\partial t}\right) &= \mu_0\mathbf{J} + \mu_0\epsilon_0 \frac{\partial \mathbf{E}}{\partial t} \\ &= \left(2z - \mu_0\epsilon_0 e^t(x+y)\right)\mathbf{i} - \mu_0\epsilon_0 e^t(y+z)\mathbf{j} \\ &\quad + \left(2x - \mu_0\epsilon_0 e^t z\right)\mathbf{k} + \mu_0\epsilon_0 e^t\left((x+y)\mathbf{i} + (y+z)\mathbf{j} + z\mathbf{k}\right) \\ &= 2z\mathbf{i} + 2x\mathbf{k} \end{aligned}$$

imply that equation (8.15) is satisfied. Clearly,

$$div\,\mathbf{B} = div\,(e^t\mathbf{i} + (x^2 - z^2)\mathbf{j} + (e^t + 1)\mathbf{k}) = 0,$$

so (8.16) is satisfied, and we are done.

7. Let E_r, E_θ and E_z be the components of $\mathbf{E}$, and let B_r, B_θ and B_z be the components of $\mathbf{B}$ in cylindrical coordinates. From the formula in Section 4.3 for *curl* in cylindrical coordinates, we get

$$curl\,\mathbf{E} = \frac{1}{r}\begin{vmatrix} \mathbf{e}_r & r\mathbf{e}_\theta & \mathbf{e}_z \\ \frac{\partial}{\partial r} & \frac{\partial}{\partial \theta} & \frac{\partial}{\partial z} \\ E_r & rE_\theta & E_z \end{vmatrix}$$

and so the formula $curl\,\mathbf{E} = -\partial\mathbf{B}/\partial t$ reads (compare components of $\mathbf{e}_r$, $\mathbf{e}_\theta$ and $\mathbf{e}_z$)

$$\frac{1}{r}\left(\frac{\partial E_z}{\partial \theta} - \frac{\partial(rE_\theta)}{\partial z}\right) = \frac{1}{r}\left(\frac{\partial E_z}{\partial \theta} - r\frac{\partial E_\theta}{\partial z}\right) = -\frac{\partial B_r}{\partial t}$$

$$\frac{1}{r}\left(-\frac{\partial E_z}{\partial r} + \frac{\partial E_r}{\partial z}\right) = -\frac{\partial B_\theta}{\partial t}$$

$$\frac{1}{r}\left(\frac{\partial(rE_\theta)}{\partial r} - \frac{\partial E_r}{\partial \theta}\right) = \frac{1}{r}\left(E_\theta + r\frac{\partial E_\theta}{\partial r} - \frac{\partial E_r}{\partial \theta}\right) = -\frac{\partial B_z}{\partial t}.$$

9. Since $\mathbf{E} = \sin x \sin t\mathbf{j}$, it follows that $div\,\mathbf{E} = (\partial/\partial y)(\sin x \sin t) = 0$. From $\mathbf{B} = \sin x \cos t\mathbf{k}$ it follows that $div\,\mathbf{B} = (\partial/\partial z)(\sin x \cos t) = 0$. From

$$curl\,\mathbf{E} = \begin{vmatrix} \mathbf{i} & \mathbf{j} & \mathbf{k} \\ \partial/\partial x & \partial/\partial y & \partial/\partial z \\ 0 & \sin x \sin t & 0 \end{vmatrix} = \cos x \sin t\mathbf{k}$$

and

$$-\frac{\partial \mathbf{B}}{\partial t} = -\sin x(-\sin t)\mathbf{k}$$

it follows that (8.18) is satisfied. However,

$$curl\,\mathbf{B} = \begin{vmatrix} \mathbf{i} & \mathbf{j} & \mathbf{k} \\ \partial/\partial x & \partial/\partial y & \partial/\partial z \\ 0 & 0 & \sin x \cos t \end{vmatrix} = -\cos x \cos t\mathbf{j},$$

whereas

$$\frac{\partial \mathbf{E}}{\partial t} = \sin x \cos t\mathbf{j};$$

so (8.19) does not hold.

8.5 Chapter Review

Solution to Computer Project

1. (a) Parametrize $\mathbf{c}$ as $\mathbf{c}(t) = (\cos t, \sin t, 1)$, $t \in [0, 2\pi]$. Then

$$\mathbf{F}(\mathbf{c}(t)) = (\cos t \sin^2 t - 2, e^{-1}\cos t, \sin t - \cos t - 2),$$

$\mathbf{c}'(t) = (-\sin t, \cos t, 0)$ and so

$$\int_{\mathbf{c}} \mathbf{F} \cdot d\mathbf{s} = \int_0^{2\pi} \mathbf{F}(\mathbf{c}(t)) \cdot \mathbf{c}'(t)\, dt$$
$$= \int_0^{2\pi} (-\cos t \sin^3 t + 2\sin t + e^{-1}\cos^2 t)\, dt.$$

Using a computer, table of integrals or substitution (and double-angle formulas), we get

$$\int_{\mathbf{c}} \mathbf{F} \cdot d\mathbf{s} = \frac{\pi}{e}.$$

(b) From $(1-a)(x^2+y^2)+a=1$ it follows that the circle $x^2+y^2=1$, $z=1$, from (a) is a common boundary of all S_a, $a<1$.

The *curl* of $\mathbf{F}$ is computed to be

$$\operatorname{curl}\mathbf{F} = \begin{vmatrix} \mathbf{i} & \mathbf{j} & \mathbf{k} \\ \partial/\partial x & \partial/\partial y & \partial/\partial z \\ xy^2-2 & xe^{-x^2-y^2} & (y-x-2) \end{vmatrix} = (1, 1, e^{-x^2-y^2}(1-2x^2) - 2xy).$$

Parametrize S_a as

$$\mathbf{r}(u,v) = (u, v, (1-a)(u^2+v^2)+a), \qquad (u,v) \in D,$$

where D is the disk $u^2+v^2 \le 1$ in the uv-plane. Thus

$$\mathbf{N}_a = \mathbf{T}_u \times \mathbf{T}_v = \begin{vmatrix} \mathbf{i} & \mathbf{j} & \mathbf{k} \\ 1 & 0 & 2(1-a)u \\ 0 & 1 & 2(1-a)v \end{vmatrix} = -2(1-a)u\mathbf{i} - 2(1-a)v\mathbf{j} + \mathbf{k},$$

and so

$$\iint_{S_a} \operatorname{curl}\mathbf{F} \cdot d\mathbf{S} = \iint_D \left(-2(1-a)(u+v) + e^{-u^2-v^2}(1-2u^2) - 2uv\right) dA.$$

Passing to polar coordinates, we get

$$\iint_{S_a} \operatorname{curl}\mathbf{F} \cdot d\mathbf{S}$$
$$= \int_0^{2\pi} \left(\int_0^1 [-2r(1-a)(\cos\theta + \sin\theta) + e^{-r^2}(1-2r^2\cos^2\theta) - 2r^2 \sin\theta\cos\theta] dr \right) d\theta$$
$$\approx 1.15573 + 0.1 \cdot 10^{-9} a.$$

(This result was obtained using Maple. Now $\pi/e \approx 1.15573$, and that is the value of the path integral in (a); the fact that the results in (a) and (b) must be the same is due to Stokes' Theorem. The extra term $0.1 \cdot 10^{-9}a$ was produced by an error in computation (when asked for a 35-digit precision, a computer returns $1.15573... + 0.1 \cdot 10^{-34}a$). No term that contains a can be a part of the solution — that would imply that $\iint_{S_a} \operatorname{curl}\mathbf{F} \cdot d\mathbf{S}$ depends on the surface S_a, which contradicts Stokes' Theorem.)

(c) Substituting $z=1$ into $z = x^2+y^2+a$ we get $x^2+y^2 = 1-a$, $z=1$; so the circle $\mathbf{c}_a$ is the boundary of S_a, for $a<1$.

Parametrize $\mathbf{c}_a$ as

$$\mathbf{c}_a(t) = (\sqrt{1-a}\,\cos t, \sqrt{1-a}\,\sin t, 1), \qquad t \in [0, 2\pi];$$

then $\mathbf{c}_a'(t) = (-\sqrt{1-a}\,\sin t, \sqrt{1-a}\,\cos t, 0)$,

$$\mathbf{F}(\mathbf{c}_a(t)) = ((1-a)^{3/2}\cos t \sin^2 t - 2, (1-a)^{1/2}e^{1-a}\cos t, (1-a)^{1/2}(\sin t - \cos t) - 2),$$

and so

$$\int_{\mathbf{c}_a} \mathbf{F}\cdot d\mathbf{s} = \int_0^{2\pi} [-(1-a)^2 \cos t \sin^3 t + 2(1-a)^{1/2}\sin t + (1-a)e^{-(1-a)}\cos^2 t]\, dt.$$

Using a computer (or changing to polar coordinates), we get

$$\int_{\mathbf{c}_a} \mathbf{F}\cdot d\mathbf{s} = e^{-(1-a)}\pi(1-a).$$

The particular values are: $\int_{\mathbf{c}_{1/2}} \mathbf{F}\cdot d\mathbf{s} = e^{-1/2}\pi/2$, $\int_{\mathbf{c}_0} \mathbf{F}\cdot d\mathbf{s} = e^{-1}\pi$, $\int_{\mathbf{c}_{-1}} \mathbf{F}\cdot d\mathbf{s} = 2e^{-2}\pi$ and $\int_{\mathbf{c}_{-10}} \mathbf{F}\cdot d\mathbf{s} = 11e^{-11}\pi$.

Parametrize S_a as

$$\mathbf{r}_a(u,v) = (u, v, u^2+v^2+a), \qquad (u,v) \in D,$$

where D is the disk $u^2+v^2 \le 1-a$ in the uv-plane. Then

$$\mathbf{N} = \mathbf{T}_u \times \mathbf{T}_v = \begin{vmatrix} \mathbf{i} & \mathbf{j} & \mathbf{k} \\ 1 & 0 & 2u \\ 0 & 1 & 2v \end{vmatrix} = -2u\mathbf{i} - 2v\mathbf{j} + \mathbf{k}$$

and

$$\iint_{S_a} curl\,\mathbf{F}\cdot d\mathbf{S} = \iint_D [-2(u+v) + e^{-u^2-v^2}(1-2u^2) - 2uv]\, dA.$$

Passing to polar coordinates, we get

$$\iint_{S_a} curl\,\mathbf{F}\cdot d\mathbf{S}$$

$$= \int_0^{2\pi}\left(\int_0^{\sqrt{1-a}} [-2r(\cos\theta+\sin\theta) + e^{-r^2}(1-2r^2\cos^2\theta) - 2r^2\sin\theta\cos\theta]dr\right)d\theta.$$

Using a computer, we obtain the following numerical values (that coincide with the results for the corresponding path integrals): $\int_{S_{1/2}} curl\,\mathbf{F}\cdot d\mathbf{S} = e^{-1/2}\pi/2$, $\int_{S_0} curl\,\mathbf{F}\cdot d\mathbf{S} = e^{-1}\pi$, $\int_{S_{-1}} curl\,\mathbf{F}\cdot d\mathbf{S} = 2e^{-2}\pi$ and $\int_{S_{-10}} curl\,\mathbf{F}\cdot d\mathbf{S} = 11e^{-11}\pi$.

Solutions to Further Explorations

1. There is no need to compute path integral(s)! The region D bounded by $\mathbf{c}$ consists of the disk $x^2+y^2 \le 16$ from which two disks of radius 1 (with centers at $(-2,0)$ and $(2,0)$) have been removed. In Section 8.1 we showed that $\int_{\mathbf{c}} y dx = -A(D)$, where $A(D)$ is the area of D. So

$$\int_{\mathbf{c}} y dx = -(16\pi - \pi - \pi) = -14\pi.$$

3. Since $x^2+y^2=\epsilon^2$, it follows that

$$\int_{\mathbf{c}_\epsilon} \alpha = \int_0^{2\pi} \left(\frac{-\epsilon \sin t}{1-\epsilon^2}(-\epsilon \sin t) + \frac{\epsilon \cos t}{1-\epsilon^2}\epsilon \cos t \right) dt$$
$$= \int_0^{2\pi} \frac{\epsilon^2}{1-\epsilon^2}\, dt = \frac{2\pi\epsilon^2}{1-\epsilon^2}.$$

So, as $\epsilon \to 0$, $\iint_{\mathbf{c}_\epsilon} \alpha \to 0$; as $\epsilon \to 1^+$, $\iint_{\mathbf{c}_\epsilon} \alpha \to -\infty$; as $\epsilon \to 1^-$, $\iint_{\mathbf{c}_\epsilon} \alpha \to \infty$; and as $\epsilon \to \infty$, $\iint_{\mathbf{c}_\epsilon} \alpha \to -2\pi$.

5. Since $div\, \mathbf{F} = 3x^2+3z^2+3y^2$, it follows from the Divergence Theorem that

$$\iint_S \mathbf{F}\cdot d\mathbf{S} = \iiint_V div\, \mathbf{F}\, dV = 3\iiint_V (x^2+y^2+z^2)\, dV,$$

where V is the three-dimensional solid bounded by the two spheres. Passing to spherical coordinates, we get

$$3\iiint_V (x^2+y^2+z^2)\, dV = 3\int_0^{2\pi} \left(\int_0^{\pi} \left(\int_1^3 \rho^2\, \rho^2 \sin\phi d\rho \right) d\phi \right) d\theta$$
$$= 3\int_0^{2\pi} \left(\int_0^{\pi} \tfrac{242}{5} \sin\phi d\phi \right) d\theta$$
$$= \tfrac{726}{5} \left(\int_0^{2\pi} d\theta \right) \left(\int_0^{\pi} \sin\phi\, d\phi \right)$$
$$= \tfrac{726}{5}(2\pi)(2) = \tfrac{2904}{5}\pi \approx 1824.637$$

7. (a) Since

$$\frac{\partial}{\partial x}\left(\frac{x}{x^2+y^2} \right) = \frac{x^2+y^2-2x^2}{(x^2+y^2)^2} = \frac{y^2-x^2}{(x^2+y^2)^2}$$

and (by symmetry — interchange x and y)

$$\frac{\partial}{\partial y}\left(\frac{y}{x^2+y^2} \right) = \frac{x^2-y^2}{(x^2+y^2)^2},$$

it follows that $div\, \mathbf{F} = 0$.

(b) Parametrize $\mathbf{c}_\epsilon$ by $\mathbf{c}_\epsilon(t) = (\epsilon \cos t, \epsilon \sin t)$, $t \in [0, 2\pi]$; then $\mathbf{c}'_\epsilon(t) = (-\epsilon \sin t, \epsilon \cos t)$ and the outward unit normal is $\mathbf{n}(t) = (\cos t, \sin t)$ (i.e., $\mathbf{n}(t)$ is of unit length, perpendicular to $\mathbf{c}'_\epsilon(t)$ and points away from the origin). It follows that

$$\mathbf{F}(\mathbf{c}_\epsilon(t)) \cdot \mathbf{n}(t) = \left(\frac{\epsilon \cos t}{\epsilon^2}, \frac{\epsilon \sin t}{\epsilon^2} \right) \cdot (\cos t, \sin t) = \frac{1}{\epsilon}$$

and therefore

$$\int_{\mathbf{c}_\epsilon} \mathbf{F}\cdot\mathbf{n}\, ds = \int_0^{2\pi} \frac{1}{\epsilon}\, \|\mathbf{c}'_\epsilon(t)\|\, dt = \int_0^{2\pi} dt = 2\pi.$$

(c) No matter what $\epsilon > 0$ is taken, the region enclosed by $\mathbf{c}_\epsilon$ contains the origin; therefore, $\mathbf{F}$ is not C^1 on that region, and the Divergence Theorem does not apply.

9. See Figure 8.5.9 for the orientation of $\mathbf{c}$ and the surface normal $\mathbf{N} = \mathbf{T}_u \times \mathbf{T}_v$. Since

$$curl\,\mathbf{F} = \begin{vmatrix} \mathbf{i} & \mathbf{j} & \mathbf{k} \\ \partial/\partial x & \partial/\partial y & \partial/\partial z \\ e^x yz & -e^y & e^z \end{vmatrix} = e^x y\mathbf{j} - e^x z\mathbf{k},$$

it follows by Stokes' Theorem that the circulation of $\mathbf{F}$ along $\mathbf{c}$ is

$$\int_{\mathbf{c}} \mathbf{F} \cdot d\mathbf{s} = \int_{\mathbf{c}} (e^x yz\mathbf{i} - e^y\mathbf{j} + e^z\mathbf{k}) \cdot d\mathbf{s} = \iint_S (e^x y\mathbf{j} - e^x z\mathbf{k}) \cdot d\mathbf{S}.$$

The equation of the plane going through the given points is $x/1 + y/(-2) + z/1 = 1$; i.e., $z = 1 - x + y/2$. Parametrize the planar region bounded by $\mathbf{c}$ as $\mathbf{r}(u,v) = (u, v, 1 - u + v/2)$, where $(u,v) \in D$, and D is the triangular region in the uv-plane defined by $u = 0$, $v = 0$ and $u - v/2 = 1$. The surface normal $\mathbf{N}$ is

$$\mathbf{N} = \mathbf{T}_u \times \mathbf{T}_v = \begin{vmatrix} \mathbf{i} & \mathbf{j} & \mathbf{k} \\ 1 & 0 & -1 \\ 0 & 1 & 1/2 \end{vmatrix} = \mathbf{i} - \mathbf{j}/2 + \mathbf{k}$$

(which is the right direction!). Hence

$$\begin{aligned}
\iint_S (e^x y\mathbf{j} - e^x z\mathbf{k}) \cdot d\mathbf{S} &= \iint_D (e^u v\mathbf{j} - e^u(1 - u + v/2)\mathbf{k}) \cdot (\mathbf{i} - \mathbf{j}/2 + \mathbf{k})\, dA \\
&= \iint_D (-e^u + ue^u - ve^u)\, dA \\
&= \int_0^1 \left(\int_0^{2u-2} ((u-1)e^u - e^u v) dv \right) du \\
&= \int_0^1 \left((u-1)e^u v - \tfrac{1}{2}e^u v^2\right) \Big|_0^{2u-2} du \\
&= \int_0^1 \left((u-1)e^u(2)(u-1) - e^u \tfrac{1}{2}(4)(u-1)^2\right) du \\
&= \int_0^1 0\, du = 0.
\end{aligned}$$

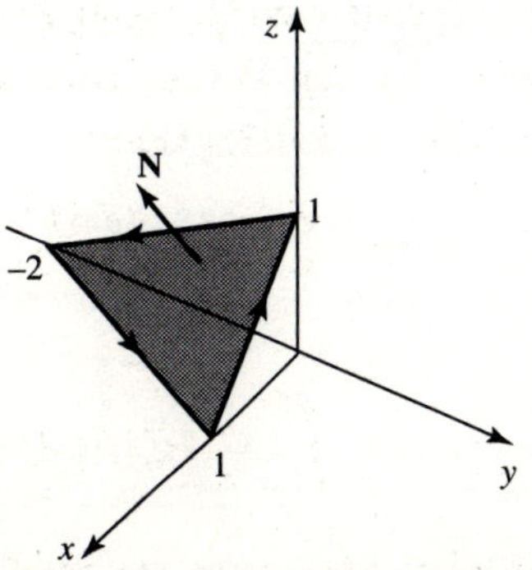

Figure 8.5.9

11. (a) Parametrize D by $\mathbf{r}(u,v) = (u,v,0)$, where (u,v) in D; thus $\mathbf{N} = \mathbf{T}_u \times \mathbf{T}_v = \mathbf{i}\times\mathbf{j} = \mathbf{k}$. Next,

$$curl\,(-ye^{x^2}\mathbf{i}) = \begin{vmatrix} \mathbf{i} & \mathbf{j} & \mathbf{k} \\ \partial/\partial x & \partial/\partial y & \partial/\partial z \\ -ye^{x^2} & 0 & 0 \end{vmatrix} = e^{x^2}\mathbf{k} = \mathbf{F}.$$

(b) By Stokes' Theorem,

$$\iint_D curl\,(-ye^{x^2}\mathbf{i}) \cdot \mathbf{N}\,dA = \int_{\mathbf{c}} (-ye^{x^2}\mathbf{i}) \cdot d\mathbf{s},$$

where $\mathbf{c}$ is the boundary of D oriented counterclockwise.

Let us evaluate the path integral. Parametrize the segment from $(0,0)$ to $(1,0)$ by $\mathbf{c}_1(t) = (t,0)$, $t \in [0,1]$. Then $\mathbf{c}_1'(t) = (1,0)$ and

$$\int_{\mathbf{c}_1} (-ye^{x^2}\mathbf{i}) \cdot d\mathbf{s} = \int_0^1 0\,dt = 0.$$

Parametrize the segment from $(1,0)$ to $(1,1)$ by $\mathbf{c}_2(t) = (1,t)$, $t \in [0,1]$. Then $\mathbf{c}_2'(t) = (0,1)$ and

$$\int_{\mathbf{c}_2} (-ye^{x^2}\mathbf{i}) \cdot d\mathbf{s} = \int_0^1 (-te,0)\cdot(0,1)\,dt = 0.$$

Finally, parametrize the segment from $(1,1)$ to $(0,0)$ by $\mathbf{c}_3(t) = (1-t,1-t)$, $t \in [0,1]$. Then $\mathbf{c}_3'(t) = (-1,-1)$ and

$$\begin{aligned}\int_{\mathbf{c}_3} (-ye^{x^2}\mathbf{i}) \cdot d\mathbf{s} &= \int_0^1 (-(1-t)e^{(1-t)^2},0)\cdot(-1,-1)\,dt \\ &= \int_0^1 (1-t)e^{(1-t)^2}\,dt = -\tfrac{1}{2}e^{(1-t)^2}\Big|_0^1 = \tfrac{1}{2}(e-1).\end{aligned}$$

Adding up the three integrals we get

$$\int_{\mathbf{c}} (-ye^{x^2}\mathbf{i}) \cdot d\mathbf{s} = \tfrac{1}{2}(e-1).$$

(c) The double integral in (b) is computed to be

$$\int_0^1 \left(\int_0^x e^{x^2}\,dy\right) dx = \int_0^1 xe^{x^2}\,dx = \tfrac{1}{2}e^{x^2}\Big|_0^1 = \tfrac{1}{2}(e-1).$$

13. (a) By definition,

$$\begin{aligned}*\alpha &= *(x^2tdxdy + zt^2dtdz - xy^2ztdtdy) \\ &= x^2tdtdz - c^{-2}zt^2dxdy + c^{-2}xy^2ztdzdx.\end{aligned}$$

(b) From (a),

$$\begin{aligned}*(*\alpha) &= *(x^2tdtdz - c^{-2}zt^2dxdy + c^{-2}xy^2ztdzdx) \\ &= -x^2tc^{-2}dxdy - c^{-2}zt^2dtdz + c^{-2}xy^2ztdtdy \\ &= -c^{-2}(x^2tdxdy + zt^2dtdz - xy^2ztdtdy) = -c^{-2}\alpha.\end{aligned}$$

(c) Using (a) again,

$$\begin{aligned}\alpha \wedge (*\alpha) &= (x^2tdxdy + zt^2dtdz - xy^2ztdtdy) \wedge (x^2tdtdz - c^{-2}zt^2dxdy + c^{-2}xy^2ztdzdx)\\ &= x^4t^2dxdydtdz - c^{-2}z^2t^4dtdzdxdy - c^{-2}x^2y^4z^2t^2dtdydzdx\\ &= (x^4t^2 - c^{-2}z^2t^4 - c^{-2}x^2y^4z^2t^2)dxdydzdt.\end{aligned}$$

15. By Stokes' Theorem,

$$\int_{\mathbf{c}} (f\nabla g + g\nabla f) \cdot d\mathbf{s} = \iint_S curl\,(f\nabla g + g\nabla f) \cdot d\mathbf{S}.$$

By a product rule for *curl*, we get

$$curl\,(f\nabla g) = f curl\,(\nabla g) + \nabla f \times \nabla g = \nabla f \times \nabla g,$$

since $curl\,(\nabla g) = \mathbf{0}$. Similarly,

$$curl\,(g\nabla f) = \nabla g \times \nabla f,$$

and so $curl\,(f\nabla g + g\nabla f) = \mathbf{0}$ and

$$\iint_S curl\,(f\nabla g + g\nabla f) \cdot d\mathbf{S} = 0.$$